BIFURCATIONS AND CHAOS IN PIECEWISE-SMOOTH DYNAMICAL SYSTEMS

WORLD SCIENTIFIC SERIES ON NONLINEAR SCIENCE

Editor: Leon O. Chua
University of California, Berkeley

Series A. MONOGRAPHS AND TREATISES

Volume 27: The Thermomechanics of Nonlinear Irreversible Behaviors — An Introduction
G. A. Maugin

Volume 28: Applied Nonlinear Dynamics & Chaos of Mechanical Systems with Discontinuities
Edited by M. Wiercigroch & B. de Kraker

Volume 29: Nonlinear & Parametric Phenomena*
V. Damgov

Volume 30: Quasi-Conservative Systems: Cycles, Resonances and Chaos
A. D. Morozov

Volume 31: CNN: A Paradigm for Complexity
L. O. Chua

Volume 32: From Order to Chaos II
L. P. Kadanoff

Volume 33: Lectures in Synergetics
V. I. Sugakov

Volume 34: Introduction to Nonlinear Dynamics*
L. Kocarev & M. P. Kennedy

Volume 35: Introduction to Control of Oscillations and Chaos
A. L. Fradkov & A. Yu. Pogromsky

Volume 36: Chaotic Mechanics in Systems with Impacts & Friction
B. Blazejczyk-Okolewska, K. Czolczynski, T. Kapitaniak & J. Wojewoda

Volume 37: Invariant Sets for Windows — Resonance Structures, Attractors, Fractals and Patterns
A. D. Morozov, T. N. Dragunov, S. A. Boykova & O. V. Malysheva

Volume 38: Nonlinear Noninteger Order Circuits & Systems — An Introduction
P. Arena, R. Caponetto, L. Fortuna & D. Porto

Volume 39: The Chaos Avant-Garde: Memories of the Early Days of Chaos Theory
Edited by Ralph Abraham & Yoshisuke Ueda

Volume 40: Advanced Topics in Nonlinear Control Systems
Edited by T. P. Leung & H. S. Qin

Volume 41: Synchronization in Coupled Chaotic Circuits and Systems
C. W. Wu

Volume 42: Chaotic Synchronization: Applications to Living Systems
E. Mosekilde, Y. Maistrenko & D. Postnov

Volume 43: Universality and Emergent Computation in Cellular Neural Networks
R. Dogaru

*Forthcoming

WORLD SCIENTIFIC SERIES ON NONLINEAR SCIENCE Series A Vol. 44
Series Editor: Leon O. Chua

BIFURCATIONS AND CHAOS IN PIECEWISE-SMOOTH DYNAMICAL SYSTEMS

Zhanybai T. Zhusubaliyev
Kursk State Technical University, Russia

Erik Mosekilde
The Technical University of Denmark

World Scientific
New Jersey • London • Singapore • Hong Kong

Published by

World Scientific Publishing Co. Pte. Ltd.

5 Toh Tuck Link, Singapore 596224

USA office: Suite 202, 1060 Main Street, River Edge, NJ 07661

UK office: 57 Shelton Street, Covent Garden, London WC2H 9HE

British Library Cataloguing-in-Publication Data
A catalogue record for this book is available from the British Library.

BIFURCATIONS AND CHAOS IN PIECEWISE-SMOOTH DYNAMICAL SYSTEMS

ISBN 981-238-420-0

This book is printed on acid-free paper.

Printed in Singapore by Mainland Press

PREFACE

This book is prepared in order to present a number of important new phenomena related to the description of nonlinear dynamical systems whose behavior is controlled by piecewise-smooth differential equations.

The practical significance of these phenomena is illustrated through a series of well-documented and realistic applications to switching power converters, relay systems, and different types of pulse-width modulated control systems. Other examples are derived from mechanical engineering, digital electronics, and economic business cycles theory.

Besides providing an overview of the bifurcation scenarios and transitions to chaos observed in piecewise-smooth dynamical systems, the book may also serve as a textbook in nonlinear dynamics for professionally engaged control and power engineers. With this purpose, the book includes two introductory chapters, that can help readers from these fields get accustomed with the concepts of modern nonlinear dynamics. To our knowledge, no other monograph with this focus is available.

Three basic bifurcation scenarios are known for the transition to chaos in smooth dissipative systems: The infinite cascade of period-doubling bifurcations of a cycle (Feigenbaum scenario), the transition through various types of intermittency as a stable cycle ceases to exist (Pomeau-Manneville scenario), and the transition via different forms of torus destruction (Ruelle-Takens-Newhouse scenario). However, these scenarios do not exhaust the possible mechanisms. There is a much broader class of bifurcation phenomena in systems described by non-smooth differential equations that still lack a detailed description and whose theoretical and numerical analysis presents an extremely difficult problem. At the same time, such systems are of significant practical interest, particularly in engineering.

All problems in mechanical engineering involving collisions, finite clearances, and stick-slip phenomena violate the requirements of smoothness to the equations of motion. A simple example is the so-called impact oscillator that is restricted in its motion by collisions with a fixed wall. This type of phenomenon arises, for instance, when the displacement of a vibrating steam pipe become large enough to engender collisions with other construction parts, in connection with pendulum type centrifugal vibration absorbers, in railroad dynamics, and for rotating machines with clearance in the bearings. Another well-known example is the mooring of an oiltanker at sea. In this case the lack of smoothness is related to a drastic change in the force as the mooring cable is stretched.

In microeconomics and management science, human decisions often involve logical functions: Depending on the relative costs of two different options, either one or the other will be chosen. In simulations of managerial systems with repeated decisions, these logical functions play the role of non-smooth restrictions, and we observe similar types of complex nonlinear dynamic phenomena as in other systems with non-smooth equations of motion. This includes, for instance, finite cascades of period-doublings, period-tripling bifurcations, new types of direct transitions to chaos, and resonance tongues delineated by border-collision bifurcations (rather than by saddle-node bifurcations as in smooth systems). An example from this realm is considered in Chapter 7.

However, the main part of the book is dedicated to the analysis of bifurcations and chaotic oscillations in switching power converters, and in relay and pulse-width modulated control systems. Systems of this type have a broad range of applications for power control and temperature regulation and they are also used to obtain highly stabilized electric and magnetic fields. Examples are power supplies for radio-electronics, computer equipment and spacecrafts, test stands for investigations of high- or low-temperature superconductivity, electron microscopes, and nuclear magnetic resonance tomographs.

In Chapter 3 we introduce a system of autonomous differential equations with discontinuous right-hand sides that can serve as a model of relay systems. We illustrate how investigation of this model can be reduced to the study of the properties of a multi-dimensional piecewise-smooth mapping, and describe a general approach to determining the periodic solutions and analyzing their local stability. The problem of determining the limit cycles is reduced to solving a system of transcendental equations with respect to the switching instants of the relay element. This approach allows us to obtain not only the stable but also the unstable periodic orbits.

The second part of Chapter 3 contains two practically important examples of relay control systems. As a first example we consider a DC/DC converter with relay control whose behavior is described by a four-dimensional set of piecewise-linear autonomous differential equations. As a second example we consider a system applied to regulate the motor torque of a DC electric drive. The mathematical model of this system is represented by a four-dimensional set of piecewise-smooth autonomous differential equations.

A separate section is devoted to a detailed description of the technical aspects of the implementation of relay control systems.

In Chapter 4 we perform a detailed analysis of complex behavior and transitions to chaos for two different types of relay control systems that one can meet in real-world technical applications. The aim is to investigate the structure of the two-parameter diagram of dynamical modes and to examine the regularities in the occurrence of periodic cycles. We also determine the main bifurcations and analyze the nature of the transitions to chaotic dynamics.

We show that relay systems can display situations where several locally stable limit cycles with different dynamic characteristics coexist for a wide range of parameters. These cycles typically arise in hard transitions, for example through saddle-node bifurcations, and with changing parameters they can undergo either a finite or an infinite sequence of period-doubling bifurcations, resulting in the transition to chaos. Moreover, the transition to chaos can take place through the appearance and subsequent destruction of a quasiperiodic motion on a two-dimensional torus. As a result we have parameter domains wherein, alongside with locally stable limit cycles, there are co-existing modes of chaotic or quasiperiodic oscillations. Under such conditions the action of external noise, even of low intensity, can induce a sudden transition from one dynamic state to another and, in particular, from regular to chaotic dynamics.

In Chapter 5 we investigate the main bifurcations and routes to chaos for pulse-modulated control systems. As a specific example we consider a DC/DC converter with pulse-width modulation of the second kind. The behavior of such a converter is described by a three-dimensional system of piecewise-smooth differential equations with an external periodic action.

The chapter contains a detailed analysis of border-collision bifurcations and transitions to chaos exhibited by multi-dimensional piecewise-smooth dynamical systems. We show that transitions from periodic to chaotic oscillations can occur through a sequence of border-collision bifurcations including period-doubling, -tripling, -quadrupling, quintupling, etc., bifurcations. More complex bifurcational transitions with fundamentally new properties are cited as examples, for instance, finite sequences of period-tripling bifurcations and soft transitions from one family of cycles to another family of cycles with multiple periods.

Chapter 6 contains the results of an investigation of a number of new phenomena that can arise in piecewise-smooth dynamical systems whose motion involves two (or more) periodic components. When two self-oscillatory systems interact (or a self-oscillatory system is subjected to an external periodic forcing), the total motion can be viewed as occurring on the surface of a two-dimensional torus and, in the absence of resonances, the motion is said to be quasiperiodic. One problem is here associated with the role that border-collision bifurcations play in the synchronization of the two modes, and in the transition to chaos when the torus finally breaks down.

As a specific example of a piecewise-smooth dynamical system with a quasiperiodic route to chaos we consider a three-dimensional piecewise smooth mapping describing the behavior of a DC/DC converter with pulse-width modulation of the first kind. One of the main ideas of Chapter 6 is to show that the transitions to chaos through two-frequency quasiperiodicity in piecewise smooth systems can differ fundamentally from the mechanisms described in the existing literature. At the same time we show that the boundaries of the Arnol'd tongues for non-smooth systems are formed by border-collisions bifurcations, rather than by the saddle-node bifurcations known from smooth systems. Moreover, the internal organization of the tongues is also quite different from the case of smooth systems.

In spite of its orientation towards engineering problems, the book addresses theoretical and numerical problems in connection with border-collision bifurcations and other peculiarities of piecewise-smooth systems in sufficient detail to be of interest to nonlinear scientists in general.

We would like to express our gratitude to our students and collaborators who have played a significant role in the often extremely demanding process of unraveling the complicated bifurcation structure of a nonlinear dynamic system. These thanks are primarily devoted to Vadim Rudakov, Sergei Pinaev, Elena Emelyanova, Mikael Togeby, Erik Reimer Larsen, Jakob Laugesen, Olga Sosnovtseva, Lasse Christiansen, Tue Lehn-Schiøler, Carsten Knudsen, Jesper Skovhus Thomsen, John D. Sterman, Maciej Szymkat, and Hiroshi Matsushita. Yuri and Vladimir Maistrenko, Laura Gardini, and Irina Sushko have contributed through a number of enlightening discussions of nonlinear dynamic phenomena in general and border-collision bifurcations in particular.

Most of all we would like to express our sincere appreciation to Evgeniy Soukhoterin who has been strongly involved in our recent work on study of quasiperiodic route to chaos in pulse-width modulated control systems. Evgeniy Soukhoterin has also contributed invaluably to the preparation of the present book. Without his engagement and care, the book might never have been published.

Zh. Zhusubaliyev acknowledges support of The Danish Natural Science Research Council (SNF).

The book is dedicated to our wives Inna and Margit.

Kursk and Lyngby, March 2003
Zhanybai Zhusubaliyev and Erik Mosekilde

CONTENTS

Chapter 1

ON THE DYNAMICS OF NONLINEAR SYSTEMS

1.1 Unpredictability and Chaos

In a most surprising manner, modern theory of nonlinear dynamics has disclosed a number of misconceptions in our understanding of the laws of motion. The notion we have about predictability of classical mechanical systems, for instance, is not correct in general, but is founded on a few exceptional cases where the equations of motion can be solved analytically. For most conservative systems, the space of initial conditions is scattered with regions in which the slightest distortion can lead to a completely different trajectory [1–3]. Moreover, in these regions the systems show forms of behavior that differ qualitatively from the simple motions we know for integrable systems [4, 5].

Dissipative systems, i.e., systems with friction, can also show unpredictability and chaos, provided that their motion is maintained through a sufficient supply of energy or through strong external forcing [6–9]. Nonlinear systems can also exhibit two or more simultaneous solutions such that varying initial conditions can cause a system to choose one stationary solution rather than another. Moreover, the boundary between the conditions that lead to the different stationary states may be fractal and show structure down to the smallest detail [10, 11]. Close to such boundaries, the uncertainty in the final outcome will diminish more slowly than does the uncertainty in the initial conditions [12]. If

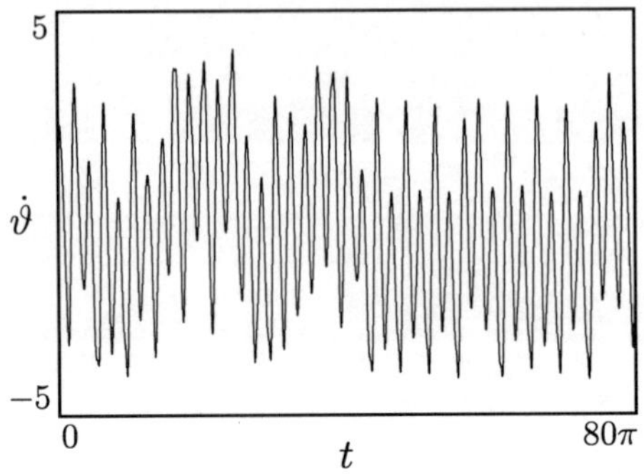

Figure 1.1. Temporal variation of the forced pendulum (1.1) for $\alpha = 0.2$ and $B = 2.4$. Note the unusually complex dynamics generated by the simple deterministic system. The smallest change in the initial conditions will, after a while, cause the system to follow a very different trajectory.

the initial conditions are specified with twice the accuracy, our ability to predict the resulting stationary state will only improve by a factor less than two.

Figure 1.1 shows an example of a chaotic motion generated by the forced pendulum equation [13, 14]

$$\frac{\mathrm{d}^2\vartheta}{\mathrm{d}t^2} + \alpha\frac{\mathrm{d}\vartheta}{\mathrm{d}t} + \sin\vartheta = B\cos t \tag{1.1}$$

where the nonlinearity is associated with the sine-function. $\alpha = 0.2$ is the damping coefficient and $B = 2.4$ is the amplitude of the forcing. Even though the dynamics is produced by a simple and fully deterministic system it displays several similarities with the outcome of a random process. In particular, one cannot use an observed variation in the magnitude of the excursions to predict the magnitude of the next excursion. This resembles the assertion in the theory of stochastic systems that one cannot use the past outcomes of tosses with a (fair) coin to predict the next outcome.

It is important to stress, however, that nonlinear systems with chaotic dynamics and linear stochastic systems behave quite differently. This is revealed, for instance, in their responses to external perturbations. For a nonlinear system the response is strongly dependent on its actual state. While the smallest distortion under certain conditions can produce a dramatic change in behavior, even a major blow may, at other stages, seem to have negligible effect. For a linear system, on the other hand, the response to an external perturbation will always be the same.

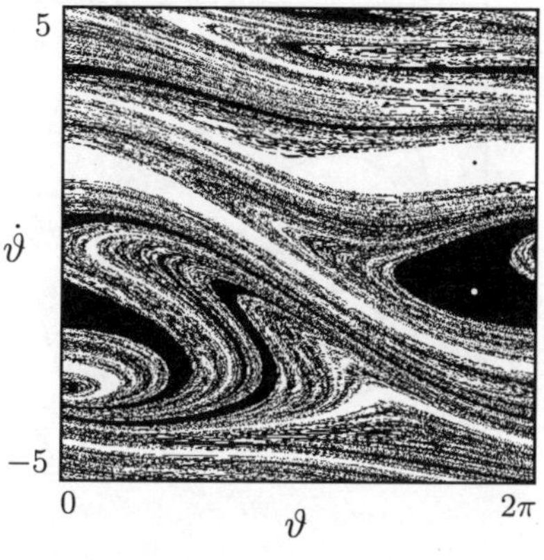

Figure 1.2. Fractal basin boundary for the forced pendulum (1.1) with $\alpha = 0.1$ and $B = 2.0$. Black points denote initial conditions that lead to one of the two coexisting stable periodic states. White points lead to the other solution.

Figure 1.2 shows an example of the complicated *basins of attraction* that one can find in nonlinear dynamical systems with two (or more) coexisting stable states. The basin of attraction is the set of initial conditions that lead to a particular stationary state. For $\alpha = 0.1$ and $B = 2.0$, the forced pendulum (1.1) displays two different stable solutions (indicated in the figure by black and white dots). Somewhere between these stable states one can find a so-called *saddle cycle*, i.e., a periodic orbit that is attracting from certain directions and repelling in others. The *stable manifold* to the saddle cycle, i.e., the set of initial conditions from which the system asymptotically approaches the unstable periodic orbit, represents the boundary of the basins of attraction for the two stable solutions. This is illustrated schematically in Fig. 1.3.

Figure 1.2 is a so-called *stroboscopic map*, i.e., a description of the state $(\vartheta, \dot{\vartheta})$ of the system each time the forcing term $B \cos t$ has completed a full cycle. Besides the role that the stable manifold to the saddle cycle plays in delineating the boundary between the two basins, the schematic illustration in Fig. 1.3 also shows how the branches of the *unstable manifold* to the saddle cycle lead to each of the stable cycles. These cycles are depicted as focus points. This implies that they have a pair of complex conjugate eigenvalues so that trajectories will approach the attracting states in a spiralling manner.

In Chapter 2, Fig. 2.10(a) shows an example of a similar structure. Here we consider the basins of attraction for the coexisting period-1 and period-3 cycles

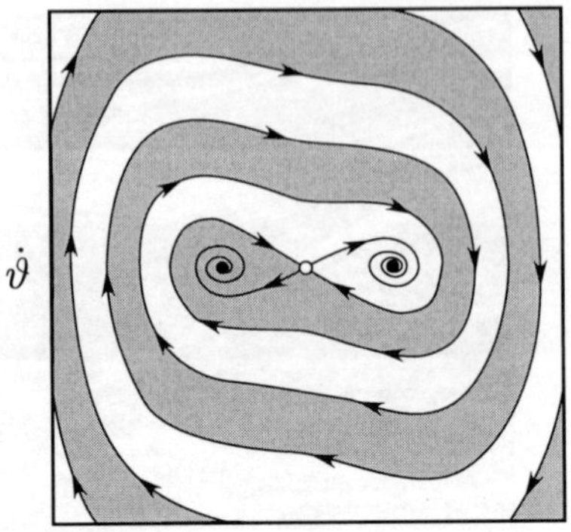

Figure 1.3. Basins of attraction for two coexisting stable periodic solutions (full circles). The stable manifold to the saddle point (open circle) delineates the basin boundary.

of the Hénon map. The period-1 cycle is a node with real eigenvalues while the period-3 cycle is a focus. The stable manifold of a period-3 saddle cycle defines the boundary between the two basins of attraction. Fractal basin boundaries as illustrated in Fig. 1.2 for the forced pendulum arise in connection with the formation of a so-called *Smale horseshoe* [15, 16] when the stable and unstable manifolds of the saddle cycle start to intersect. This will typically happen when the damping becomes relatively low. Figures 2.11 and 2.12 in Chapter 2 show examples of fractal basin boundaries for the Hénon map.

The forced pendulum is obviously an extremely simple system, and the complicated dynamic phenomena that this system displays are truly astonishing. If the forcing amplitude is low enough, the pendulum exhibits a single stable oscillatory state, resembling in the limit the regular oscillations of a harmonic oscillator. The transition between the two types of behavior occurs via a so-called *pitchfork bifurcation* in which the single stable oscillatory state becomes unstable and yields to the formation of two mutually symmetric oscillatory states.

Another example of a simple nonlinear oscillator is the Duffing system first studied in detail by Ueda [17–19]

$$\frac{\mathrm{d}^2 x}{\mathrm{d}t^2} + \alpha \frac{\mathrm{d}x}{\mathrm{d}t} + x^3 = B \cos t. \tag{1.2}$$

This system may represent a nonlinear mechanical oscillator for which the restoring force varies as x^3 rather than linearly with the displacement. Al-

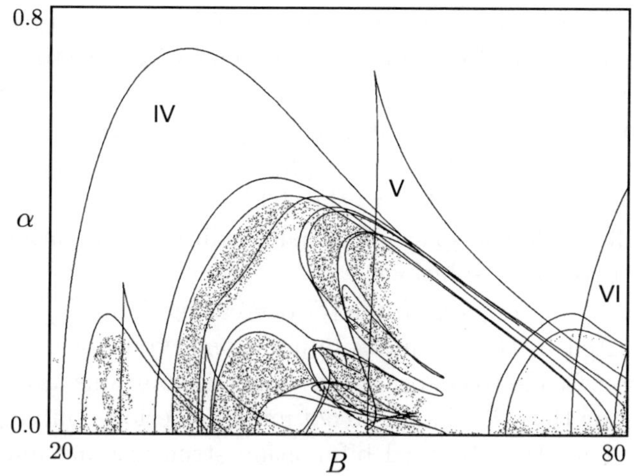

Figure 1.4. Phase diagram illustrating the local bifurcations for the Duffing oscillator (1.2). In regions IV and VI, delineated by pitchfork bifurcation curves, the oscillations are associated with even resonances. Odd resonances occur in region V. Dotted areas are regions where chaotic solutions can be found.

ternatively, the Duffing oscillator may serve as a model of an electronic resonance circuit with a nonlinear magnetization characteristics for the inductive component.

Figure 1.4 reproduces part of the phase diagram for the Duffing oscillator [20, 21]. This diagram presents the number and type of the solutions to (1.2) as a function of the forcing amplitude B and the damping coefficient α. Each time the system crosses one of the fully drawn lines in the figure, a local bifurcation takes place, and the character of the solution changes. In the coarsest interpretation the phase diagram may be described in terms of a series of resonances in which the nonlinear oscillator completes $m = 1, 2, 3, \ldots$ cycles for each period of the external forcing. In these resonances, the odd modes ($m = 1, 3, 5, \ldots$) are symmetric whereas the even modes ($m = 2, 4, 6, \ldots$) always involve a pair of mutually symmetric solutions. Hence, symmetry-breaking pitchfork bifurcations delineate the regions in parameter space where even resonance modes are found.

Regions in parameter space where large-amplitude odd solutions occur are bounded by curves of *saddle-node bifurcations*. In such a bifurcation, a stable (node) and an unstable (saddle) cycle of the same periodicity arise (or disappear) simultaneously. For a nonlinear oscillator, the apparent resonance frequency depends on the forcing amplitude. This gives rise to a so-called *tilted resonance curve* [7, 22]. Hence, in a certain region, delineated by the saddle-node bifurcation curves, the oscillator will display a stable high amplitude (resonant) solution together with a low amplitude (nonresonant) solution. As described above, the stable manifold to the saddle cycle arising in the saddle-node bifurcations will divide the space of initial conditions into basins of attraction for each of the two stable solutions.

With increasing forcing amplitude (or decreasing damping), the asymmetric solutions generated in the pitchfork bifurcations typically proceed through a period-doubling cascade to chaos. The symmetric solutions associated with the odd resonances, on the other hand, do not undergo period-doubling bifurcations. To fully account for the observed bifurcation structure one must also discuss various subharmonic and ultra-subharmonic resonances as well as the global bifurcations in which, for instance, a chaotic attractor collides with its basin boundary [19, 20].

The dotted areas in the phase diagram of Fig. 1.4 are regions where chaotic dynamics is observed. When chaos occurs, the system has developed an infinite number of unstable periodic orbits, and the trajectory moves apparently at random among these cycles without being able to settle down [17, 19]. The presence of a dense set of unstable periodic orbits and the existence of trajectories that wander around and visit the neighborhood of each of these cycles (*ergodicity* or transitivity) are two of the main hallmarks of deterministic chaos. A third characteristics is the aforementioned sensitivity to the initial conditions. The (largest) *Lyapunov exponent* [23, 24] measures the average rate of divergence of two nearby trajectories. A positive value of this exponent is generally taken as a signature of chaos.

With a background in medicine, Rössler [25, 26] was presumably first to construct a set of homogeneous equations that could generate chaotic dynamics

$$\begin{aligned} \frac{\mathrm{d}x}{\mathrm{d}t} &= -y - z; \\ \frac{\mathrm{d}y}{\mathrm{d}t} &= x + ay; \\ \frac{\mathrm{d}z}{\mathrm{d}t} &= b + z(x - c). \end{aligned} \tag{1.3}$$

To understand how these equations work one can try to imagine the different

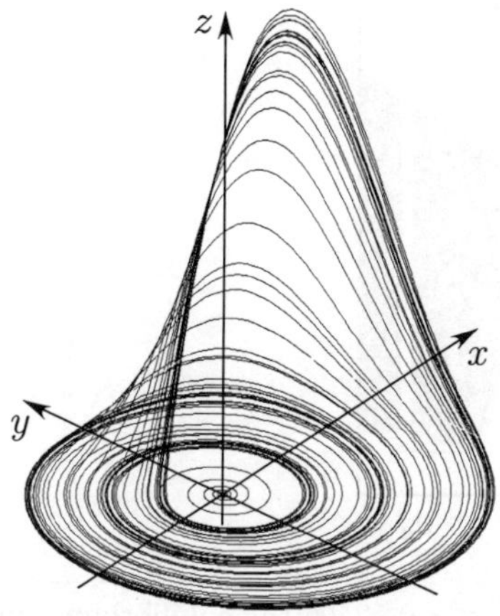

Figure 1.5. Rössler attractor. Stationary solution to the Rössler system (1.3) for $a = b = 0.2$ and $c = 5.7$. The Rössler system can be motivated in a chemical reaction where x, y, and z play the role of concentrations of various chemical species [27].

steps in the model formulation. Starting with a damped harmonic oscillator (in two dimensions), the first step could be to change the sign of the damping term in order to replace the stable focus point by an unstable one:

$$\frac{\mathrm{d}x}{\mathrm{d}t} = -y; \qquad \frac{\mathrm{d}y}{\mathrm{d}t} = x + ay$$

with $a > 0$.

Since it is not allowed that trajectories in a deterministic system intersect, containment of the dynamics requires a third dimension. The next step is, therefore, to introduce a variable that makes it possible for the system to move away from the xy-plane when the amplitude of the spiralling motion becomes sufficiently large.

$$\frac{\mathrm{d}x}{\mathrm{d}t} = -y; \qquad \frac{\mathrm{d}y}{\mathrm{d}t} = x + ay; \quad \frac{\mathrm{d}z}{\mathrm{d}t} = b + z(x - c).$$

Here, the term b serves to keep z positive. The second term gives rise to a fast growth of z when x exceeds c. For $x < c$, the second term is relatively insignificant as long as z stays small. Finally, variations in z are coupled back to the first equation to arrange for x to decrease as z grows large. This folds the trajectory back towards the unstable equilibrium point.

Figure 1.5 shows an example of the stationary solution to the Rössler system for $a = b = 0.2$ and $c = 5.7$. For $c^2 \geq 4ab$, the system has two equilibrium points

$$x_0 = \frac{c}{2} \pm \sqrt{\frac{c^2}{4} - ab}; \qquad y_0 = -\frac{x_0}{a}; \qquad z_0 = \frac{x_0}{a}$$

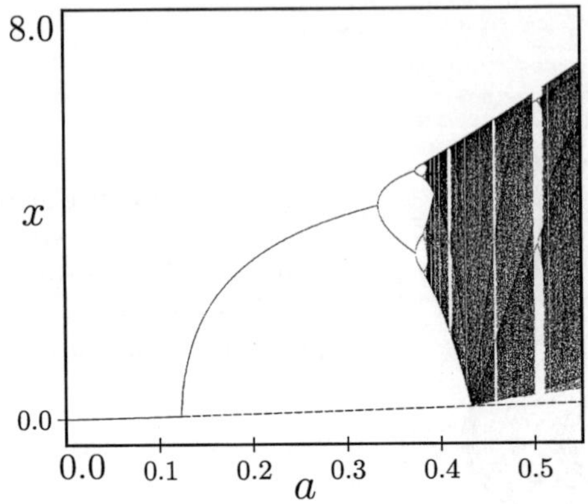

Figure 1.6. Bifurcation diagram for the Rössler system. At $a \approx 0.12496$ the equilibrium point undergoes a Hopf bifurcation, and the system starts to perform self-sustained oscillations. At $a \approx 0.33$, a period-doubling cascade is initiated, and at $a \approx 0.38$ a transition to chaos takes place. $b = 2.0$ and $c = 4.0$.

of which the point with a minus in front of the squareroot falls close to the center of phase space. For the considered parameter values, this point is a *saddle-focus* with one, numerically relatively large, negative eigenvalue and two complex conjugate eigenvalues with positive real parts. Hence, in accordance with the above discussion, the motion may be described as a spiralling dynamics near the xy-plane away from the unstable focus, a folding of the trajectory by virtue of the nonlinear term $z(x - c)$, and a new approach towards the equilibrium point along its stable direction.

For small values of a, the complex conjugate eigenvalues have negative real parts, and the equilibrium point is stable. As a increases, the eigenvalues cross the imaginary axis for $a = a_{\text{H}} \approx 0.12496$, the Rössler system undergoes a *Hopf bifurcation*, and for $a > a_{\text{H}}$ it starts to exhibit stable limit cycle behavior. The further development may be followed in the bifurcation diagram of Fig. 1.6. Here we have plotted the maximum value of x observed in each oscillation as the parameter a is slowly scanned from the Hopf bifurcation point to $a = 0.53$. The figure also shows the position of the equilibrium point.

At $a \approx 0.33$, the bifurcation curve splits into two branches, indicating that the dynamics alternates between two different maxima. This is the signature of a period-doubling bifurcation. As a is further increased, the system undergoes a cascade of bifurcations in which period-4, period-8, period-16, etc., dynamics

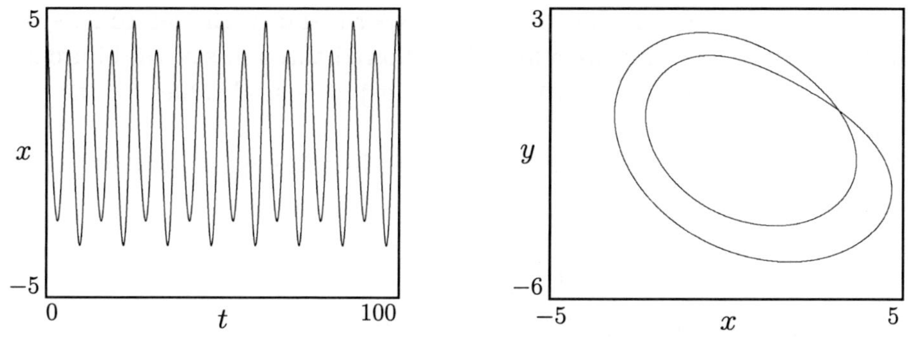

(a)

(b)

Figure 1.7. Time and phase plots for the period-2 solution that exists in the Rössler system (1.3) for $a = 0.35$. Note that the trajectory does not intersect itself. The apparent intersection is an artefact of the two-dimensional projection. $b = 2.0$ and $c = 4.0$.

arise and disappear, until at $a \approx 0.38$ the dynamics becomes chaotic.

Figure 1.7 presents time and phase plots of the period-2 solution generated by the Rössler system for $a = 0.35$. In the temporal variation we notice the characteristic alternation between high and low maxima, corresponding to the two different values of x in the bifurcation diagram. The *phase plot* shows that the system now completes two full rotations before it repeats itself. In a Fourier analysis of the temporal variation, the doubling of the period will reveal itself by the presence of *subharmonic components* at one-half the fundamental frequency. The original limit cycle continues to exist, but now as a saddle cycle with one stable and one unstable transversal direction.

In the chaotic regime $a > 0.38$ one finds a large number of *windows with periodic dynamics*. In particular, we notice the *period*-3 *window* at $a \approx 0.41$. Each window starts with a saddle-node bifurcation in which a stable (node) and an unstable (saddle) cycle of the same periodicity are born, and the windows are terminated by new period-doubling cascades. This development is universal in the sense that the same type of transitions can be observed in physical [28], chemical [29], biological [30], and electronic [31] systems. Moreover, the universality is quantitative in the sense that the period-doubling cascades can be characterized by two parameters, the *Feigenbaum constants* $\alpha_{\mathrm{F}} \approx 2.50290787\ldots$ and $\delta_{\mathrm{F}} \approx 4.6692016\ldots$ [32, 33], that are the same for a wide class of systems. The order in which the periodic windows arise in the chaotic regime also follow universal rules [34, 35]. Because of their low noise levels and the accuracy with which the parameters can be controlled, experiments with electronic systems allow the period-doubling cascade to be followed through several bifurcations. Hence it is possible to obtain (approximate) experimental values for the scaling constants [31].

The basic model in the study of the period-doubling transition to chaos is the *logistic map*

$$\begin{aligned} x_k &= f_\lambda(x_{k-1}); \\ f_\lambda(x) &= \lambda x(1-x). \end{aligned} \tag{1.4}$$

We shall give a short review of some of the most pertinent results obtained for this map in connection with our discussion of simple one-dimensional maps in Sec. 1.3. More detailed discussions can be found in standard textbooks in nonlinear dynamics [6–9, 16].

Figure 1.8 illustrates how the behavior of the Rössler oscillator can be related to the dynamics of a simple non-invertible, one-dimensional map such as the logistic map. Following Poincaré [36] we start by introducing a plane of intersection with the attractor. Figure 1.8(a) shows the mutual arrangement of the Rössler attractor and the cutting plane, and Fig. 1.8(b) displays the projection of the points of intersection between the attractor and the plane onto the xz-plane. Since the considered motion is deterministic it is clear that the position of a given point of intersection is determined by the position of the previous point. Hence, besides the positions of the various points of intersection we also introduce a numbering of subsequent intersections.

In this way, the Poincaré map in Fig. 1.8(b) provides a one-to-one mapping of the intersection plane onto itself. For the considered attractor this mapping appears to take place along a one-dimensional curve. In reality, the "curve" is not one-dimensional, but it is a fractal structure of infinitely many layers folded on top of one another. Only, the contraction of the system is so strong that the folding is nearly complete. Assuming now that we can treat the *Poincaré section* as a one-dimensional curve, we may use the value of the x-coordinate x_k of point k in this section as a discrete-time variable. When x_k is depicted as a function of x_{k-1} we finally arrive at the *return map* shown in Fig. 1.8(c). The construction of such a return map represents an enormous simplification of the problem. It should be noted, however, that by contrast to real physical systems, the one-dimensional return map is non-invertible. This implies that we cannot use the map in a backward direction to uniquely determine where the system came from.

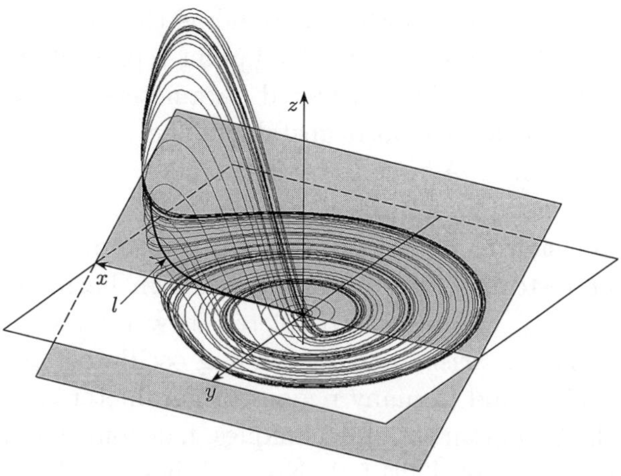

(a)

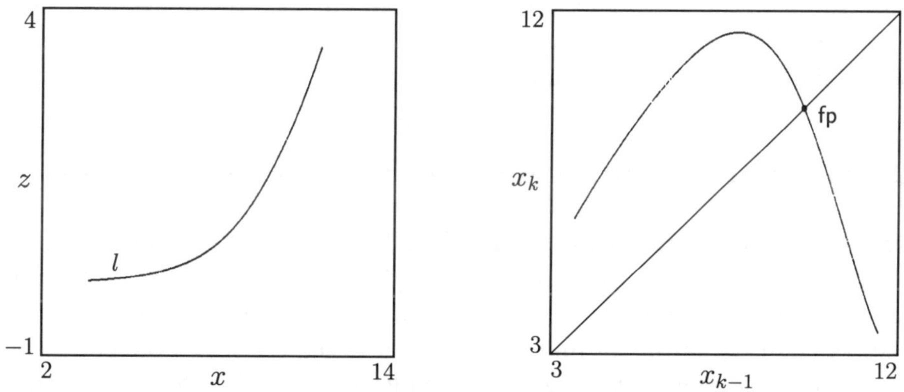

(b) (c)

Figure 1.8. Construction of a return map for the Rössler oscillator. (a) Intersection of the attractor with a Poincaré plane (shaded gray). The set of intersection points is denoted l. (b) Distribution of intersection points (projection of the set l onto xz-plane.). (c) Return map with the fixed point fp.

It is obvious that the return map in Fig. 1.8(c) is qualitatively similar to a parabola, i.e., to the logistic map (1.4). The ascending part of the map represents the expanding dynamics close to the unstable fixed point. When operating in this region, the value of x increases from iteration to iteration. The decreasing branch of the map represents the folding in the system. The *fixed point* (denoted fp in Fig. 1.8(c)) is found as the intersection between the map and the diagonal $x_k = x_{k-1}$. The fixed point represents the period-1 cycle in the Rössler system. For the considered parameters, this cycle is unstable. The fixed point of the map is also unstable since the slope at this point is numerically larger than 1. As one iterates the map, a point close to the fixed point will therefore move away from it.

The main conclusion to be drawn from the above discussion is that as we replace the linear restoring force of the well-known harmonic oscillator with a nonlinear force term, we enter a completely new regime where surprisingly complex phenomena can arise. For the Duffing oscillator, analytical solutions are no longer possible, and in many regions of parameter space the classic approximation methods (perturbation techniques, harmonic balancing, averaging, equivalent linearization, etc. [22]) fail. None of these methods can account for the complexity of chaotic dynamics. This conclusion does not depend on the precise form of the nonlinear term. If the forcing amplitude is high enough and the damping sufficiently low, most nonlinear oscillators can be expected to produce similar complex phenomena.

1.2 Oscillations and Chaos in Engineering Systems

As illustrated for the Rössler system in Fig. 1.6, nonlinear dynamic phenomena are often initiated by a Hopf bifurcation [6–9] where the stationary equilibrium state of a dynamical system becomes unstable and yields to a state of self-sustained oscillations of finite amplitude. Since the first application of radiowaves (G. Marconi, 1901), we have made use of active devices to produce oscillations and waves of a broad spectrum of frequencies, ranging from ultrasound oscillations for crack detection in aircraft wings and scanning of pregnant women over radio- and microwaves to laser, X-ray and synchrotron radiation as used in optical communication and modern materials science.

As we strive for cheaper, faster, and more efficient solutions to a variety of technical problems we will be forced to operate closer and closer to the stability threshold. In certain systems, crossing of the threshold may be acceptable. In other systems, the emergence of self-sustained oscillations could represent a severe threat to safety or to the durability of the structure. In all cases, it will be important to understand precisely how the oscillations arise, and what happens as we cross into the region of instability. This is precisely the area of interest for nonlinear dynamics.

Transonic flutter in aircraft wings represents a typical example of a problem in contemporary engineering [37, 38]. Flutter denotes a characteristic form of self-excited oscillations that can arise through the interaction of an aerodynamic flow and the elastic modes of a mechanical structure such as, for instance, the torsion and bending modes of an aircraft wing. Spontaneous oscillations of this kind typically occur in the transonic range, i.e., when the aircraft operates close to the velocity of sound. They are thought to be related to a separation of the airflow associated with shock waves propagating along the surface of the wing.

When flutter occurs it compromises not only the long term durability of the wing structure, but also the operational safety, energy efficiency, and flight performance. Effective means for the prevention of flutter are, therefore, mandatory in the certification of new flight vehicles, and considerable effort, theoretically as well as experimentally, is devoted to the study of methods for active flutter control [39, 40].

Wind tunnel experiments [39–41] performed at the National Aerospace Laboratory (NAL) in Tokyo over a period of 4–5 years have consistently shown that the onset of flutter occurs abruptly as the dynamic pressure reaches a critical value. Moreover, as the dynamic pressure is subsequently reduced, the oscillations persist all the way down to pressures that are nearly 10 % lower than the onset pressure. Here the flutter oscillations disappear in a new abrupt transition.

With its hysteresis and abrupt transitions the above behavior is typical of a *subcritical Hopf bifurcation*. The equilibrium point of the aeroelastic system loses its stability as a pair of complex conjugated eigenvalues cross the imaginary axis. However, by contrast to the supercritical Hopf bifurcation, observed, for instance, in the Rössler system (Fig. 1.6), the transition is not associated with the emergence of a stable limit cycle. Rather, for aerodynamic pressures below the bifurcation point, an unstable (saddle) cycle coexists with the stable equilibrium point, and the loss of stability is associated with the saddle cycle closing in on this point. At aerodynamic pressures about 10 % below the Hopf bifurcation point, the saddle cycle stabilizes in a saddle-node bifurcation, and the finite amplitude stable limit cycle (the node) continues to exist for increasing aerodynamic pressures.

Experiments were performed with a 1.043 m long model of a high aspect ratio wing from an advanced, energy efficient aircraft. In order to permit ex-

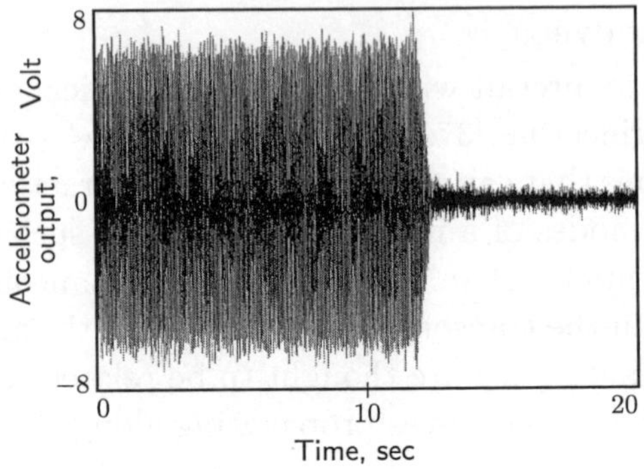

Figure 1.9. Variation of the accelerometer signal in a wind tunnel experiment with a high aspect ratio wing structure. The dynamic pressure is gradually reduced and, at $t \approx 11.5$ s, the large amplitude (flutter) oscillations in the wing disappear abruptly. The experiment was performed with an airspeed around 0.8 M [41].

periments with active flutter control [40], the model was equipped with leading and trailing edge control surfaces that could be activated by means of a pair of servo motors. During the experiments the behavior of the wing was monitored by four accelerometers and seven pairs of strain gauges.

Figure 1.9 shows the results of a particular experiment. Here, the dynamic pressure starts at a value above the nominal flutter pressure, and we follow the signal from one of the accelerometers as the air speed is gradually reduced. In the beginning, the wing structure performs self-sustained oscillations of a nearly constant amplitude. At $t \approx 11.5$ s, however, the accelerometer signal drops abruptly from approximately 6 V to less than 1 V. This indicates that the system has passed the saddle-node bifurcation in which the stable limit cycle disappears as it merges with the unstable limit cycle born in the Hopf bifurcation. The basic frequency of the flutter oscillations is about 22 Hz. However, the aeroelastic system produces strong higher harmonic peaks in the spectrum (up to the order of seven or more), indicating that nonlinear interactions are at work.

Figure 1.10 shows the experimentally determined bifurcation diagram. The nominal flutter transition occurs at a pressure $p_c \approx 27.9$ kPa. As the aeroelastic system is taken beyond this point, large amplitude oscillations arise without warning. This type of bifurcation is sometimes referred to as a *dangerous bifurcation* [42]. The amplitude of the flutter oscillations in the 1 m wing model reaches values of 12–13 mm. A linear stability analysis of the aeroelastic struc-

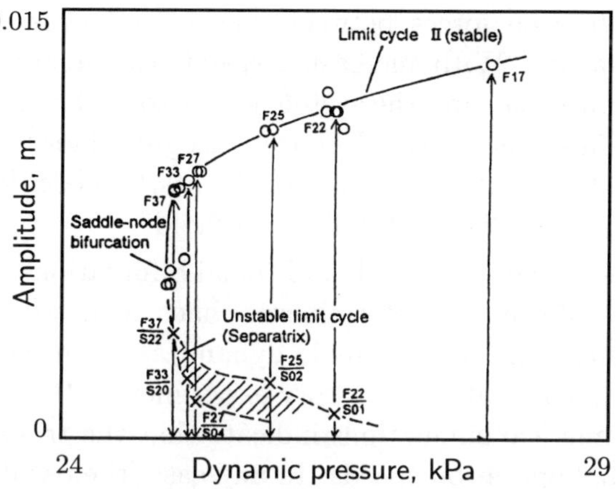

Figure 1.10. Experimentally obtained bifurcation diagram illustrating the subcritical nature of the flutter transition. The diagram was constructed from phase plots obtained with different initial conditions and six different values of the dynamic pressure [41].

ture is quite demanding [39–41]. Such an analysis can, in principle, determine the nominal flutter point. However, the analysis cannot describe the variation of the unstable limit cycle or determine the position of the saddle-node bifurcation. The unstable limit cycle with its stable manifold serves as a *separatrix* (basin boundary) between those initial conditions that lead to the stable equilibrium point, and those that lead to limit cycle oscillations.

As the aeroelastic system approaches the Hopf bifurcation point from below, noise components with frequencies close to the frequencies of the critical modes will start to grow in amplitude, and they will show up in a spectral analysis of the wing dynamics. *Noisy precursors* [43] of this type may be used in the flutter control. It is important to realize, however, that the presence of noise will shift the onset point for the flutter oscillations to lower air speeds than calculated by the linear stability analysis. For the wing model, the amplitude of the unstable limit cycle (the separatrix) is relatively small. This implies that random fluctuations in the air flow can excite high amplitude oscillations across the separatrix well before the nominal flutter point is reached [41].

Oscillations of a somewhat similar nature arise, for instance, in *railroad dynamics* [44–46] and in connection with *metal cutting processes* [47, 48].

The dynamic behavior of railroad vehicles is governed by two types of nonlinearity. There is a kinematic nonlinearity associated with the profiles of the wheels, and the friction forces between wheels and rails contribute another source of nonlinearity. With increasing speed, the steady state equilibrium tends to become unstable, and the system starts to perform self-excited oscillations (hunting). This phenomenon has been recognized for more than 100 years, and it is also well-known that the onset of the *hunting oscillations* often occurs at speeds well below the linear stability threshold.

As the speed is further increased, additional bifurcations take place, and the dynamics may transform into deterministic chaos. Due to stochastic irregularities that often mask the deterministic dynamics, these phenomena are most readily observed in controlled laboratory experiments. However, both such experiments and numerical simulations indicate that the transition to chaos can occur below normal operation speeds. In any case, the oscillations can play an important role for the riding comfort as well as for the wear of the wheels and the associated maintenance costs. Hence, it is important to develop wheel profiles and suspension systems so that one can suppress the oscillations as much as possible. Particularly, the occurrence of asymmetric modes may increase the wear.

Knudsen *et al.* [44] have considered a strongly simplified model consisting of a dicone (i.e., a pair of cones connected by a rod), suspended in the frame of a bogie and rolling on two parallel rails with circular cross sections. The wheelset can perform lateral motions in which the dicone is displaced sidewards. This will cause the cones to roll on different diameters, hence coupling the lateral motion to a rotation of the wheelset in the horizontal plane (yaw).

As the speed V of the wheelset is increased, the system first undergoes a subcritical Hopf bifurcation at $V \approx 15.7$ m/s (≈ 57 km/h). In this bifurcation an unstable limit cycle, that has existed below the bifurcation point, disappears as the equilibrium point turns unstable. Without warning the system then jumps into a state of finite amplitude oscillations. The unstable limit cycle arises together with the stable limit cycle in a saddle-node bifurcation at $V \approx 11.9$ m/s (≈ 43 km/h). Again we encounter the situation where random irregularities (in the rails or the wheels) can excite large amplitude oscillations across a separatrix at speeds well below the point of linear stability.

At $V \approx 12.3$ m/s (≈ 44 km/h) the symmetric limit cycle oscillation undergoes a symmetry-braking pitchfork bifurcation, and for a certain range of V a pair of mutually symmetric modes exist. Immediately above the subcritical Hopf bifurcation for the equilibrium point, the limit cycle oscillation undergoes a new Hopf bifurcation producing quasiperiodic motion on a two-torus [6–9]. This implies that the system exhibits oscillations at two incommensurable frequencies. In the quasiperiodic regime a variety of resonances arise. Here the motions adjust to one another in such a way that there is a rational ratio between the periods. The torus is destructed in a global bifurcation through the collision with an unstable periodic orbit, and for higher values of V the corresponding stable periodic orbit undergoes a global bifurcation.

In a couple of subsequent papers, Knudsen *et al.* [45, 46] have considered more realistic wheel profiles. With the introduction of a flanged profile, the mechanical system is no longer smooth, but abrupt changes in the forces that act on the wheelset occur whenever a flange comes in contact with the rails. The lack of smoothness produces a wide variety of new bifurcation phenomena, including direct transitions to chaos, discontinuous period-doubling bifurcations, and transitions from period-m dynamics to $2m$-band chaotic dynamics. These are precisely the phenomena that are characteristic for the type of systems we are going to discuss in the present book. A short introduction to some of these phenomena will be given in the next sections.

The vibrations (*chatter*) that can arise in *metal cutting* and other machining processes are associated with instabilities of a similar nature. By measuring the spectral distribution and correlation dimension of the vibrations produced by an orthogonal cutting machine, Grabec [47] demonstrated how the system passes through a sequence of bifurcations as the chip width or the cutting speed is increased. At relatively low intensity cutting, quasiperiodic oscillations were observed. At higher cutting speeds, the oscillations became chaotic to finally turn into strongly anharmonic oscillations at very high speeds. Random irregularities arising, for instance, from surface roughness or from the finite grain size of the workpiece play a significant role in exciting the oscillations. But the instability itself is controlled by deterministic nonlinear mechanisms.

Grabec [47, 48] also developed a model of the cutting process, accounting for elastic and plastic deformations as well as for friction forces normal and parallel to the cutting surface. Other nonlinear mechanisms that might be at work include fracture formation, thermal softening, and work hardening of the material. A detailed understanding of these mechanisms is clearly important to combine improved surface smoothness and dimensional accuracy with prolonged tool lifetimes and higher machine productivity. The development of new intelligent manufacturing machines will provide an even stronger motivation for theoretical and experimental work in this area.

The properties of the cutting force and of the forces produced by dry friction in solids are similar in several aspects, and the appearance of complex dynamic phenomena and chaos is generic to problems involving dry friction and stick-slip motion [49, 50]. A review of some of these problems may be found in the recent books by Blazejczyk-Okolewska *et al.* [51] and by Wiercigroch and de Kraker [52]. Instabilities and nonlinear dynamic phenomena also arise in connection with milling, sheet spinning, extruding, ultrasonic machining, water-jet cutting, laser welding, and electro-discharge machine processing [53]. Methods from nonlinear dynamics are obviously essential for achieving a deeper understanding and a better control of these processes. Nonlinear dynamics allows us to examine the initial instability in more detail and to follow what happens in the subsequent instabilities as the system is forced to operate at even higher speeds. The interesting result is that many of the phenomena we encounter are generic to large classes of systems. For this reason, information obtained from analysis of relatively simple models can be useful in more complicated practical situations. The detailed mathematical argumentation for this assertion is provided by the center manifold theory [16].

Some of the problems that can arise in mechanical systems with instabilities or large amplitude forcing have been dealt with from a more theoretical point of view in a series of papers on impact oscillators [54–57]. Such oscillators display a particular type of bifurcation, called *grazing bifurcation*, that occurs at the moment when the amplitude becomes large enough for the oscillator to touch a solid boundary. Technical examples of such phenomena include, for instance, a Jeffcott rotor with bearing clearance [58] and a pendulum type vibration absorber [59]. A somewhat related problem is the deterministic description of die tossing [60] or of the trajectories in a pin-ball machine [61].

A thermostatically controlled hot-water radiator represents an example of a *technical control system* with negative feedback and *hysteresis*. The behavior of such a system depends on the slope of the regulator characteristics. As long as this slope is small enough, the system has a stable equilibrium point. However, since the regulator operates with proportional control, the attained room temperature will vary with the net heat losses to the surroundings. To minimize this variation and to improve the function of the valve in general, new designs are being developed with higher regulator slope, less hysteresis, and better closing characteristics. By virtue of the delay associated with the finite heat capacities of both the radiator and the room itself, these designs tend to produce instability in the form of self-sustained opening and closing cycles with a typical period of 1–3 h. Such behavior has repeatedly been observed in full-scale tests of valves from a number of leading manufacturers.

Following Togeby *et al.* [62] we may consider a system of two such oscillating radiators heating separate air volumes, but with a certain degree of heat exchange. The two air volumes can be viewed as parts of a single room or as two separate rooms, depending on the value of the coupling parameter h_{w}. This parameter measures the heat flow between the two air volumes per degree of temperature difference and per square meter of common wall area. For simplicity, each air volume is considered to be a single compartment with complete temperature equilibration. For both subsystems the model includes a radiator, a thermostatic valve, a pipe that supplies hot water to the radiator, and a representation of the building confinement. Ventilation and outdoor temperature are taken to be constant, and the observed behavior is generated entirely by the deterministic equations of motion of the system.

To account for phase shifts and thermal delays, the supply pipes and the radiators are each divided into three segments. Similarly, each regulator is divided into two parts which represent, respectively, the sensor and the body of the valve. Altogether, the model contains 20 state variables, each representing the heat content of a particular compartment or space.

The static and dynamic properties of the valves are characterized by the throughflow of hot water in the fully open position, the gain (from sensor temperature to water flow) $-\mathrm{d}V/\mathrm{d}T_{\mathrm{s}}$, the hysteresis parameter ΔT_{hyst}, and the closing ability. Hysteresis is caused by static friction and mechanical slip in the valve construction. The sensor is characterized by its heat capacity, the effective sensor surface area and the heat-exchange coefficient h_{s}. This latter parameter is a measure of the thermal contact between the sensor and the surrounding air.

Parameter values corresponding to typical household installations were assumed throughout. The regulator gain is $-\mathrm{d}V/\mathrm{d}T_{\mathrm{s}} = 400\text{–}500\ \mathrm{Lh^{-1}K^{-1}}$, the hysteresis parameter is $\Delta T_{\mathrm{hyst}} = 0.1$ K and the sensor mass is 15 g. The two radiators differ particularly with respect to their weights and the pipe-to-air sensitivity ratios which are, respectively, 15 kg and $1:100$ for radiator 1, and 65 kg and $1:9$ for radiator 2. The set point for radiator 1 is 18.9 °C while for radiator 2 this temperature is 19.3 °C. Let us consider what happens as the pa-

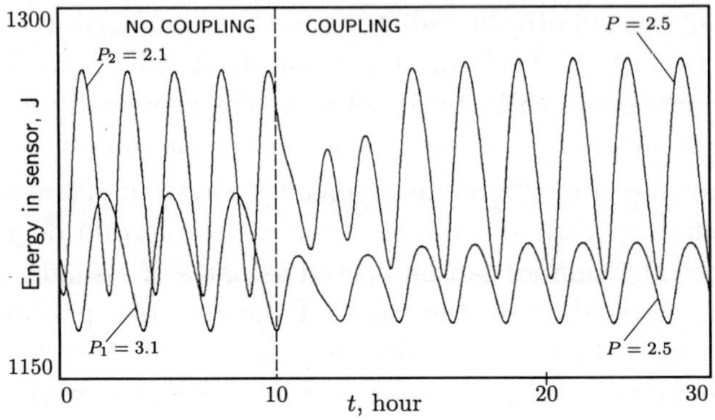

Figure 1.11. Graphs of the energy content of the thermostatic sensors of a pair of hot-water radiators. P_1 and P_2 are the periods (in hours) for the uncoupled radiators, and P is the period of the synchronized oscillations. $h_{s1} = 10.0\ \mathrm{Wm^{-2}K^{-1}}$ and $h_{s2} = 18.0\ \mathrm{Wm^{-2}K^{-1}}$.

rameters h_s for radiator 2 (h_{s2}) and h_w are changed. These parameters control the uncoupled oscillation period for radiator 2 and the strength of the coupling between two subsystems, respectively. The parameter $h_{s1} = 10.0\ \mathrm{Wm^{-2}K^{-1}}$ is kept constant.

The first interesting phenomenon to report is mode-locking [63–67]. This phenomenon is illustrated in Fig. 1.11 where the energy content in each of the two sensors is plotted as a function of time t. Until $t = 10$ h, the subsystems are assumed to be uncoupled ($h_w = 0$). Owing to their different parameters, the radiators oscillate with different periods: $P_1 = 3.1$ h and $P_2 = 2.1$ h. At $t = 10$ h, h_w is abruptly increased to $25\ \mathrm{Wm^{-2}K^{-1}}$. After approximately 5 h of transient behavior, the two radiators are seen to entrain with each other at an intermediate period of $P = 2.5$ h.

Figure 1.12(a) shows a phase plot of the stationary $1:1$ mode-locked behavior. Here, we have plotted simultaneous values of the energy contents of sensors 1 and 2. The motion is seen to be completely periodic and this $1:1$ entrainment is maintained over an interval of h_{s2} ranging up to $h_{s2} = 19.6\ \mathrm{Wm^{-2}K^{-1}}$. If h_{s2} is increased beyond this limit, qualitative shifts in the behavior of the system take place.

Mode-locking occurs in finite intervals for all rational ratios of the two periods. As an example, Fig. 1.12(c) shows the phase plot of a $3:4$ oscillation obtained for $h_{s2} = 22.0\ \mathrm{Wm^{-2}K^{-1}}$. With this value of h_{s2}, radiator 1 performs precisely three cycles each time radiator 2 performs four cycles. The $3:4$ mode-

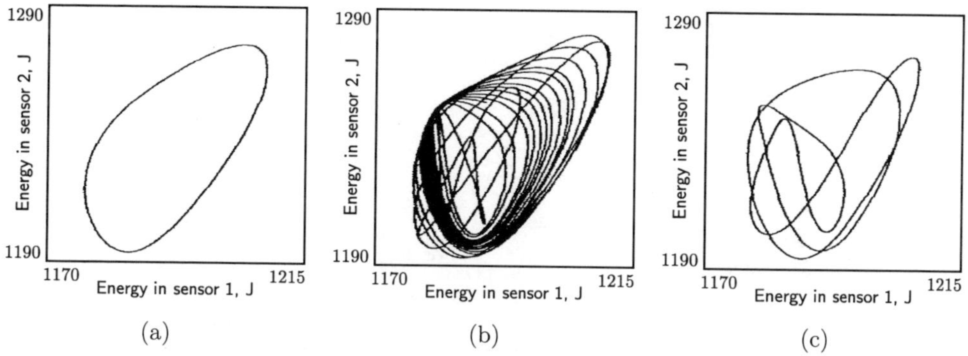

Figure 1.12. Phase plots of the energy content of sensor 2 versus the energy content of sensor 1: (a) 1 : 1 mode-locked behavior for $h_{s2} = 18.0\ \mathrm{Wm^{-2}K^{-1}}$, (b) quasiperiodic behavior for $h_{s2} = 19.7\ \mathrm{Wm^{-2}K^{-1}}$, and (c) 3 : 4 mode-locked behavior for $h_{s2} = 22.0\ \mathrm{Wm^{-2}K^{-1}}$. Quasiperiodic behavior is distinguished from deterministic chaos by not showing sensitivity to the initial conditions.

locked solution exists over an interval of h_{s2} ranging from 21.8 to 22.0 $\mathrm{Wm^{-2}K^{-1}}$. Between this interval and the 1 : 1 mode-locking interval, we find intervals with 5 : 6, 11 : 14 and 13 : 17 mode-locked behavior. Interspersed with these intervals again, there are solutions for which the two radiators oscillate with incommensurate periods. The stationary behavior is here quasiperiodic, and the winding attractor never closes to itself. This type of behavior is illustrated in Fig. 1.12(b) for $h_{s2} = 19.7\ \mathrm{Wm^{-2}K^{-1}}$.

The interval in which the particular mode-locking occurs depends on the ratio of the involved periods such that entrainment between modes with simple period ratios and with period ratios near 1 tends to be more pronounced than entrainment between modes with more complex ratios [63–67]. A more complete illustration of the entrainment phenomenon is obtained by plotting the observed mode-locking ratio as a function of h_{s2}. Figure 1.13 shows an example of such a construction for $h_w = 25\ \mathrm{Wm^{-2}K^{-1}}$. The figure shows a "devil's staircase" of mode-locked solutions. The *devil's staircase* is a fractal structure since, under repeated magnification, new mode-locked intervals continue to emerge.

In the intervals of some of the steps, mode-converting and period-doubling bifurcations can be observed. Thus, for instance, the 2 : 3 step contains intervals in which 4 : 6 and 8 : 12 solutions are found. Similarly, the 3 : 5 step contains intervals with 6 : 10, 12 : 20 and 9 : 15 solutions.

The interval in which a particular mode-locking takes place also depends on

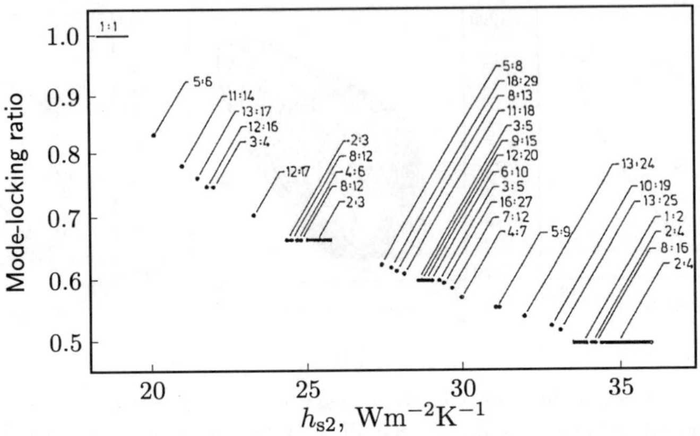

Figure 1.13. Devil's staircase obtained by plotting the period ratio of the two oscillating radiators as a function of h_{s2}. Between the $2:3$ and $5:8$ mode-locked solutions, narrow intervals of entrainment with period ratios of $7:11$, $9:14$, $12:19$, etc., can be expected.

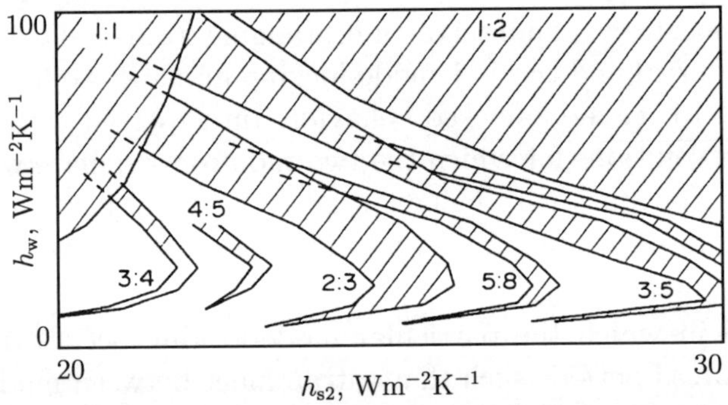

Figure 1.14. Phase diagram showing the variation of some of the principal mode-locked intervals with the coupling parameter h_w. The regions of mode-locked behavior are referred to as Arnol'd tongues.

the strength of the coupling between the two radiators. An overview of this variation is provided by the phase diagram in Fig. 1.14. Here, the contours of some of the principal mode-locking regions are plotted as functions of h_{s2} and h_w. Without coupling, i.e. for $h_w = 0$, there can, of course, be no mode-locking at all. As h_w is increased, wider and wider intervals of mode-locked behavior develop, and above a certain coupling strength the intervals start to overlap. Several different periodic attractors then exist simultaneously, and the stationary motion of the system depends on the initial conditions. In the present example, the critical coupling strength appears to be approximately $h_w = 15$ Wm^{-2}K^{-1}. With further increase of h_w, the simple mode-locked solutions ($1:1$ and $1:2$) gradually take dominance at the expense of other periodic attractors.

Above the critical coupling, quasiperiodic behavior ceases to exist and the system either locks itself into a periodic motion or performs chaotic oscillations [21, 62]. The chaotic behavior is found to develop through a cascade of period-doubling bifurcations, through *frustration* between two nearly periodic motions, or through destabilization of the two-dimensional torus associated with the quasiperiodic motion. Frustration is a special form of *intermittency* [6–9] in which the coupled oscillator system moves in an apparently regular, frequency-locked dynamics for a while, enters a turbulent phase of highly irregular dynamics, and settles for some time in another region of nearly regular, frequency-locked dynamics. However, a Poincaré section of the phase space trajectory shows that none of the nearly regular motions exists as periodic orbits, neither stable nor unstable. During the phase of nearly regular dynamics, the system "chases the ghost" of a periodic orbit that has disappeared in a saddle-node bifurcation at a nearby parameter value. The frustrated dynamics consists in a continuous, and more or less random, transition between the two ghost states.

The temporal variation of the energy content in sensor 1 corresponding to a chaotic behavior is plotted in Fig. 1.15(a). The sensitivity of the chaotic trajectory to the initial conditions is illustrated in Fig. 1.15(b). Here, we have plotted the distance between two chaotic trajectories starting with practically identical initial conditions; in one case the initial conditions are specified to 11 significant digits, and in the other to 10. Computations are performed to 15 significant digits. The distance between the two solutions (as measured in terms of its initial value) is seen to grow almost exponentially over 5–6 orders of magnitude. If the simulation is extended, this growth continues until the distance becomes comparable to the amplitudes (10^{10}) of the oscillations. From then on, the different initial conditions cause the system to follow completely different trajectories.

On-off regulation is applied in a multitude of technical control systems, including the regulation of heat pumps, refrigeration systems and oil burners. Interaction between two such systems (or periodic forcing of a single system) can produce a similar variety of complex nonlinear phenomena as we have de-

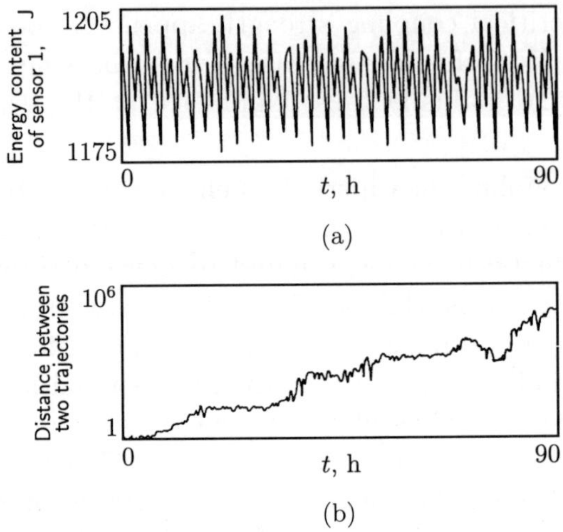

Figure 1.15. (a) Temporal variation of the energy content of sensor 1. The trajectory is here chaotic. (b) Distance between two trajectories started at almost identical initial points. The near exponential growth of this distance over many orders of magnitude represents sensitivity to the initial conditions.

scribed for the coupled radiator systems, including mode-locking, quasiperiodic behavior, period-doubling, mode-conversion, hysteresis due to coexisting periodic solutions, and deterministic chaos. Traditional analysis has tended to concentrate on the individual subsystem and to overlook the behavioral complexity that can arise through interaction. In practice, decoupling is often performed by means of various types of buffers. However, this approach introduces additional losses, and these sources of ineffictivity can be minimized if the behavioral complexity resulting from a closer coupling can be managed.

1.3 Simple One-Dimensional Maps

As noted in Sec. 1.1, the logistic map (1.4) is the basic model for discussing the period-doubling transition to chaos. We shall not consider this map in detail here, but refer the reader to standard textbooks in nonlinear dynamics [6–9, 16]. Let us recall, however, that the metric and topological properties of the logistic map have been studied in significant detail over the last two or three decades [68–70]. It is known, for instance, that $f_\lambda(x)$ for $\lambda \in [1, 4]$ has a unique attracting state that can be periodic (a cycle of points) or chaotic

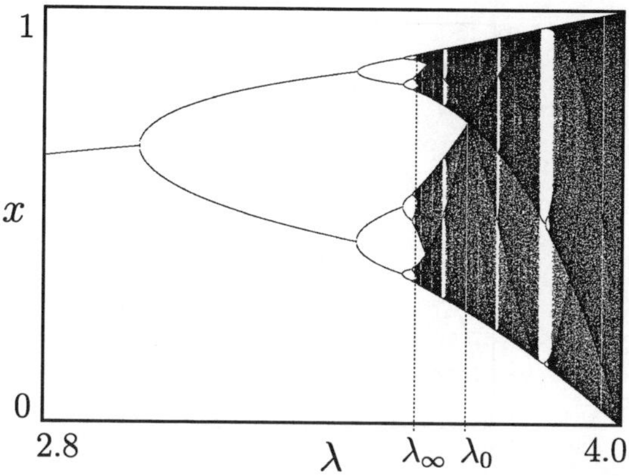

Figure 1.16. Bifurcation diagram for the logistic map (1.4). λ_∞ denotes the accumulation point for the period-doubling cascade, and λ_0 is a homoclinic bifurcation point where two chaotic bands merge into a single chaotic attractor. Note, the periodic windows in the chaotic regime. The scaling factor $\delta_F \approx 4.6692016\ldots$ describes the asymptotic rate of convergence for the period-doubling cascade in parameter space, and $\alpha_F \approx 2.50290787\ldots$ describes the rate at which the splitting in state space decreases from one period-doubling to the next.

(a cycle of chaotic intervals). The set of parameters λ for which $f_\lambda(x)$ has an attracting cycle is open and everywhere dense, and the set of parameters that produce a chaotic solution has a Cantor-like structure with a measure $\mu > 0$. In the considered parameter range, the basin of attraction is the unit interval $x \in [0, 1]$.

Feigenbaum [32, 33] applied ideas (renormalization group theory) from statistical mechanics to prove the universal properties of unimodal, one-dimensional maps with quadratic maxima. This shed new light on the significance of the scaling factors α_F and δ_F and initiated a broad interest in the study of nonlinear dynamic phenomena. (A map is unimodal when it has one extremum, and the extremum is quadratic if the second derivative of the map is finite). Other important contributions were made by Sharkovsky [34], Coullet and Tresser [71], Metropolis *et al.* [35], Singer [72], and Landford [73].

Figure 1.16 shows part of the bifurcation diagram for the logistic map with its well-known period-doubling cascade, the transition to chaos, and the characteristic periodic windows. As mentioned in Sec. 1.1, each periodic window is initiated by a saddle-node bifurcation, and the window terminates in a new period-doubling transition to chaos. The recurrent (or fractal) structure of the bifurcation diagram itself may be described in terms of the so-called box-within-

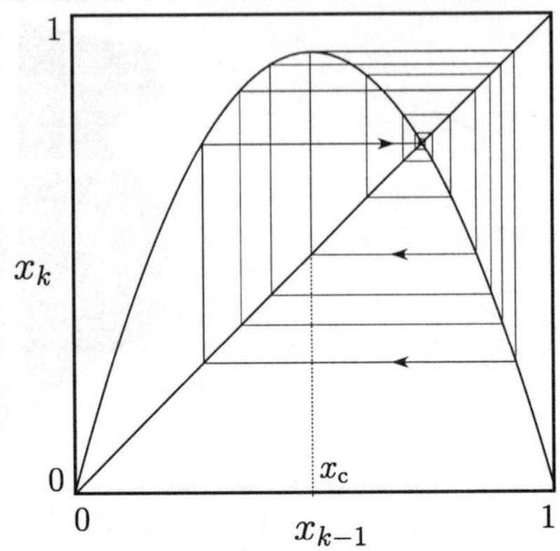

Figure 1.17. Construction of a homoclinic orbit to the unstable fixed point. A homoclinic orbit is a trajectory that starts at the fixed point and after a (finite or infinite) number of iterations returns to the same point. $\lambda \approx 3.67857$.

a-box structure proposed by Mira [74].

As the parameter λ is increased beyond the accumulation point $\lambda_\infty \approx 3.569$ for the first period-doubling cascade, one observes a reverse cascade of band-merging bifurcations. At $\lambda \approx 3.67857$, for instance, two chaotic bands merge into a single band as they collide with the unstable fixed point $x_1 = 1 - 1/\lambda$. As illustrated in Fig. 1.17 this happens when the *critical point* $x_c = 1/2$ is mapped into the unstable fixed point in three iterations of $f_\lambda(x)$. The critical point x_c here denotes the point at which the map has vanishing slope. (We shall also use the term "critical point" for specific parameter values where a bifurcation takes place). Starting from (the neighborhood of) the unstable equilibrium point one can construct a trajectory that (after a large number of iterations) maps into the critical point. Hence, the band merging takes place in a homoclinic bifurcation in which it becomes possible for a trajectory that leaves the unstable fixed point to return to this point after a (finite or infinite) number of iterations. The homoclinic bifurcation is said to be a global bifurcation because it depends on the conditions in a larger part of phase space. Such bifurcations cannot be predicted from a linear stability analysis.

Let us return for a moment to the bifurcation diagram in Fig. 1.16. This diagram displays an interesting structure of particularly dark curves. These curves obviously represent values of x that are visited particularly often during the iteration process. These values are the iterates of the critical point x_c [75].

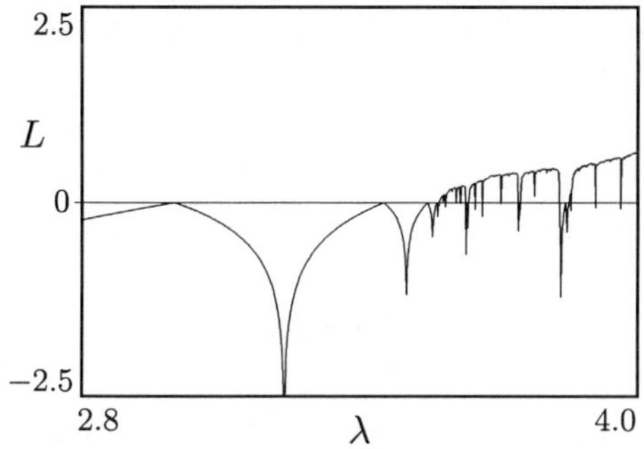

Figure 1.18. Variation of the Lyapunov exponent for the logistic map. Note the characteristic variation through the period-doubling cascades. A positive value of L is taken as indication of chaotic dynamics. From a strictly mathematical point of view, the presence of a dense set of unstable periodic orbits obviously requires some reservation with respect to the value of the numerical calculations.

Because of the vanishing slope of the map, trajectories starting from a region around x_c will all be mapped into a very narrow interval at (or just below) the maximum value $f(x_c) = \lambda/2$. Hence, the visiting frequency (or the *invariant density*) is particularly large along the upper edge of the bifurcation diagram. After this focusing process, the trajectories tend to remain relatively close to one another through several subsequent iterations. Hence, there is a pronounced lower boundary to the bifurcation diagram at $f^{(2)}(1/2) = \lambda^2(2-\lambda)/8$, and we can easily follow the curves representing $f^{(3)}(1/2)$, $f^{(4)}(1/2)$, etc. At the homoclinic bifurcation point λ_0, $f^{(3)}(1/2) = f^{(4)}(1/2) = f^{(5)}(1/2) = \cdots \approx 0.727766$, indicating that a trajectory starting from the critical point will hit the unstable fixed point after three iterations.

To characterize the stability properties of the stationary solutions one can introduce the *Lyapunov exponent* [6–9, 23, 24]

$$L = \lim_{N\to\infty} \frac{1}{N} \sum_{k=1}^{N} \ln |f'_\lambda(x_{k-1})|. \tag{1.5}$$

As previously noted, this exponent measures the average rate of divergence of two nearby trajectories. For parameter values at which the attracting state is periodic $L < 0$, while for (typical) itineraries on a chaotic set $L > 0$. Figure 1.18 shows the variation of the Lyapunov exponent for the logistic map in the interval $2.8 < \lambda < 4.0$. This is the same interval as considered in Fig. 1.16. The variation of the Lyapunov exponent clearly reveals both the period-doubling cascades and the presence of periodic window in the chaotic regime. In each bifurcation point, $L = 0$.

The *iterates of the critical point* may also be used to locate the periodic windows [75]. The stability of an m-periodic cycle is determined by the product of the slopes of the map in the points of the cycle, i.e., by the eigenvalue

$$\rho = \prod_{k=1}^{m} f'_\lambda(x_{k-1}).$$

If one of the slopes vanishes, i.e., if the cycle passes through the critical point $x_c = 1/2$, the eigenvalue ρ will also vanish. In this case the cycle is said to be *superstable*, because the transient approach to the cycle will be faster than exponential. Around the parameter value at which $\rho = 0$ there will be an interval in which $|\rho| < 1$. This is the interval where the period-m cycle is stable. Hence, to find windows of stable period-m dynamics we only have to follow the iterate $f_\lambda^{(m)}(1/2)$ under variation of λ until the curve crosses $x = 1/2$. At the point λ_m where

$$f_\lambda^{(m)}\left(\frac{1}{2}\right) = \frac{1}{2}$$

an m-cycle exists that passes through the critical point, and around this value for λ the m-cycle is stable. Figure 1.19 shows such a construction in order to find the period-3 and period-4 windows of the logistic map. The construction can easily be extended to higher values of m, and it can be used to determine the number of windows of different periodicities and their relative positions. The construction is equivalent in many ways to the analysis in terms of symbolic dynamics presented by Metropolis *et al.* [35].

Discussions of *resonances* and *synchronization phenomena* in systems of two interacting oscillators are often based on the so-called *sine-circle map* [63–67]

$$\vartheta_k = f(\vartheta_{k-1}) = \vartheta_{k-1} + \Omega - \frac{K}{2\pi} \sin 2\pi\vartheta_{k-1} \mod 1. \tag{1.6}$$

Here ϑ_{k-1} measures the phase of one of the oscillators each time the other oscillator has completed one full cycle. Ω measures the mean phase advance per cycle, and $K > 0$ is the coupling constant. Due to the presence of the coupling term, the phase advance $\vartheta_k - \vartheta_{k-1}$ in a particular iteration depends on the initial phase ϑ_{k-1}. In general terms this implies that there are phase relations between the two oscillators that the system tends to avoid or to pass relatively quickly and other phase relations at which the system tries to stay.

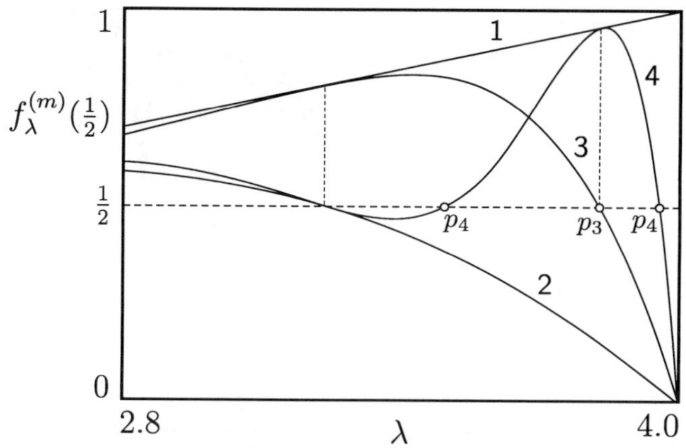

Figure 1.19. Construction to locate the periodic windows by means of the iterates of the critical point $f_\lambda^{(m)}(1/2)$. Each curve representing the variation of $f_\lambda^{(m)}(1/2)$ $(m = 1, 2, 3, 4)$ with λ is marked with the appropriate value of m; p_3 represents a superstable period-3 orbit, and p_4 gives the positions of two superstable period-4 cycles.

If the coupling $K > 2\pi|\Omega|$, the map $f(\vartheta_{k-1})$ has two points of intersection with the diagonal $\vartheta_k = \vartheta_{k-1}$. One point, in which the slope $f'(\vartheta_{k-1})$ is numerically less than 1, is a stable fixed point. This point represents the phase of the $1:1$ (or $0:1$) mode-locked dynamics. The other fixed point is a saddle point that has emerged together with the stable fixed point in the saddle-node bifurcation that delineates the edge of the *resonance zone.*

As noted in connection with our discussion of the coupled radiators in Sec. 1.2, nonlinear oscillators tend to synchronize with one another such that one oscillator completes precisely p cycles each time the other oscillator completes q cycles, with p and q being integers. If the regions of stable synchronized dynamics are determined for the sine-circle map one obtains a so-called *Arnol'd tongue diagram* [6–9]. Figure 1.20 is a simplified version of such a diagram. Here, the numbers p/q associated with the various tongues represent the mode-locking ratio, i.e., the ratio of the number of cycles completed by each of the two oscillators within a given time interval. In the parameter region corresponding to the 2/5 resonance tongue, for instance, one of the oscillators complete 5 cycles each time the other completes 2 cycles. For high values of the coupling parameter $(K > 1)$, the tongues start to overlap, and their internal structure becomes quite complicated. We shall return to this problem in Chapter 2 in connection with a discussion of the so-called *swallow-tail structure.*

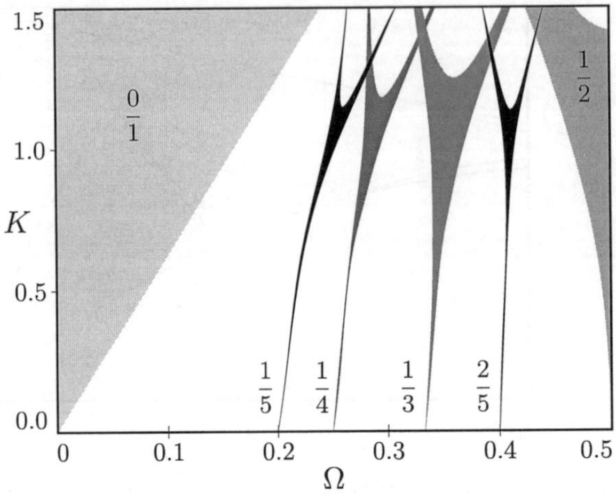

Figure 1.20. Simplified Arnol'd tongue diagram for the sine-circle map. The diagram shows the main regions of mode-locked behavior for two coupled oscillators. A more detailed diagram is shown in Fig. 2.29.

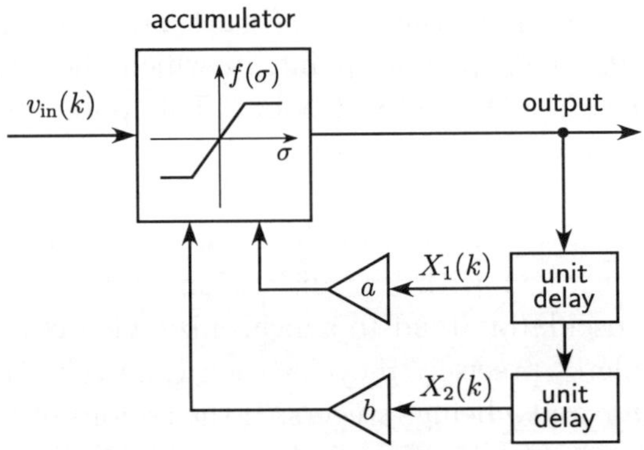

Figure 1.21. Block diagram for a 2nd order digital filter with saturation arithmetic.

Piecewise linear maps arise, for instance, in connection with the description of signal processing in digital electronic systems [76, 77]. Figure 1.21 shows the block diagram of a second order digital filter, known as a $\Sigma\Delta$-modulator. In conjunction with a quantizer that produces short wordlength (e.g., one bit) codes, this type of filter may be used as a noise shaper at the input to the digital/analog converter in disk recorders and portable phones. The idea is to reduce the requirements to the D/A-converter by cutting down the wordlength of the digital code while increasing the sampling frequency. The quantizing noise introduced in this way is spectrally shaped by the filter in such a manner that the major noise components are shifted away from the signal band.

As illustrated by the block diagram, delayed and amplified versions, $aX_1(k)$ and $bX_2(k)$, of the output signal are fed back to the input, and a nonlinear function $f(\sigma)$ limits the signal in order to maintain it within the dynamical range of the filter. Mathematically, this can be expressed in terms of the two-dimensional discrete mapping

$$\begin{aligned} X_1(k+1) &= f\{v_{\text{in}}(k) + aX_1(k) + bX_2(k)\}; \\ X_2(k+1) &= X_1(k), \end{aligned} \tag{1.7}$$

where k denotes the discrete time variable.

Several different choices are possible for $f(\sigma)$. The function

$$f(\sigma) = \begin{cases} -1 & \sigma < -1; \\ \sigma & -1 \leq \sigma \leq 1; \\ 1 & \sigma > 1 \end{cases}$$

that we have sketched in the block diagram represents so-called saturation arithmetic. An alternative choice, referred to as two-component arithmetic, uses a modulus-2 constraint to maintain $f(\sigma)$ within the dynamical range from -1 to $+1$.

Investigation of the stability of the fixed point $(X_1, X_2) = (0, 0)$ for the autonomous system ($v_{\text{in}}(k) = 0$) shows that the *sigma-delta modulator* can generate a variety of nonlinear dynamic phenomena, depending on the choice of the gain factors in the feedback. For $a + b = 1$, the fixed point undergoes a pitchfork bifurcation producing two mutually symmetric fixed points $(1, 1)$ and $(-1, -1)$. For $a - b = -1$, the original fixed point undergoes a period-doubling bifurcation leading to the formation of a stable period-2 cycle $(1, -1) \rightarrow (-1, 1)$. Finally, for $b = -1$, the fixed point undergoes a Hopf bifurcation producing quasiperiodicity and various forms of mode-locked dynamics. In analogy with the synchronization phenomena for the coupled radiator system mentioned above, this may be discussed in terms of a devil's staircase of synchronization intervals if the rotation in the phase plane per iteration of the map is plotted as a function of a. With two-component arithmetic, the filter can display chaotic dynamics [76].

Piecewise linear maps also arise in the description of *human decision-making behavior* [78, 79]. Chapter 7 provides a more detailed analysis of such a problem. Here we shall discuss some of the bifurcations that can occur in a discrete time model of a production-distribution chain, the so-called *beer model.* This model was originally developed at the Sloan School of Management in order to illustrate some of the mechanisms involved in the generation of macroeconomic business cycles [80, 81]. At the same time, the model can be used to examine how different order policies influence the stability of a distribution chain. The piecewise linear mapping arises from logical considerations. Some decisions may be optimal under certain conditions. In other situations, qualitatively different decisions are better. Much theory in management science is formulated around such *situations of choice.*

An interesting aspect of piecewise linear maps is that they display bifurcation phenomena that differ essentially from the bifurcations that one can observe in smooth systems. In smooth maps, such as the logistic map (1.4), the form and character of the bifurcations are controlled by the curvature of the map. A period-doubling takes place when the slope of the map $f'(x)$ evaluated at the fixed point becomes equal to -1. However, the second and third order derivatives of the map $f''(x)$ and $f'''(x)$ determine what happens in the neighborhood of the bifurcation point. If the *Schwarzian derivative* [82]

$$Sf(x) \equiv \frac{f'''(x)}{f'(x)} - \frac{3}{2}\left(\frac{f''(x)}{f'(x)}\right)^2 \tag{1.8}$$

is negative, the period-doubling will be supercritical, i.e., a stable period-2 cycle will arise in the bifurcation. In the opposite case where $Sf(x) > 0$, the bifurcation is subcritical, and an unstable period-2 cycle that exists before the transition disappears in the bifurcation.

For the logistic map $f''(x) = -2\lambda$ and $f'''(x) = 0$. $Sf(x)$ is therefore negative, and the first period-doubling bifurcation is supercritical (as are all subsequent period-doubling bifurcations). Moreover, the curvature of the logistic map restrains the emerging period-2 cycle and ensures that its amplitude (at least in the beginning) grows as $\sqrt{\lambda - 3}$, with $\lambda = 3$ being the bifurcation point. For piecewise linear maps, on the other hand, the second and higher order derivatives vanish and, if it stabilizes at all, the period-2 cycle that emerges in a period-doubling bifurcation will jump abruptly to a finite amplitude. This is illustrated in Fig. 1.22. Here we have shown how a period-doubling bifurcation appears in a so-called skew tent map as the slope of the descending

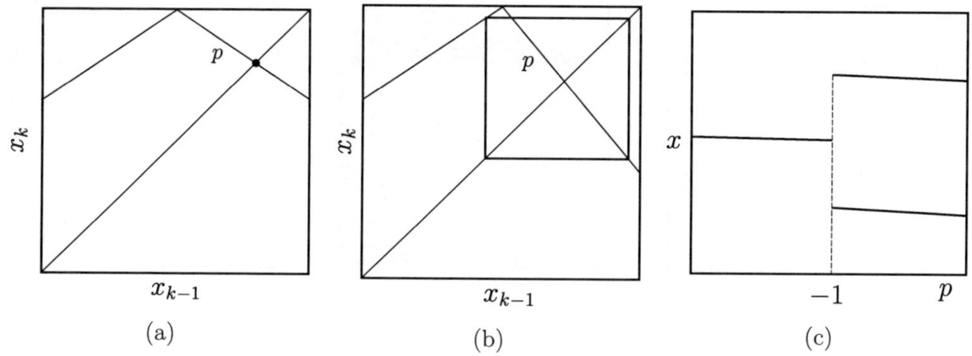

Figure 1.22. Period-doubling in the skew tent map. p is the slope of the descending line segment. (a) $p > -1$, (b) $p < -1$, and (c) sketch of the corresponding bifurcation diagram. Note the abrupt formation of a period-2 cycle of finite amplitude.

leg passes through -1. We may recall how the pitchfork and period-doubling bifurcations discussed in connection with the sigma-delta modulator displayed similar abrupt jumps.

For the logistic map, the transition to chaos takes place through a cascade of period-doubling bifurcations starting with the first period-doubling at $\lambda = 3$ and accumulating at the Feigenbaum point $\lambda_\infty \approx 3.571$. As illustrated in Fig. 1.18, the presence of this period-doubling cascade is clearly revealed in the variation of the Lyapunov exponent with the nonlinearity parameter λ. If the logistic map is replaced by the tent map [83]

$$x_k = T(x_{k-1}) = \begin{cases} rx_{k-1}, & 0 \le x_{k-1} < \frac{1}{2}; \\ r(1 - x_{k-1}), & \frac{1}{2} \le x_{k-1} < 1, \end{cases} \tag{1.9}$$

with $0 < r < 2$, the nature of the transition becomes very different. Since the slope of $T(x)$ is numerically the same in all points, calculation of the Lyapunov exponent immediately gives

$$L = \lim_{N \to \infty} \frac{1}{N} \sum_{k=1}^{N} \ln |T'(x_{k-1})| = \ln r.$$

L varies monotonously and smoothly from negative values for $r < 1$ to positive values for $r > 1$. Hence, for the tent map we observe a direct transition from the stable fixed point $(x_{k-1}, x_k) = (0, 0)$ to a chaotic attractor at $r = 1$ where the Lyapunov exponent changes sign.

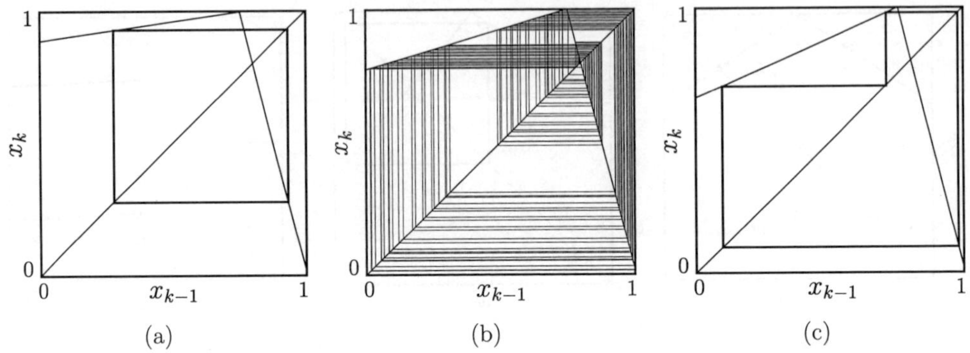

Figure 1.23. Stationary solutions to the skew tent map for different values of the slope of the ascending line segment. (a) period-2 cycle for $l = 0.15$, (b) chaotic attractor for $l = 0.30$, and (c) period-3 cycle for $l = 0.45$. $p = -4$.

Let us pursue some of the peculiarities of piecewise linear maps a little further by considering the skew tent map [84, 85]

$$x_k = T_{l,p}(x_{k-1}) = \begin{cases} l x_{k-1} + c, & 0 \le x_{k-1} \le d; \\ p\,(x_{k-1} - 1), & d \le x_{k-1} \le 1 \end{cases} \tag{1.10}$$

with $l > 0$ and $p < 0$ being the slopes of the two line segments. To ensure that the map is continuous and that the top point falls in $(x_{k-1}, x_k) = (d, 1)$, the following conditions

$$c = 1 - l\left(1 + \frac{1}{p}\right) \quad \text{and} \quad d = 1 + \frac{1}{p}$$

are imposed on the remaining parameters.

In Fig. 1.23 we have sketched the form of different stationary solutions for the map $T_{l,p}$ when the slope l of the ascending line segment is increased while the slope of the descending line segment is maintained at $p = -4$. In Fig. 1.23(a) $l = 0.15$, and the skew tent map exhibits a stable period-2 cycle. The chaotic solution shown in Fig. 1.23(b) is obtained for $l = 0.30$, and the stable period-3 cycle of Fig. 1.23(c) is observed for $l = 0.45$. It is interesting to note how the chaotic attractor for $l = 0.30$ avoids the region around the unstable period-2 solution.

In order to illustrate the transitions between the various stationary solutions in more detail, Fig. 1.24 displays a one-dimensional bifurcation diagram obtained by gradually increasing l from 0.15 to 0.65. Again $p = -4$. The bifurcation diagram shows how the stable period-2 cycle that exists for $l = 0.15$

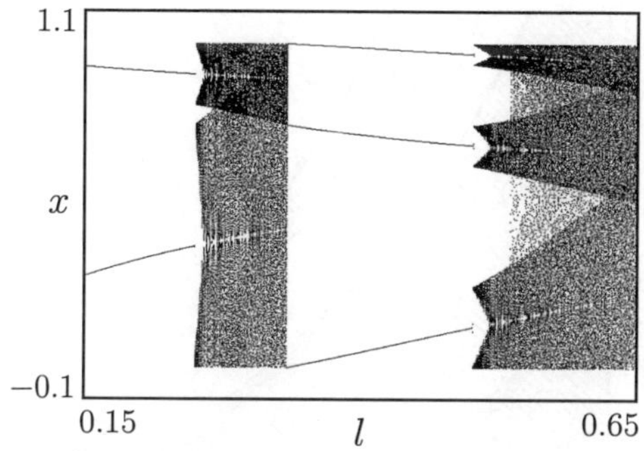

Figure 1.24. Bifurcation diagram for the skew tent map with $p = -4$. Note the abrupt transitions from a period-2 cycle to 4-band chaotic dynamics and from a period-3 cycle to 6-band chaos.

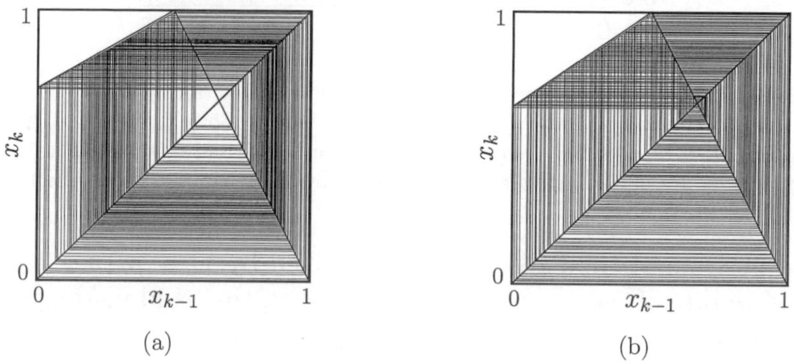

Figure 1.25. Band merging bifurcation for $p = -2$. (a) Two-band chaotic attractor for $l = 0.58 < 2/3$ and (b) one-band attractor for $l = 0.7 > 2/3$.

abruptly bifurcates into a 4-band chaotic attractor at $l \approx 0.25$. With further increase of l, the four chaotic bands merge, first into two chaotic bands in a homoclinic bifurcation at $l \approx 0.258$ and subsequently into a single chaotic band at $l \approx 0.267$. At $l \approx 0.333$ a stable period-3 cycle abruptly arises, and the window in which this cycle exists is terminated at $l = 0.5$ with the formation of a 6-band chaotic attractor. As l continues to increase we can again observe how the 6-band attractor is transformed first into a 3-band chaotic attractor and then again into a 1-band attractor.

As illustrated in Fig. 1.25 the homoclinic bifurcation in which the two-band chaotic attractor transforms into a one-band attractor occurs when the second iterate of the top point of the map falls precisely in the unstable fixed point.

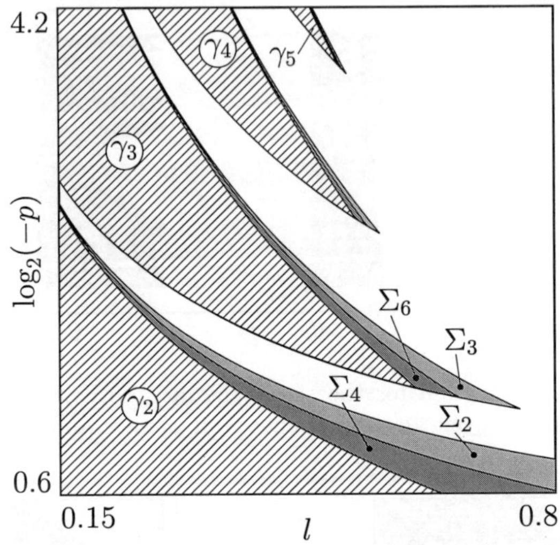

Figure 1.26. Two-dimensional bifurcation diagram for the skew tent map. γ_n denote regions of stable period-n dynamics, and Σ_m are regions of m-band chaotic dynamics. The multiband chaotic regions with $m = 8$ and 4, respectively with $m = 10$ and 5, that are adjacent to the domains γ_4 and γ_5 are quite narrow and, hence, are not explicitly marked in the figure.

With the position of the fixed point being given by $x^* = p/(p-1)$ and with the second iterate of the top point falling in $(0, c)$, the band merging transition takes place for $c = p/(p-1)$ or, for $p = -2$, $l = 2/3$.

The two-dimensional bifurcation diagram in Fig. 1.26 provides an overview of the dynamics of the skew tent map when both l and p are varied. Note that the vertical axis displays $-p$ on a logarithmic scale. γ_n denotes a region with stable period-n dynamics and Σ_m is a region with m-band chaotic dynamics. For $\log_2(-p) = 2$ we recover the one-dimensional bifurcation sequence of Fig. 1.24. In the vertical direction one observes a sequence of regions with period-2, period-3, period-4, etc., dynamics. This sequence can be related to the so-called period-adding sequence for the logistic map [35], i.e., the sequence of last appearances as stable solutions for the various periodic orbits.

One of the most striking aspects of the bifurcation diagrams in Fig. 1.24 and 1.26 is the absence of the period-doubling sequences we have seen for smooth maps. At the same time one observes how the reverse cascades of band merging bifurcations are truncated into a pair of transitions. To understand why the period-doubling cascades are missing we note that the period-2 cycle has a point on each of the two line segments. The eigenvalue of this cycle is the product of the two slopes, i.e., $\vartheta = lp$, and the period-2 orbit is stable as long as $|lp| < 1$. With increasing l the orbit becomes unstable for $lp = -1$. This value of the eigenvalue indicates the occurrence of a period-doubling bifurcation. However, the period-4 orbit that arises in this bifurcation has two points on each of the line segments. Hence, its eigenvalue is l^2p^2, and this eigenvalue exceeds 1 as soon as $|lp| > 1$. We conclude that the period-4 orbit is unstable from the moment it is born. The same argument is true for the period-8, period-16, etc., cycles. All the period-doubling bifurcations therefore take place at the same point, and the system makes a direct transition to chaotic dynamics.

The period-3 cycle has two points on the line segment with the (numerically) lower slope, and its eigenvalue is l^2p. The period-3 cycle is born in a so-called border-collision bifurcation (to be explained in the next section) when $c = d$ or $p = -(1+l)/l$. It disappears in a transition to 6-band chaotic dynamics when $l^2p = -1$. The period-4 cycle has three points on the line segment with the lower slope and one point on the other line segment. Its region of existence can be determined in a similar way as for the period-3 orbit.

Let us conclude this section by noting that due to a kind of degeneracy, associated with the absence of second and higher order derivatives, many bifurcations that for smooth maps occur in separate bifurcation points coincide for piecewise linear maps. Hence we observe a truncation of the bifurcation diagram. The purpose of the next section is to provide a first introduction to the new type of bifurcations, the so-called *border-collision bifurcations*, that are characteristic for nonsmooth systems.

1.4 Border-Collision Bifurcations

The above examples have illustrated some of the peculiar bifurcation phenomena that arise in piecewise linear maps. For piecewise-smooth dynamical systems one can in general distinguish between two types of bifurcations. The first type is similar to the bifurcations we know for smooth dynamical systems. These include the local bifurcations (saddle-node, pitchfork, period-doubling, and Hopf bifurcations), where a periodic orbit loses its stability as one of its eigenvalues (or a pair of eigenvalues) crosses out through the unit circle, and the global bifurcations (homoclinic bifurcations) where a connection is established from an unstable solution and back to the same solution along one of its stable directions. The band merging bifurcations considered in Sec. 1.3 for the logistic map and for the skew tent map are examples of homoclinic bifurcations.

The second type of bifurcations, referred to as *border-collision bifurcations*, are connected with situations where the trajectory starts to intersect one of the so-called sewing surfaces, i.e., surfaces that divide the phase space into domains of different dynamics. Within each such domain the system is smooth, but the equations of motion change abruptly from one domain to the next. This type of bifurcation, which typically involves abrupt jumps in the eigenvalues of the orbit, cannot occur in smooth dynamical systems.

A simple type of border-collision bifurcation consists in the continuous transformation of a solution from one type into another. However, more complicated phenomena are also possible, including new types of direct transition to chaos, and the merging and disappearance of solutions of different types. The direct transition from a stable fixed point to a chaotic attractor that we demonstrated for the simple tent map in Sec. 1.3 is an example of a border-collision bifurcation. The extraordinary complexity and large variety of nonlinear dynamic phenomena that one can observe in mechanical systems with discontinuities are associated with the interplay between local and homoclinic bifurcations on one side and different forms of border-collision bifurcations on the other. It is important to realize, however, that similar phenomena can arise in electrotechnical, chemical, biological, and economic systems. In particular, as we shall discuss in the following chapters, they can play a significant role in relay and pulse-width modulated control systems.

The study of border-collision bifurcations has a long history in the Russian literature [86–88]. Here, they are commonly referred to as *C-bifurcations*, derived from the Russian word *shivanije* meaning "sewing." Periodic orbits for dynamical systems with discontinuous equations of motion are characterized by a particular sequence of passings of the trajectory through the domains of local smoothness. At the borders of these domains the trajectory is sewed together through some type of continuity conditions. A forced mechanical oscillator that operates in regions with a different restraining force is a typical example. With the abrupt change in the force constant at the domain border, the acceleration will also change abruptly. However, as discussed for the mooring buoy in the next section, both the position and the velocity of the oscillator will vary continuously across the border.

In his early work [86], Feigin obtained analytical conditions for a period-doubling bifurcation to occur and for two cycles of different types to merge and disappear. Feigin subsequently developed conditions for the emergence of complex subharmonic oscillations and of families of unstable periodic orbits [87] and he established a relation between the analytical conditions for border-collision bifurcations and the eigenvalue spectrum of the local point maps [88].

The initial analyses were performed for a two-dimensional piecewise linear set of differential equations describing a nonautonomous mechanical oscillator. However, similar methods were applied to the study of piecewise linear one- and two-dimensional maps. For one-dimensional maps Feigin illustrated how chaotic oscillations can arise via border-collision bifurcations, and for two-dimensional maps he obtained the pattern of dynamical regimes and studied the structure of the bifurcation boundaries. An overview of these results may be found in Feigin's book on Forced Oscillations in Systems with Discontinuities [89]. In this book one can also find a classification of border-collision bifurcations of codimension-1. The term *codimension* is a measure of the number of active parameters for a dynamical system. When wee say that a bifurcation is of codimension-1 it implies that the bifurcation point is specified by a single condition among the parameters.

In the Western literature, one of the first explicit studies of border-collision bifurcations seems to be a paper by Nusse and Yorke [90]. These authors also seem to have coined the term "border-collision bifurcation". Based primarily on numerical simulations of two-dimensional piecewise linear maps, Nusse and Yorke showed bifurcation diagrams with period-2 to period-3, period-2 to period-5, and period-2 to period-11 transitions. Nusse and Yorke also observed direct transitions from period-2 dynamics to chaotic behavior. The same line of research was continued in a couple of subsequent papers [91, 92] in which more detailed explanations of the observed bifurcation phenomena were given and applications to physical and economic systems were discussed. Particular attention was paid to the analysis of border-collision bifurcations in which cycles of chaotic intervals (i.e., multiband chaotic dynamics) softly arise from a stable periodic cycle.

Maistrenko *et al.* [84, 85, 93] have performed a more mathematical oriented study of border-collision bifurcations in piecewise linear maps in connection with an analysis of the bifurcation structure of the Chua circuit [94, 95]. We have already presented some of their results for the skew tent map in Sec. 1.3. Different families of two-dimensional piecewise linear noninvertible maps have also been considered by Gumowsky and Mira [96], by Gardini [97], and by Mira *et al.* [98].

di Bernardo *et al.* [99, 100] have recently provided an overview of some of Feigin's results, and these authors have also discussed the relation between border-collision bifurcations and grazing bifurcations that occur when a mechanical oscillator bounces off of a hard wall. Finally we should mention that, approximately at the same time as the first paper by Nusse and Yorke, Uherka [101] wrote a contribution in which he presented so-called "tongue art." This topic is of interest in connection with our discussion of resonance phenomena and border-collision bifurcations on a two-dimensional torus in Chapters 5, 6, and 7.

The purpose of the present section is to discuss some of the above results in order to provide a clearer picture of the characteristics of border-collision bifurcations. Let us start by considering a slightly different version of the skew tent map [90–92]

$$x_k = T_\mu(x_{k-1}) = \begin{cases} ax_{k-1} + \mu, & x_{k-1} \leq 0; \\ bx_{k-1} + \mu, & x_{k-1} > 0 \end{cases} \tag{1.11}$$

and let us examine the different types of bifurcations that this map can produce for $a = 0.5$ and different values of $b < 0$. The position μ of the top point of the map is used as the bifurcation parameter.

For $\mu < 0$, the map (1.11) displays a stable fixed point with $x^* = 2\mu$. For $|b| < 1$, the map also displays a stable fixed point for positive values of μ, but the fixed point coordinate is now $x^* = 1/(1-b)$. At $\mu = 0$, the system undergoes a border-collision bifurcation as the fixed point crosses from the ascending to the descending part of the map. In this bifurcation the eigenvalue of the fixed point abruptly changes from a to b.

For $b < -1$, the fixed point on the right-hand part of the map is no longer stable. Figure 1.27 shows a series of one-dimensional bifurcation diagrams obtained by varying μ for different values of b. In Fig. 1.27(a) $b = -1.5$, and the map displays stable period-2 behavior for positive values of μ. For $b = -3.5$, the border-collision bifurcation at $\mu = 0$ produces stable period-3 dynamics (b), for $b = -4.15$ it produces 6-band chaotic dynamics (c), for $b = -4.4$ three-band chaotic dynamics (d), and for $b = -5.5$ one-band chaos (e). These results are in full concordance with the scenario we observed for the skew tent map in Sec. 1.3. Important aspects of the observed bifurcation structure are (i) that the fixed point can bifurcate directly into many different periodic and chaotic solutions,

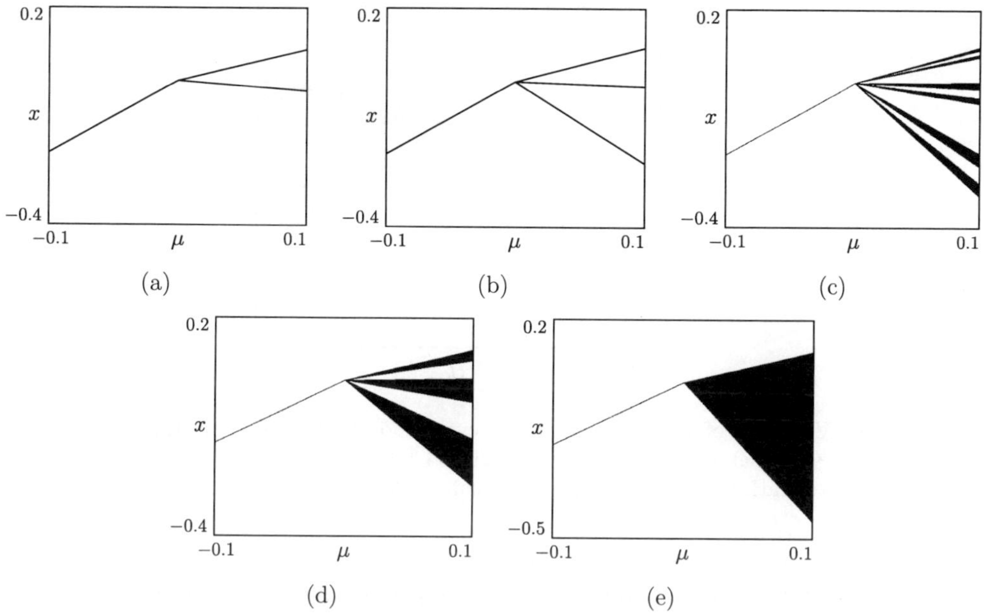

Figure 1.27. Series of one-dimensional bifurcation diagrams for the skew tent map (1.11). The slope of the descending part of the map is $b = -1.5$ (a), $b = -3.5$ (b), $b = -4.15$ (c), $b = -4.4$ (d), and $b = -5.5$ (e).

(ii) that these solutions emerge directly from the fixed point as it crosses the border between the two regions of the map, and (iii) that the solutions grow linearly in amplitude with the distance to the bifurcation point (rather than in the parabolic manner known for bifurcations in smooth maps).

It should be stressed that, besides stable periodic (or chaotic) solutions, the border-collision bifurcation at $\mu = 0$ produces a great variety of unstable periodic orbits. This set often includes several cycles with the same periodicity. For $b = -3.5$, for instance, along with a stable period-3 cycle, the border-collision bifurcation gives rise to an unstable fixed point, an unstable period-2 cycle, an unstable period-3 cycle, an unstable period-4 cycle, two unstable period-5 cycles, two unstable period-6 cycles, four unstable period-7 cycles, five unstable period-8 cycles, etc. Altogether we have identified 2490 unstable periodic orbits with a periodicity $m \leq 50$. Figure 1.28 shows the stable period-3 cycle and the set of 12 unstable cycles with periodicity $m \leq 7$ that arise in the border-collision bifurcation for $b = -3.5$.

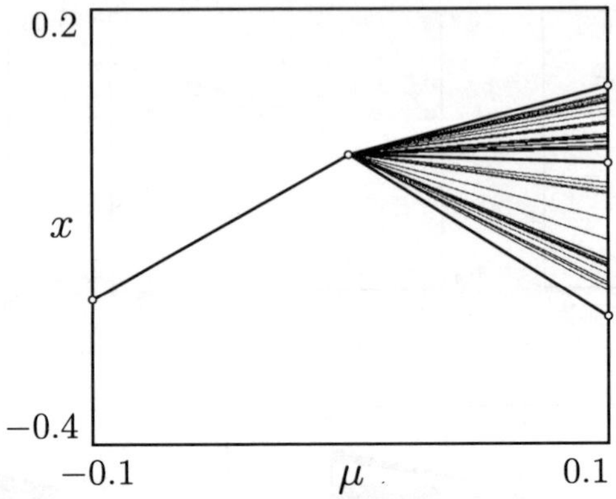

Figure 1.28. Variation of the stable period-3 cycle and the set of 12 unstable cycles with periodicity $m \leq 7$ that arise in the border-collision bifurcation for $a = 0.5$ and $b = -3.5$. Stable cycles are marked with thick lines and small circles.

Let us finally examine how the structure of the Arnol'd tongue diagram in Fig. 1.20 is changed when the sine-circle map is replaced by a piecewise linear map. Following Uherka [101] we may consider the map

$$\vartheta_k = R(\vartheta_{k-1}) = \Omega + \begin{cases} 2\vartheta_{k-1}, & 0 \leq \vartheta_{k-1} < b; \\ \dfrac{(1-4b)\,\vartheta_{k-1} + b}{1-2b}, & b \leq \vartheta_{k-1} < 1-b; \\ 2\vartheta_{k-1} - 1, & 1-b \leq \vartheta_{k-1} < 1. \end{cases} \tag{1.12}$$

Again the map is given modulus 1.

This map, referred to as a *rotating sawtooth map*, is illustrated in Fig. 1.29(a), and the corresponding mode-locking diagram is shown in Fig. 1.29(b). We note how the tongues now have assumed a sausage-like structure. Moreover, the edges of the resonance regions are now constituted by border-collision bifurcations.

Somewhat similar pictures were obtained by Yang and Hao [102, 103] in a study of a slightly different version of the sawtooth map. These authors obtained explicit expressions for the position of the tongue edges, determined the coordinates to the so-called *shrinking points* where the width of the tongues reduces to zero, and derived formulae for the (infinite set of) curves joining the shrinking points for different tongues. One such curve is plotted as a dotted line in Fig. 1.29(b). Alstrøm [104] proved that the dimension of the fractal set of

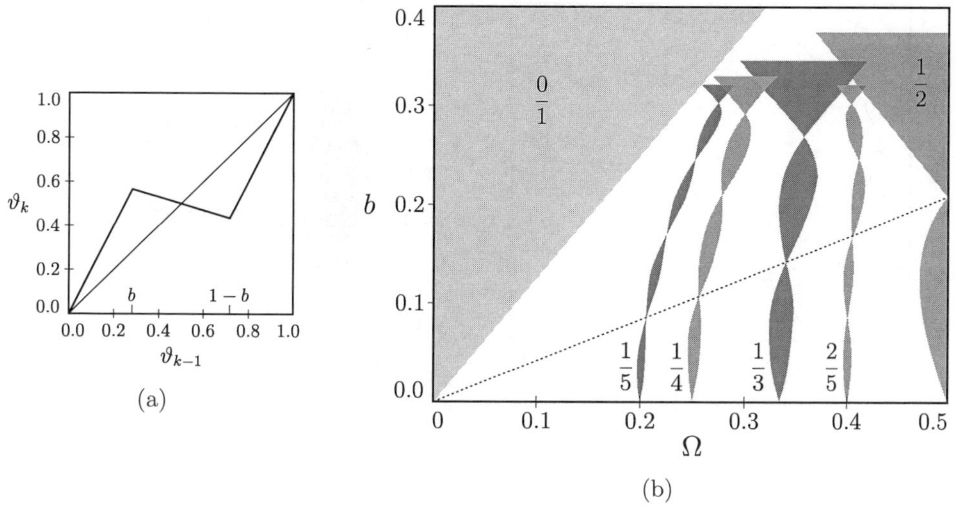

Figure 1.29. Rotating sawtooth map for $\Omega = 0$ (a). Corresponding resonance tongue diagram (b). Notice the characteristic sausage-like structure of the tongues. Compare with the Arnol'd tongue diagram for the sine-circle map in Fig. 1.20.

quasiperiodic motions vanishes at the critical point where the map becomes non-monotonous. For the rotating sawtooth map in Fig. 1.29(a) loss of monotonicity occurs for $b = 0.25$.

1.5 The Sewing Approach

An illustrative example of the sewing approach is the analysis of a mooring buoy subjected to a regular forcing from the waves of the ocean [7]. The buoy may be described as an inverted pendulum that, by virtue of buoyancy, maintains an equilibrium state in the upright position. For small excursions, the one-dimensional representation of this system may be cast into the form

$$\frac{\mathrm{d}^2 x}{\mathrm{d}t^2} + \alpha \frac{\mathrm{d}x}{\mathrm{d}t} + F(x) = B \cos t. \tag{1.13}$$

As before, α denotes the damping coefficient, and B is the forcing amplitude. $F(x)$ represents the restoring force. This force has a contribution $h_1 x$ from the buoyancy. However, if an oil tanker or a large freight ship is moored to the buoy there is also a contribution to $F(x)$ from the forces transmitted through the mooring cable. As the crudest possible approach we may take this force to

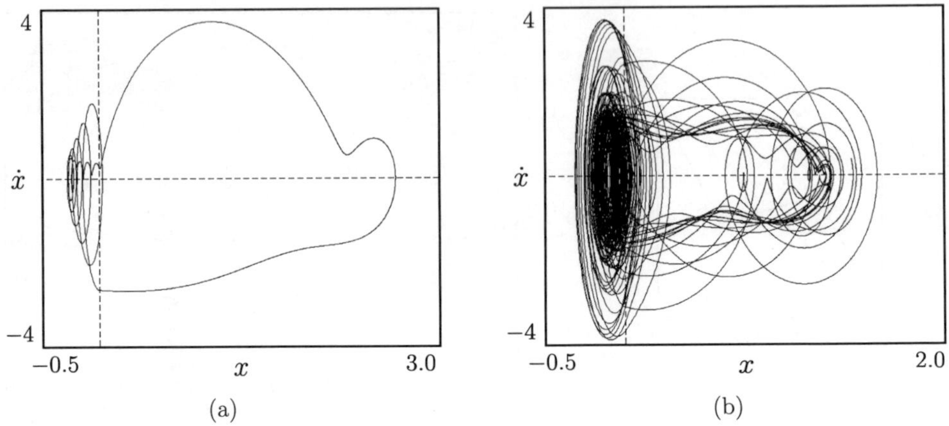

Figure 1.30. Phase projections for a simple model of a mooring buoy. The trajectory is sewed together across the surface $x = 0$. The mooring cable is stretched for $x < 0$. (a) Periodic solution for $\alpha = 2$, $h_1 = 40$, $h_2 = 360$, and $B = 100$. (b) Chaotic solution for $\alpha = 0.1$, $h_1 = 40$, $h_2 = 360$, and $B = 50$.

vanish when the cable is slackened and to contribute an elastic term $h_2 x$ when the cable is stretched. In total this gives a piecewise linear restoring force

$$F(x) = \begin{cases} (h_1 + h_2)x, & x \leq 0; \\ h_1 x, & x > 0 \end{cases}$$

with h_1 and h_2 being constants.

On both sides of the stretching point $x = 0$ the system is linear, and the equation of motion can be solved analytically. To obtain the global solution we must connect the partial solutions across the sewing surface. For the considered mechanical system, the sewing conditions are that the position x and the velocity $\dot{x}$ must vary continuously across $x = 0$.

Figure 1.30 shows the results of such a sewing procedure. For certain values of the parameters it is possible to connect the two solutions in such a way that the total dynamics becomes periodic (Fig. 1.30(a)). For other parameter values, this type of connection is not possible, and the piecewise linear system displays chaotic dynamics.

A more detailed illustration of the sewing approach is the global bifurcation analysis of a simplified *economic long-wave* model performed by Szymkat and Mosekilde [105]. The economic long wave is a complex phenomenon that is supposed to manifest itself in the form of alternating phases of growth and recession in the World economy with a typical period of 50–60 years [106–108]. Historical depression periods are considered to be the 1830's and 40's, the 1870's through late 1880's, the 1920's and 30's, and the period from the mid 1970's to the present time [107].

The long wave has ramifications to a broad range of economic, social and political phenomena, including interest rates, building activity, family values, lengths of skirts, homicide rates, etc. Many scholars [109, 110] associate the growth phases with the replacement of old production technologies, transportation means and energy sources with qualitative new technologies, and a significant number of empirical studies lend support to a relation between the economic long wave and variations in infrastructure investments (i.e., construction of railroads, subways, bridges, etc.).

Sterman [111] was first to formulate a model of the economic long wave. In his description, the rhythmic expansions and contractions of the economy arise as a self-sustained oscillation in the ordering and production of capital. An increase in the demand for capital leads to further increase through so-called capital self-ordering, i.e., by the fact that the capital sector in the aggregate depends on its own output to expand its production capacity. Once a capital expansion gets underway, self-reinforcing processes therefore sustain it beyond its long-term equilibrium until, at the end, nonlinear restraints to the rate of capital expansion allow production to catch up with orders. At this point, the economy is overcapitalized, and the feedback loops reverse, forcing capital production to remain below replacement levels until the excess capacity has been fully discarded, and room for a new expansion has been created. The lack of replacement orders from the capital sector during the downswing phase causes the system to shoot below its long term equilibrium level, triggering the next upswing to start.

Rasmussen *et al.* [112] considered a slightly simplified version of Sterman's model. Besides investigating the Hopf bifurcation that causes the spontaneous oscillations to arise, these authors also examined certain aspects of the global dynamics of the model, particularly the conditions under which the system is unable to recover from a depression period. Subsequent studies [113] have examined the interaction between the economic long wave and other modes (business cycles) in the economy. The effects of a disaggregation of the model into sectors that produce different types of capital have also been considered [114]. A survey of some of these results may be found in Topics in Nonlinear Dynamics [21].

The simplified long-wave model follows the flow of capital through the macroeconomic system. This flow is governed by two coupled ordinary differential equations

$$\dot{x} = \frac{x}{c} f_1(z) - 1 - \frac{x}{a}; \tag{1.14}$$

$$\dot{y} = \frac{x}{a} f_2(z) - \frac{x}{c} f_1(z) + 1, \tag{1.15}$$

one controlling the rate of change for the stock of active capital x, and the other describing variations in the supply line of capital y, i.e., in the stock of capital under production. (The term capital is used in the meaning of physical production capital, i.e., buildings, machinery, transport means, etc.). The economy is scaled relative to the desired production of goods (for consumption) which is taken to be constant and equal to 1. This implies that the two state variables x and y are measured in years.

c is the capital/output ratio and a the average capital lifetime. Typical values for these parameters are 4 and 20 years, respectively. In the expression for $\dot{x}$, x/c represents the production capacity, also referred to as the potential output. $f_1(z)$ is the capacity utilization factor, and x/a is the capital depreciation rate. $(x/c)f_1(z)$ thus represents the total production, and by subtracting the production of goods, i.e. 1, we obtain the production of capital. The auxiliary variable

$$z = \frac{c}{x}\left(\frac{y}{d} + 1\right) \tag{1.16}$$

is the ratio of the desired production $y/d + 1$ to the potential output x/c. Here, $d = 3$ years denotes the normal delivery delay for capital. The desired production is the production of goods plus the desired production of capital y/d.

The nonlinear function $f_1(z)$ describes how the utilization of the available production capacity varies with z. Economic considerations indicate that

$$f_1'(z) \geq 0, \; f_1(0) = 0, \; f_1(1) = 1, \text{ and } f_1(z) \to b_1 \text{ for } z \gg 1, \tag{1.17}$$

i.e., capital utilization is a growing function of z that falls to zero if no production is required and reaches a saturation level $b_1 > 1$ for high values of desired production over potential output. By the applied normalization, capital utilization is unity when desired production equals potential output. The function $f_2(z)$ describes the variation in order placements relative to capital discards.

Economic considerations again lead to

$$f_2'(z) \geq 0, \; f_2(0) = 0, \; f_2(1) = 1, \text{ and } f_2(z) \to b_2 \text{ for } z \gg 1 \tag{1.18}$$

with $b_2 > 1$.

With the above assumptions, the long wave model has a unique equilibrium for $z = 1$ at

$$x_{\mathrm{e}} = \frac{ca}{a-c} \qquad \text{and} \qquad y_{\mathrm{e}} = \frac{cd}{a-c}. \tag{1.19}$$

If e_1 and e_2 denote the slopes of the functions $f_1(z)$ and $f_2(z)$ at $z = 1$, the equilibrium point is stable for

$$e_2 < \frac{ad}{c}\left(\frac{1}{a} + \frac{e_1}{d} - \frac{1-e_1}{c}\right). \tag{1.20}$$

At this value of the slope, the system undergoes a bifurcation similar to a Hopf bifurcation, and at higher values of e_2 it displays self-sustained oscillations. At the Hopf bifurcation point the oscillation period is [112]

$$T = \frac{2\pi a}{\sqrt{e_2(a-c)/d}}. \tag{1.21}$$

As illustrated in Fig. 1.31, to define the nonlinear functions we divide the positive half-line into four intervals: $I_1 = [0, 1-w_1[$, $I_2 = [1-w_1, 1+w_2[$, $I_3 = [1+w_2, 2[$, and $I_4 = [2, \infty[$, where w_l are constants satisfying $0 < w_l < 1$, for $l = 1, 2$. In each of these intervals the functions $f_1(z)$ and $f_2(z)$ are assumed to be piecewise linear, i.e.,

$$f_i(z) = h_{ik} + g_{ik} z, \quad \text{for } z \in I_k \tag{1.22}$$

for $i = 1, 2$ and $k = 1, 2, 3, 4$. The coefficients g_{ik} and h_{ik} may be expressed in terms of w_l and the slopes e_i at the equilibrium point. For $k = 1$, for instance, we have $h_{i1} = 0$ and

$$g_{i1} = (1 - e_i w_1)/(1 - w_1). \tag{1.23}$$

We also note that $g_{12} = e_1$ and $g_{22} = e_2$. By a proper choice of the various constants, we can obtain functions that fulfill the assumptions (1.17) and (1.18) (except at points where their first derivatives are discontinuous). In the following analysis we shall confine our investigation to a class of functions that preserve the convexity properties shown in Fig. 1.31, i.e., we shall assume that

$$0 < e_1 < 1, \; e_2 > 1 \qquad \text{and} \qquad b_i < 1 + e_i, \; i = 1, 2. \tag{1.24}$$

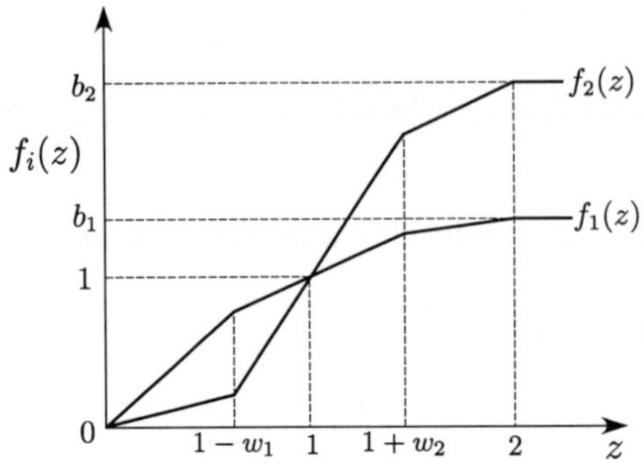

Figure 1.31. Piecewise linearization of the nonlinear relations. b_1 and b_2 are the saturation levels of $f_1(z)$ and $f_2(z)$, respectively.

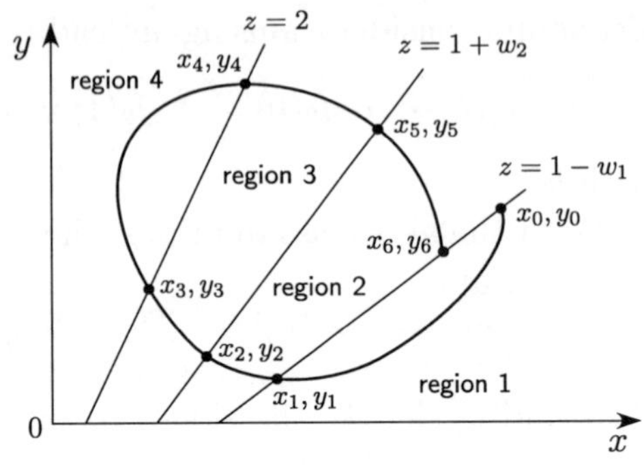

Figure 1.32. Division of the (x, y)-plane into four regions. The slanting lines represent the sewing surfaces.

The above definitions lead to a division of the positive quadrant of the phase plane into four regions corresponding to the four intervals I_k. This simplification arises from the fact that the two nonlinear functions depend on the same intermediate variable z. Trajectories intersect the dividing lines (or sewing surfaces) at subsequent points. These points may be numbered consecutively as shown in Fig. 1.32.

The piecewise linearization of the nonlinear functions that we are using here is common in economic and managerial models from the so-called Forrester school [115]. This approach is recommended on the grounds that, since a detailed theoretic foundation of the nonlinear relations is seldom available, it would be an overspecification to make use of particular analytic functions.

In each of the regions of Fig. 1.32 the dynamics of the model is described by a specific set of linear differential equations. It is convenient to express the solutions to these equations in terms of local coordinates around the equilibrium point even when this point lies outside the specific region. For the regions 1, 2 and 3 the unique equilibria are given by

$$x_{ek} = \frac{ac}{a(h_{1k} + g_{1k}z_{ek}) - c} \quad \text{and} \quad y_{ek} = d\left[\frac{z_{ek}(a + x_{ek})}{a(h_{1k} + g_{1k}z_{ek})} - 1\right] \tag{1.25}$$

with $z_{ek} = (1 - h_{2k})/g_{2k}$. Note that $(x_{e2}, y_{e2}) = (x_{\mathrm{e}}, y_{\mathrm{e}})$. There is no equilibrium point for the dynamics in region 4. However, $\dot{x} = 0$ for

$$x = x_{e4} = \frac{ac}{b_1 a - c}. \tag{1.26}$$

The equations for x and y in region k can now be written in the form

$$\begin{pmatrix} \dot{x} \\ \dot{y} \end{pmatrix} = M_k \begin{pmatrix} x - x_{ek} \\ y - y_{ek} \end{pmatrix} \tag{1.27}$$

where

$$M_k = \begin{pmatrix} h_{1k}/c - 1/a & g_{1k}/d \\ h_{2k}/a - h_{1k}/c & cg_{2k}/(ad) - g_{1k}/d \end{pmatrix} \tag{1.28}$$

The matrix M_1 has two real eigenvalues: $-1/a$ and $-1/(y_{e1} + d)$. Eliminating the time variable we get the equation for the trajectory in region 1

$$\begin{aligned} F_1(x_0, y_0, x, y) = &- \left[x_0 - x_{e1} + A(y_0 - y_{e1})\right]^a (y - y_{e1})^{y_{e1}+d} \\ &+ \left[x - x_{e1} + A(y - y_{e1})\right]^a (y_0 - y_{e1})^{y_{e1}+d} = 0 \end{aligned} \tag{1.29}$$

where

$$A = \frac{g_{11}a(y_{e1} + d)}{d(a - y_{e1} - d)}.$$

As illustrated in Fig. 1.32, x_0 and y_0 denote the coordinates of the point in which the trajectory enters region 1. For usual parameter values, the matrices M_k for $k = 2, 3$ have conjugate pairs of complex eigenvalues with a real part δ_k and an imaginary part $\pm\omega_k$. δ_k and ω_k can easily be found from the relations $2\delta_k = \operatorname{tr} M_k$ and $\delta_k^2 + \omega_k^2 = \det M_k$. Introducing the following change of variables

$$\begin{aligned} u_k(x, y) &= \omega_k \left[\left(\frac{h_{2k}}{a} - \frac{h_{1k}}{c}\right)(x - x_{ek}) - \left(\frac{h_{1k}}{c} - \frac{1}{a}\right)(y - y_{ek})\right]; \\ v_k(x, y) &= -\frac{\delta_k u_k(x, y)}{\omega_k} - \det M_k (y - y_{ek}); \\ r_k(x, y) &= \sqrt{u_k^2(x, y) + v_k^2(x, y)} \end{aligned} \tag{1.30}$$

and

$$\varphi_k(x, y) = 2 \arctan \left\{ \frac{v_k(x, y)}{u_k(x, y) + r_k(x, y)} \right\}, \tag{1.31}$$

we obtain the trajectory equations for the two regions in a common form

$$\begin{aligned} F_k(x_j, y_j, x, y) = &-r_k(x_j, y_j) \exp\left[\frac{\delta_k}{\omega_k} \varphi_k(x_j, y_j)\right] \\ &+ r_k(x, y) \exp\left[\frac{\delta_k}{\omega_k} \varphi_k(x, y)\right] = 0. \end{aligned} \tag{1.32}$$

x_j and y_j now denote the coordinates to the points at which the trajectory enters the respective regions. $\arctan(\cdot) = \tan^{-1}(\cdot)$.

The eigenvalues of the matrix M_4 are 0 and $1/x_{e4}$. The trajectories in region 4 are thus given by

$$F_4(x_3, y_3, x, y) = -x_3 - y_3 + B\left[x_3 + x_{e4}\ln(x_3 - x_{e4})\right] \\ + x + y - B\left[x + x_{e4}\ln(x - x_{e4})\right] = 0 \quad (1.33)$$

where

$$B = \frac{b_2 - 1}{a} x_{e4}$$

and x_{e4} was defined in (1.26). Again x_3 and y_3 determine the point at which the trajectory enters the considered region.

In order to find the complete trajectory we also need the equations for the sewing lines $z = z_1 = 1 - w_1$, $z = z_2 = 1 + w_2$ and $z = z_3 = 2$. From the relation (1.19) we see that they can be written in the following general form for $i = 1, 2, 3$

$$y = l_i(x) = \frac{p_i}{q_i} x - d$$

where

$$p_i = \frac{z_i d}{\sqrt{c^2 + z_i^2 d^2}} \quad \text{and} \quad q_i = \frac{c p_i}{z_i d}. \tag{1.34}$$

Let us assume that $y_0 = l_1(x_0)$. The numerical process of following the trajectory starting at (x_0, y_0) consists of the computation of points (x_i, y_i) (see Fig. 1.32) satisfying subsequent systems of equations. First we find (x_1, y_1) by solving $y_1 = l_1(x_1)$ and $F_1(x_0, y_0, x_1, y_1) = 0$ then (x_2, y_2) satisfying $y_2 = l_2(x_2)$ and $F_2(x_1, y_1, x_2, y_2) = 0$ and so on.

In this way we can define the mappings

$$(x_i, y_i) \to (x_{i+1}, y_{i+1}) = P_{i+1}(x_i, y_i) = \big(Q_{i+1}(x_i, y_i), R_{i+1}(x_i, y_i)\big). \tag{1.35}$$

Unfortunately, explicit formulae for this mapping do not exist. A major advantage of the present approach, however, is that the accuracy can be kept at the same level along the whole trajectory. Let us also show how the approach can provide answers to a number of concrete questions.

1.5.1 Existence of a periodic solution

The problem of the existence of a periodic solution for the long-wave model can be treated by means of classical methods for 2-dimensional systems, particularly

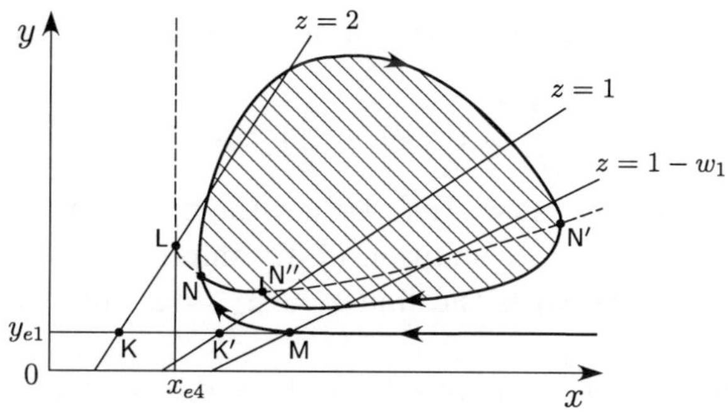

Figure 1.33. Invariant set containing a periodic solution.

the Poincaré-Bendixson theorem [16, 116]. Let us try to establish the sufficient conditions for the existence of a bounded positive invariant set of phase space containing as the only equilibrium the point (x_e, y_e) defined in (1.19). We shall assume that the capital/output ratio $c > c_0$ where the Hopf bifurcation point c_0 is given by (1.20). Hence, the equilibrium is unstable.

The geometry of the flow produced by (1.14) and (1.15) can be characterized by the use of *isoclines*, i.e., by curves joining points with a constant ratio $\dot{y}/\dot{x}$. The ∞-isocline, for instance, represents all points satisfying $\dot{x} = 0$. This isocline is illustrated by the broken curve in Fig. 1.33.

In order to construct the bounded *invariant set* we can follow the trajectory starting in region 1 with $y = y_{e1}$. We want to show that this trajectory intersects the ∞-isocline at some point N. Since there is no equilibrium in region 4 and the unique equilibrium in region 3 is unstable, the only way to avoid this intersection is to cross one of the line segments KM or KL in Fig. 1.33. To prevent the first possibility we must prove that $\dot{y} > 0$ on K′M where K′ denotes the point of intersection of the segment KM and the line $z = 1$. This follows from (1.15) that shows that for $z > 1$, y increases as long as $\dot{x} < 0$, provided that

$$\left[f_2(z)\frac{c}{a} - f_1(z)\right]\left(\frac{y_{e1}}{d} + 1\right) + z > 0 \tag{1.36}$$

for $z \in]1 - w_1, 1]$. Because the left hand side of (1.36) is linear with respect to z and equals 0 for $z = 1 - w_1$, we only need to check if the condition is fulfilled for $z = 1$. Substituting (1.25) into (1.36) we obtain

$$c(1 - e_2) < a(1 - e_1),$$

a condition that is satisfied for all positive c and a due to the assumptions (1.24).

To exclude the case where the trajectory crosses the segment KL we must show that

$$Q_3\left[P_2\left(\frac{q_1}{p_1}(y_{e1}+d), y_{e1}\right)\right] - x_{e4} = G > 0 \tag{1.37}$$

if the left hand side makes sense. Here we use the notation introduced in (1.35) for the mappings P and Q. Condition (1.37) implies that the possible intersection of the trajectory with the line $z = z_3 = 2$ falls to the right of the point L.

Due to the structure of the matrix M_4 we have a special dynamics in region 4. We can deduce from (1.33) that the x-coordinate of all trajectories starting from (x_3, y_3) satisfying $y_3 = l_3(x_3)$ and $x_3 > x_{e4}$ has to increase (we assume that $x_{e4} > 0$). From (1.33) we see that the slope of the trajectories in region 4 is a decreasing function of x. So if the trajectory tends to infinity, its x-coordinate also has to diverge and the values of z corresponding to the points of such a trajectory remain bounded. The same is true of course for all trajectories tending to infinity in other regions.

Differentiating (1.16) with respect to time and combining (1.14), (1.15) and (1.22) we obtain

$$\dot{z} = H(z) + \frac{1}{x}\left(\frac{c}{d} + z\right)$$

where

$$H(z) = \left(\frac{1}{a} - \frac{h_{1k}}{c} - \frac{g_{1k}}{c}z\right)\left(\frac{c}{d} + z\right) + \frac{c}{ad}(h_{2k} - 1 + g_{2k}z) \tag{1.38}$$

for $z \in I_k$, $k = 1, 2, 3, 4$. The next provision is that

$$H(z) < 0 \text{ for } z \in]f_1^{-1}(c/a), 2]. \tag{1.39}$$

This condition implies that $H(z)$ is negative also for $z > 2$, because for such values of z, H is a linear function with a slope $-1/x_{e4}$. It follows from (1.39) that any trajectory with increasing x-coordinate that remains in a region with bounded z eventually has to reach the line $z = f_1^{-1}(c/a)$ and therefore cross the ∞-isocline. This means that there exists a point N′ at which our trajectory intersects the ∞-isocline for the second time and then a third point N″ positioned to the right of N (see Fig. 1.33). Let us now define the set in phase space with a boundary consisting of the trajectory joining N and N″ and part of the ∞-isocline between the same two points. Due to the uniqueness of the equilibrium and by virtue of the continuity of the vector field no trajectory can leave

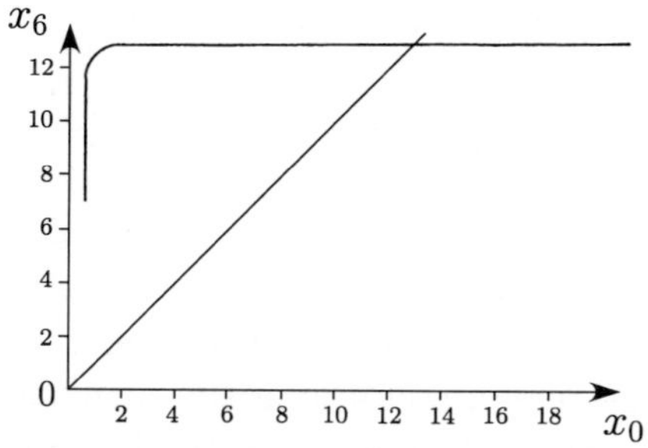

Figure 1.34. Poincaré map for the long-wave model obtained for $c = 7$ y, $a = 20$ y, $d = 3$ y, $e_1 = 0.7$, $b_1 = 1.4$, $e_2 = 3.125$, $b_2 = 4$, and $w_1 = w_2 = 0.2$. The slope of the map is very close to zero at the fixed point. Hence, this fixed point is stable.

this set. Hence, we conclude that the defined set (shaded area in Fig. 1.33), as an invariant set different from the equilibrium point, has to contain a periodic orbit.

The method applied to compute the values of the mapping introduced in (1.35) can be used to construct a rapidly converging procedure for finding the limit cycle. Having calculated the whole sequence of points up to (x_6, y_6), we have the value of the Poincaré first return map for (x_0, y_0) with the line $z = z_1$ taken as a transversal section. As an example of this map, the x-coordinate is shown in Fig. 1.34.

In order to find the fixed point of the Poincaré map we have iteratively repeated the whole process with subsequent starting points computed by the use of Newton's method. After a few iterations, this procedure generated the sequence of points (x_0, y_0), (x_1, y_1), (x_2, y_2), (x_3, y_3), (x_4, y_4), (x_5, y_5) and $(x_6, y_6) = (x_0, y_0)$ situated on the periodic orbit.

Having obtained the whole sequence of points we can now perform a more detailed analysis of the periodic orbit. The period can be computed using the formula

$$T = \sum_{i=1}^{6} t_i, \tag{1.40}$$

where

$$t_1 = (y_{e1} + d) \ln \frac{y_0 - y_{e1}}{y_1 - y_{e1}} \quad \text{and} \quad t_4 = x_{e4} \ln \frac{x_4 - x_{e4}}{x_3 - x_{e4}}.$$

The two remaining transit times are given by

$$t_i = \left[\varphi_{k_i}(x_j, y_j) - \varphi_{k_i}(x_i, y_i)\right]\frac{1}{\omega_{k_i}}$$

with φ_k as defined in (1.31). $i = 2, 3, 5, 6$, $k_2 = k_6 = 2$, and $k_3 = k_5 = 3$.

We can take the value of $x_6 - x_2$ as a measure of the size of the periodic orbit, i.e., the difference between the x-coordinates of the two points at which the limit cycle leaves the region containing the unstable equilibrium. In order to obtain a quantitative description of the rate of attraction to the limit cycle, we can calculate its Lyapunov exponents, i.e., the logarithm of the modules of the linearized Poincaré map eigenvalue. Let us assume that

$$(x_i + s_i q_i, y_i + s_i p_i) = P_i(x_{i-1} + s_{i-1}q_{i-1}, y_{i-1} + s_{i-1}p_{i-1}) \tag{1.41}$$

where (x_i, y_i) are the points on a periodic orbit, P_i are defined by (1.35) for $i = 1, 2, \ldots, 6$, the coefficients p_i and q_i are given by (1.34) for $i = 1, 2, 3$, $p_0 = -p_1$, $q_0 = -q_1$ and $p_i = -p_{7-i}$, $q_i = -q_{7-i}$ for $i = 4, 5, 6$. The Lyapunov exponent of the limit cycle can then be expressed as

$$L = \sum_{i=1}^{6} \ln \left|\frac{\mathrm{d}s_i}{\mathrm{d}s_{i-1}}\right|_{s_{i-1}=0}.$$

1.5.2 Global bifurcations

When the capital/output ratio c passes the value c_0 ($= 4$ y) for our base case parameters), a periodic orbit arises in a process similar to a Hopf bifurcation. This process cannot be treated with standard methods because of the nondifferentiability of the vector field. Using the explicit relations (1.40) and (1.41), however, we can obtain exact values, instead of local approximations, for the period and stability of the limit cycle. Let us now consider the behavior of the periodic orbit when certain parameters are varied.

We start by checking if the condition (1.37) can be satisfied for large values of c. Since, in this case, the matrix M_3 can have real roots, we apply (1.30) or the transformation

$$u_3(x, y) = \left(\frac{h_{1k}}{c} - \frac{h_{2k}}{a}\right)(x - x_{e3}) + \left(\frac{h_{1k}}{c} - \frac{1}{a} - \delta_{31}\right)(y - y_{e3})$$

and

$$v_3(x, y) = \left(\frac{g_{2k}c}{ad} - \frac{g_{1k}}{d} - \delta_{32}\right)(x - x_{e3}) - \frac{g_{1k}}{d}(y - y_{e3})$$

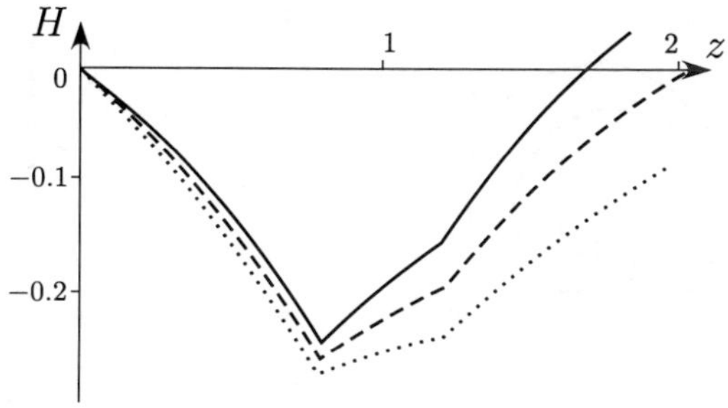

Figure 1.35. Function H for $c = 8.75$ y (dotted line), $c = 9.75$ y (broken line) and $c = 10.75$ y (continuous line). All other parameters are the same as in Fig. 1.34.

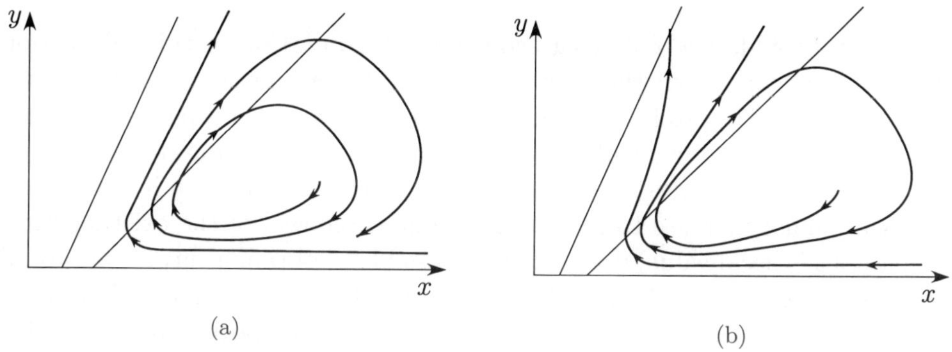

Figure 1.36. Global bifurcation in which the limit cycle disappears. Structure of the flow for $c = c^*$ (a) and for $c > c^*$ (b).

where δ_{31} and δ_{32} are the real eigenvalues of M_3. The equation for the trajectories in region 3 for the case of real eigenvalues is

$$F_3(x_j, y_j, x, y) = -[u(x_j, y_j)]^{\delta_{32}}[v(x, y)]^{\delta_{31}} + [u(x, y)]^{\delta_{32}}[v(x_j, y_j)]^{\delta_{31}} = 0. \quad (1.42)$$

Using the above procedure, the left hand side of (1.37) can be evaluated, and it turns out that the condition $G > 0$ is satisfied. Next we check the condition (1.39). Graphs of $H(z)$ for three different values of c are plotted in Fig. 1.35.

Inspection of the figure shows that a critical value c^* for c exists so that $H(z) = 0$. For $c > c^*$, condition (1.39) is violated. As we approach the critical value c^* from below we have that the period $T \to \infty$ and the Lyapunov exponent $L \to 0$. For $c = c^*$ the periodic orbit disappears in a global bifurcation. Figure 1.36 presents the structure of the flow for $c = c^*$ and $c > c^*$.

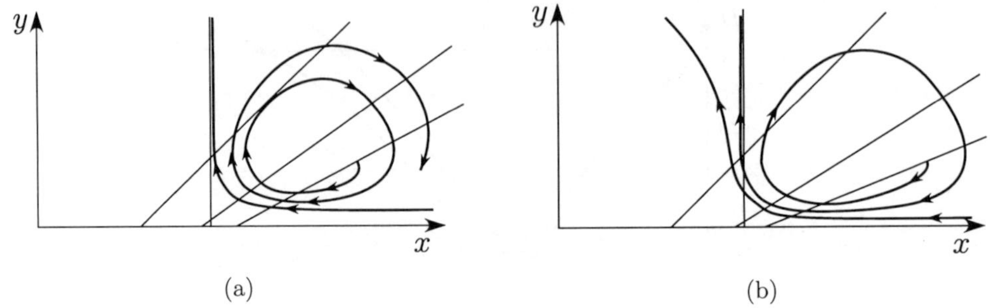

Figure 1.37. Another example of a global bifurcation that destroys the limit cycle. Structure of the flow for $w_1 = w_1^*$ (a), and $w_1 > w_1^*$ (b).

We may provide the following economic interpretation of this bifurcation: The parameter c measures the efficiency of capital reproduction, i.e., the amount of capital required to produce one unit of capital per year. This parameter represents a major factor in controlling the amplification in the capital self-ordering. As c increases, it becomes harder and harder to produce new capital and order placements must increase to secure the required renewal of the existing stock. As c passes a critical value, it is no longer possible to obtain a surplus of capital. In fact, the system is no longer able to get out of the saturation regime where the desired production is more than twice as large as the potential output. The larger c is, the more capital must be installed before the orders for capital start to grow slower than the capital itself. For $c > c^*$ the orders will continue to grow with the rate asymptotically proportional to the capital growth rate.

The second case to be discussed here concerns the slope at $z = 0$ of the function $f_2(z)$. Using (1.23) the value of this slope can be expressed in terms of the slope e_2 at the equilibrium point and the position w_1 of the sewing surface. Our analysis shows [105] that the condition (1.39) is now observed but (1.37) may be violated. This implies that there exists a critical value w_1^* for w_1 such that $G > 0$ for $w_1 > w_1^*$. The structural changes in the flow of the dynamical system (1.14) and (1.15) that occur when w_1 passes through w_1^* are illustrated in Fig. 1.37.

Both the considered cases lead to situations that are similar from a topological point of view. When the parameter under consideration is equal to its critical value, the trajectory investigated in the proof of the existence of the periodic orbit follows the horizontal line in region 1 (the 0-isocline), then breaks away tending to infinity along the straight line (the eigenray in region 3 for the first case and the ∞-isocline in the second case). When completing our phase plane with the points at infinity we see that a heteroclinic trajectory joining two saddle points at infinity appears when we pass the critical value of the parameter. This special trajectory "absorbs" the limit cycle.

The economic conditions that result from the two types of global bifurcation are quite different, however. Whereas after the first bifurcation, the economy ends up in a situation of permanent scarcity, but eternal growth, the second bifurcation describes the collapse of the economy. The values of $f_2(z)$ for $z < 1 - w_1$ express the willingness (propensity) of the society to continue to invest in new production capacity, even though the utilization of the available capacity is small. If these values of $f_2(z)$ are low, very few orders for new capital are placed during the downswing phase, and at the end of the downswing the production capacity is too small to initiate a new upswing. This result, of course, to a significant extent rests on the model's assumption that goods production takes priority over the production of capital units. This assumption is not maintained in more elaborate versions of the model [113, 114]. Nonetheless, the result lends support to the idea of starting a new upswing by massive investments in infrastructure capital.

Bibliography

[1] J. Moser, Stable and Random Motions in Dynamical Systems (Princeton Univ. Press, NJ, 1973).

[2] V. I. Arnol'd, Mathematical Methods in Classical Mechanics (Springer–Verlag, Heidelberg, 1978).

[3] R. H. G. Helleman, *Self-Generated Chaotic Behaviour in Nonlinear Mechanics*, Fund. Prob. Stat. Mech. **5**, 165–233 (1980).

[4] R. S. MacKay and J. D. Meiss (Editors), Hamiltonian Dynamical Systems: A Reprint Selection (Adam Hilger, Bristol, 1987).

[5] M. V. Berry, I. C. Percival, and N. O. Weiss (Editors), Dynamical Chaos (Proc. Roy. Soc. London, 1987).

[6] F. C. Moon, Chaotic Vibrations. An Introduction for Applied Scientists and Engineers (Wiley, New York, 1987).

[7] J. M. T. Thompson and H. B. Stewart, Nonlinear Dynamics and Chaos (Wiley, New York, 1986).

[8] H. G. Schuster, Deterministic Chaos — An Introduction (VCH Verlagsgesellschaft, Weinheim, 1989).

[9] E. Ott, Chaos in Dynamical Systems (Cambridge Univ. Press, UK, 1993).

[10] C. Grebogi, S. W. McDonald, E. Ott, and J. A. Yorke, *Final State Sensitivity: An Obstruction to Predictability*, Phys. Lett. A **99**, 415–418 (1983).

[11] F. C. Moon, *Fractal Boundary for Chaos in a Two-State Mechanical Oscillator*, Phys. Rev. Lett. **53**, 1962–1964 (1984).

[12] C. Grebogi, E. Ott, and J. A. Yorke, *Fractal Basin Boundaries, Long-Lived Chaotic Transients, and Unstable–Unstable Pair Bifurcations*, Phys. Rev. Lett. **50**, 935–938 (1983).

[13] E. G. Gwinn and R. M. Westervelt, *Horseshoes in the Driven, Damped Pendulum*, Physica D **23**, 396–401 (1986).

[14] C. Grebogi, E. Ott, and J. A. Yorke, *Basin Boundary Metamorphoses: Changes in Accessible Boundary Orbits*, Physica D **24**, 243–262 (1987).

[15] S. Smale, *Differentiable Dynamical Systems*, Bull. Amer. Math. Soc. **73**, 747–817 (1967).

[16] J. Guckenheimer and P. J. Holmes, Nonlinear Oscillations, Dynamical Systems, and Bifurcations of Vector Fields (Springer, New York, 3rd printing, 1990).

[17] Y. Ueda, *Random Transitional Phenomena in the System Governed by Duffing's Equation*, J. Stat. Phys. **20**, 181–196 (1979).

[18] Y. Ueda, *Survey of Irregular and Chaotic Phenomena in the Forced Duffing Oscillator*, Chaos, Solitons and Fractals **1**, 199–231 (1991).

[19] Y. Ueda, The Road to Chaos (Aerial Press, Santa Cruz, 1992).

[20] Y. Ueda, J. S. Thomsen, J. Rasmussen, and E. Mosekilde, *Behavior of the Solution to Duffing's Equation for Large Forcing Amplitudes*, Math. Res. **72**, 149–166 (1993).

[21] E. Mosekilde, Topics in Nonlinear Dynamics. Applications to Physics, Biology and Economic Systems (World Scientific, Singapore, 1996).

[22] D. W. Jordan and P. Smith, Nonlinear Ordinary Differential Equations (Clarendon Press, Oxford, 1987).

[23] A. Wolf, *Quantifying Chaos with Lyapunov Exponents*, in Nonlinear Science, Theory and Applications, edited by A. V. Holden (Manchester University Press, England, 1984), pp. 273–290.

[24] H. Kantz and T. Schreiber, Nonlinear Time Series Analysis (Cambridge University Press, England, 1997).

[25] O. E. Rössler, *An Equation for Continuous Chaos*, Phys. Lett. A **57**, 397–398 (1976).

[26] O. E. Rössler, *The Chaotic Hierarchy*, Z. Naturforsch. **38a**, 788–801 (1983).

[27] G. Baier and M. Klein, A Chaotic Hierarchy (World Scientific, Singapore, 1991).

[28] M. Giglio, S. Musazzi, and U. Perini, *Transition to Chaotic Behavior via a Reproducible Sequence of Period-Doubling Bifurcations*, Phys. Rev. Lett. **47**, 243–246 (1981).

[29] J. L. Hudson and J. C. Mankin, *Chaos in the Belousov-Zhabotinskii Reaction*, J. Chem. Phys. **74**, 6171–6177 (1981).

[30] K. S. Jensen, E. Mosekilde, and N.-H. Holstein-Rathlou, *Self-Sustained Oscillations and Chaotic Behavior in Kidney Pressure Regulation*, Mondes en Develop. **54/55**, 91–109 (1986).

[31] J. Testa, J. Pérez, and C. Jeffries, *Evidence of Universal Chaotic Behavior of a Driven Nonlinear Oscillator*, Phys. Rev. Lett. **48**, 714–717 (1982).

[32] M. J. Feigenbaum, *Universal Behavior in Nonlinear Systems*, Los Alamos Science **1**, 4–27 (1980).

[33] M. J. Feigenbaum, *The Universal Metric Properties of Nonlinear Transformations*, J. Stat. Phys. **21**, 669–706 (1979).

[34] A. N. Sharkovsky, *Coexistence of Cycles of a Continuous Map of the Line Onto Itself*, Ukr. Mat. Z. **16**, 61–71 (1964), in Russian.

[35] N. Metropolis, M. L. Stein, and P. R. Stein, *On Finite Limit Sets for Transformations on the Unit Interval*, J. Comb. Theory **15**, 25–44 (1973).

[36] H. Poincaré, Les Methodes Nouvelles de la Mécanique Celeste, Vol. 1–3 (Gauthier–Villar, Paris, 1899).

[37] D. Nixon (Editor), Unsteady Transonic Aerodynamics (Progress in Astronautics and Aeronautics, Vol. 120) (American Institute of Aeronautics and Astronautics, 1989).

[38] O. O. Bendiksen and G.-Y. Hwang, *Nonlinear Flutter Calculations for Transonic Wings*, in Proc. CEAS Int. Forum on Aeroelastity and Structural Dynamics, Rome, Italy (1997), p. 105.

[39] H. Matsushita, M. Hashidate, K. Saitoh, Y. Ando, K. Fujii, K. Suzuki, and D. H. Baldelli, *Transonic Flutter Control of a High Aspect Ratio Wing: Mathematical Modeling, Control Law Design and Wind Tunnel Tests*, in Proc. 19th ICAS Cong., ICAS-94-5.6.2, Anaheim, CA (1994).

[40] H. Matsushita, K. Saitoh, and P. Gránásy, *Research on Active Control of Aeroelastic Systems at National Aerospace Laboratory in Japan*, Int. J. Intelligent Mechatron.; Design Prod. **3**, 34 (1998).

[41] L. E. Christiansen, T. Lehn-Schiøler, E. Mosekilde, P. Gránásy, and H. Matsushita, *Nonlinear Characteristics of Randomly Excited Transonic Flutter*, Math. Comp. Sim. **58**, 385–405 (2002).

[42] J. M. T. Thompson, H. B. Stewart, and Y. Ueda, *Safe, Explosive, and Dangerous Bifurcations in Dissipative Systems*, Phys. Rev. E **49**, 1019–1027 (1994).

[43] K. Wiesenfeld, *Noisy Precursors of Nonlinear Instabilities*, J. Stat. Phys. **38**, 1071–1097 (1985).

[44] C. Knudsen, R. Feldberg, and A. Jaschinski, *Non-Linear Dynamic Phenomena in the Behaviour of a Railway Wheelset Model*, Nonl. Dyn. **2**, 389–404 (1991).

[45] C. Knudsen, R. Feldberg, and H. True, *Bifurcations and Chaos in a Model of a Rolling Railway Wheelset*, Phil. Trans. Roy. Soc. London A **338**, 455–469 (1992).

[46] C. Knudsen, E. Slivsgaard, M. Rose, H. True, and R. Feldberg, *Dynamics of a Model of a Railway Wheelset*, Nonl. Dyn. **6**, 215–236 (1994).

[47] I. Grabec, *Chaos Generated by the Cutting Process*, Phys. Lett. A **117**, 384–387 (1986).

[48] I. Grabec, *Chaotic Dynamics of the Cutting Process*, Int. J. Mach. Tool. Manufact. **28**, 10–32 (1988).

[49] B. Feeny and F. C. Moon, *Chaos in a Forced Dry-Friction Oscillator: Experiments and Numerical Modeling*, J. Sound Vibr. **170**, 303–323 (1994).

[50] U. Galvanetto and C. Knudsen, *Event Maps in a Stick-Slip System*, Nonl. Dyn. **13**, 99–115 (1997).

[51] B. Blazejczyk-Okolewska, K. Czolczynski, T. Kapitaniak, and J. Wojewoda, Chaotic Mechanics in Systems with Impacts and Friction (World Scientific, Singapore, 1999).

[52] M. Wiercigroch and B. de Kraker, Applied Nonlinear Dynamics and Chaos of Mechanical Systems with Discontinuities (World Scientific, Singapore, 2000).

[53] *Nonlinear Dynamics in Mechanical Processing*, COST Action P4, Research Directorate General, European Commission, internal reports.

[54] G. S. Whiston, *Global Dynamics of a Vibro Impacting Linear Oscillator*, J. Sound Vibr. **118**, 395–429 (1987).

[55] A. B. Nordmark, *Non-Periodic Motion Caused by Grazing Incidence in an Impact Oscillator*, J. Sound Vibr. **145**, 279–297 (1991).

[56] G. S. Whiston, *Singularities in Vibro-Impact Dynamics*, J. Sound Vibr. **152**, 427–460 (1992).

[57] S. Foale, *Analytical Determination of Bifurcations in an Impact Oscillator*, Phil. Trans. Roy. Soc. Lond. A **347**, 353–364 (1994).

[58] S.-K. Choi and S. T. Noah, *Mode-Locking and Chaos in a Modified Jeffcott Rotor with Bearing Clearance*, Nonl. Vibr. **144**, 21–28 (1992).

[59] S. W. Shaw and S. Wiggins, *Chaotic Motions of a Torsional Vibration Absorber*, J. Appl. Mech. **55**, 952–958 (1988).

[60] R. Feldberg, M. Szymkat, C. Knudsen, and E. Mosekilde, *Iterated-Map Approach to Die Tossing*, Phys. Rev. A **42**, 4493–4502 (1990).

[61] L.-U. W. Hansen, M. Christensen, and E. Mosekilde, *Deterministic Analysis of the Probability Machine*, Phys. Scr. **51**, 35–45 (1995).

[62] M. Togeby, E. Mosekilde, and J. Sturis, *Frequency-Locking in a Model of Two Coupled Thermostatically Controlled Radiators*, in Proc. 1988 Winter Annual ASME-Meeting, Chicago, paper 88WA/DSC-14 .

[63] V. I. Arnol'd, *Small Denominators, I: Mappings of the Circumference into Itself*, AMS Trans. Series 2 **46**, 213–284 (1965).

[64] M. R. Guevara, L. Glass, and A. Shrier, *Phase Locking, Period-Doubling Bifurcations, and Irregular Dynamics in Periodically Stimulated Cardiac Cells*, Science **214**, 1350–1353 (1981).

[65] D. A. Rand, S. Ostlund, J. Sethna, and E. D. Siggia, *Universal Transition from Quasiperiodicity to Chaos in Dissipative Systems*, Phys. Rev. Lett. **49**, 132–135 (1982).

[66] P. Bak, T. Bohr, and M. H. Jensen, *Mode-Locking and the Transition to Chaos in Dissipative Systems*, Phys. Scr. **T9**, 50–58 (1984).

[67] A. Pikovsky, M. Rosenblum, and J. Kurths, Synchronization — A Universal Concept in Nonlinear Sciences (Cambridge Univ. Press, England, 2001).

[68] P. J. Myrberg, *Iteration von Quadratwurzeloperationen*, Ann. Acad. Sci. Fenn., Ser. A **256**, 1–10 (1958); *ibid.* **268**, 1–10 (1959); *ibid.* **336**, 1–10 (1963).

[69] P. Collet and J.-P. Eckmann, Iterated Maps of the Interval as Dynamical Systems (Birkhäuser, Boston, 1980).

[70] W. De Melo and S. van Strien, One-Dimensional Dynamics (Springer, New York, 1993).

[71] P. Coullet and C. Tresser, *Critical Transition to Stochasticity for Some Dynamical Systems*, J. de Phys. Lett. **41**, L255–L258 (1980).

[72] D. Singer, *Stable Orbits and Bifurcations of Maps of the Interval*, SIAM J. Appl. Math. **35**, 260–267 (1978).

[73] O. E. Landford, *A Computer-Assisted Proof of the Feigenbaum Conjectures*, Bull. Am. Math. Soc. **6**, 427–434 (1982).

[74] C. Mira, *Chaotic Hierarchy from the One-Dimensional Box-within-a-Box Bifurcation Structure Properties*, in A Chaotic Hierarchy, edited by C. Baier and M. Klein (World Scientific, Singapore, 1991), p. 285.

[75] R. V. Jensen and C. R. Myers, *Images of the Critical Points of Nonlinear Maps*, Phys. Rev. A **32**, 1222–1224 (1985).

[76] L. O. Chua and T. Lin, *Chaos in Digital Filters*, IEEE Trans. Circuits Syst. **35**, 648–658 (1988).

[77] E. F. Stikvoort, *Some Remarks on the Stability and Performance of the Noise Shaper or Sigma-Delta Modulator*, IEEE Trans. Comm. **36**, 1157–1162 (1988).

[78] E. Mosekilde, E. R. Larsen, and J. D. Sterman, *Coping with Complexity: Deterministic Chaos in Human Decisionmaking Behavior*, in Beyond Belief: Randomness, Prediction, and Explanation in Science, edited by J. L. Casti and A. Karlqvist (CRC Press, Boca Raton, 1991), pp. 199–229.

[79] J. S. Thomsen, E. Mosekilde, and J. D. Sterman, *Hyperchaotic Phenomena in Dynamic Decision Making*, J. Syst. Anal. Mod. Sim. **9**, 137–156 (1992).

[80] W. E. Jarmain, Problems in Industrial Dynamics (MIT Press, Cambridge, MA, 1963).

[81] J. D. Sterman, *Instructions for Running the Beer Distribution Game*, Working Paper D-3679, Sloan School of Management, MIT, Cambridge, MA (1984).

[82] R. L. Devaney, An Introduction To Chaotic Dynamical Systems (Addison-Wesley, 1989).

[83] B. Derrida, A. Gervois, and Y. Pomeau, *Iteration of Endomorphisms on the Real Axis and Representation of Numbers*, Ann. Inst. Henri Poincaré **29A**, 305–356 (1978).

[84] Yu. L. Maistrenko, V. L. Maistrenko, and L. O. Chua, *Cycles of Chaotic Intervals in a Time-Delayed Chua's Circuit*, Int. J. Bifurcation and Chaos **3**, 1557–1572 (1993).

[85] Yu. L. Maistrenko, V. L. Maistrenko, S. I. Vikul, and L. O. Chua, *Bifurcations of Attracting Cycles from Time-Delayed Chua's Circuit*, Int. J. Bifurcation and Chaos **5**, 653–671 (1995).

[86] M. I. Feigin, *Doubling of the Oscillation Period with C-Bifurcations in Piecewise Continuous Systems*, Prikl. Mat. Mekh. **34**, 861–869 (1970), in Russian.

[87] M. I. Feigin, *On the Generation of Sets of Subharmonic Modes in a Piecewise Continuous System*, Prikl. Mat. Mekh. **38**, 810–818 (1974), in Russian.

[88] M. I. Feigin, *On the Structure of C-Bifurcation Boundaries of Piecewise Continuous Systems*, Prikl. Mat. Mekh. **42**, 820–829 (1978), in Russian.

[89] M. I. Feigin, Forced Oscillations in Systems with Discontinuous Nonlinearities (Nauka Publ., Moscow, 1994), in Russian.

[90] H. E. Nusse and J. A. Yorke, *Border-Collision Bifurcations Including "Period Two to Period Three" for Piecewise Smooth Systems*, Physica D **57**, 39–57 (1992).

[91] H. E. Nusse, E. Ott, and J. A. Yorke, *Border-Collision Bifurcations: An Explanation for Observed Bifurcation Phenomena*, Phys. Rev. E **49**, 1073–1076 (1994).

[92] H. E. Nusse and J. A. Yorke, *Border-Collision Bifurcation for Piecewise Smooth One-Dimensional Maps*, Int. J. Bifurcation and Chaos **5**, 189–207 (1995).

[93] Yu. L. Maistrenko, V. L. Maistrenko, and S. I. Vikul, *On Period-Adding Sequence of Attracting Cycles in Piecewise Linear Maps*, Chaos, Solitons and Fractals **9**, 67–75 (1998).

[94] T. Matsumoto, L. O. Chua, and M. Komuro, *The Double Scroll*, IEEE Trans. Circuits Syst. **32**, 798–818 (1985).

[95] L. O. Chua, *Global Unfolding of Chua's Circuit*, IEICE **E76-A**, 704–734 (1993).

[96] I. Gumowski and C. Mira, Dynamique Chaotique (Cepadues Editions, Toulouse, 1980), in French.

[97] L. Gardini, *Some Global Bifurcations of Two-Dimensional Endomorphisms by Use of Critical Lines*, Nonlinear Anal. Theor. Appl. **18**, 361–399 (1992).

[98] C. Mira, C. Rauzy, Yu. L. Maistrenko, and I. M. Sushko, *Some Properties of a Two-Dimensional Piecewise Linear Noninvertible Map*, Int. J. Bifurcation and Chaos **6**, 2299–2319 (1996).

[99] M. di Bernardo, M. I. Feigin, S. J. Hogan, and M. E. Homer, *Local Analysis of C-Bifurcations in n-Dimensional Piecewise-Smooth Dynamical Systems*, Chaos, Solitons and Fractals **10**, 1881–1908 (1999).

[100] M. di Bernardo, C. J. Budd, and A. R. Champneys, *Grazing and Border-Collision in Piecewise-Smooth Systems: a Unified Analytical Framework*, Phys. Rev. Lett. **86**, 2553–2556 (2001).

[101] D. J. Uherka, *Tongue Art*, CNLS newsletter **78**, 1–23 (1992).

[102] W.-M. Yang and B.-L. Hao, *How the Arnol'd Tongues Become Sausages in a Piecewise Linear Circle Map*, Commun. Theor. Phys. **8**, 1–15 (1987).

[103] B. L. Hao, Elementary Symbolic Dynamics and Chaos in Dissipative Systems (World Scientific, Singapore, 1989).

[104] P. Alstrøm, *Map Dependence of the Fractal Dimension Deduced from Iterations of the Circle Map*, Commun. Math. Phys. **104**, 581 (1986).

[105] M. Szymkat and E. Mosekilde, *Global Bifurcation Analysis of an Economic Long-Wave Model*, in Proc. 2nd European Simulation Conf., Antwerp (Simulation Councils Inc., San Diego, 1986), pp. 139–145.

[106] N. D. Kondratieff, *The Long Wave in Economic Life*, Rev. Econ. Stat. **17**, 105–115 (1935).

[107] J. van Duijn, The Long Wave in Economic Life (Allan and Unwin, London, 1963).

[108] W. W. Rostow, The World Economy. History and Prospects (MacMillan Press, London, 1978).

[109] E. Mansfield, *Long Waves in Technological Innovation*, Am. Econ. Ass. Papers and Proceedings **73**, 141–145 (1983).

[110] E. Mosekilde and S. Rasmussen, *Technical Economic Succession and the Economic Long Wave*, Eur. J. Operational Res. **25**, 27–38 (1986).

[111] J. D. Sterman, *A Behavioral Model of the Economic Long Wave*, J. Econ. Beh. Org. **6**, 17–53 (1985).

[112] S. Rasmussen, E. Mosekilde, and J. D. Sterman, *Bifurcations and Chaotic Behavior in a Simple Model of the Economic Long Wave*, Syst. Dyn. Rev. **1**, 92–108 (1985).

[113] E. Mosekilde, E. R. Larsen, J. D. Sterman, and J. S. Thomsen, *Non-Linear Mode-Interaction in the Macroeconomy*, Ann. Operations Res. **37**, 185–215 (1992).

[114] C. Haxholdt, C. Kampmann, E. Mosekilde, and J. D. Sterman, *Entrainment in a Disaggregated Economic Long Wave Model*, Syst. Dyn. Rev. **11**, 177–198 (1995).

[115] J. W. Forrester, Industrial Dynamics (MIT Press, Cambridge, MA, 1961).

[116] S. Lefschetz, Differential Equations — Geometric Theory (Wiley, New York, 1963).

Chapter 2
BASIC CONCEPTS AND METHODS

2.1 Bifurcation Analysis for the Hénon Map

Nonlinear dynamics makes use of a number of archetype models. Within the area of dissipative nonlinear systems the logistic map, the sine-circle map, the tent map, Duffing's equation, and the Rössler system discussed in Chapter 1 are among the most well-known models. Other important models are the Hénon map, that we shall consider in the present chapter, the Lorenz system, the impact oscillator, the Chua circuit and, particularly in the Russian literature, various models of autonomous and non-autonomous generators with inertial nonlinearity [1]. For Hamiltonian systems and in the area of quantum chaos, other models, such as the twist map and the stadium model, play the role of archetypes. Some of these models derive directly from practical problems in science and technology, while other models have been constructed as the simplest possible set of equations required to demonstrate a particular dynamic phenomenon and isolate it from the parasitic influence of a more complicated structure. The logistic model, for instance, is the basic model for the description of the classic period-doubling route to chaos, and the sine-circle map is a starting point for discussions of phase locking (or synchronization) of interacting oscillators.

In recent years, coupled logistic maps and coupled Rössler systems have been used to study riddled basins of attraction, on-off intermittency, and different forms of *chaotic synchronization* [2]. Significant advances have also been made in the study of wave propagation and *spatio-temporal chaos* in lattices of chaotic oscillators and of various forms of *clustering* in systems of globally (i.e., all-to-all) coupled oscillators. These types of behavior may be particularly relevant to living systems where large numbers of cells or functional units, each displaying complicated nonlinear phenomena, cooperate to produce a coherent response on a higher organizational level. The smooth muscle cells in the walls of our arterioles, for instance, must switch into a state of synchronization to produce a contraction of the vessel, thus reducing the blood flow. It is also possible that changes between different states of synchronization among the functional units of the kidney play a role in the regulation of our blood pressure and salt balance. In these cases, the synchronization transitions are associated with normal physiological function. On the other hand, it is well-known that synchronization of the firing activity for larger groups of cells in the brain may be an underlying cause of epilepsia and certain forms of tremor.

The *Hénon map* [3] is a two-dimensional mapping of the form

$$
\begin{aligned}
&X_k = F(X_{k-1}),\ k = 1, 2, \ldots \qquad (2.1)\\
&X = \begin{pmatrix} x_1 \\ x_2 \end{pmatrix};\ F = \begin{pmatrix} f_1 \\ f_2 \end{pmatrix};\\
&f_1 = 1 - \alpha x_1^2 + x_2;\\
&f_2 = \beta x_1.
\end{aligned}
$$

Here α represents the nonlinearity in the system and β ($|\beta| < 1$) is a measure of the dissipation. For $\beta = 0$, the Hénon map reduces to the logistic map in the form $x_k = 1 - \alpha x_{k-1}^2$, and for sufficiently small values of β (i.e. in the limit of strong dissipation) one expects the bifurcation structure of the Hénon map to be similar to that of the logistic map.

The mapping (2.1) was originally proposed by Hénon and Pomeau [4] as a simplified version of the Poincaré map for a three-dimensional flow system. Since then it has been extensively studied both numerically and from a more theoretical point of view (for references see, e.g., [1, 5–9]). Aiming at a better understanding of strange attractors and chaotic dynamics in two-dimensional (invertible) maps, Hénon and Pomeau specified the parameter values to be $\alpha = 1.4$ and $\beta = 0.3$, and similar values have been used in a number of subsequent studies. It is worth mentioning, however, that for positive values of β the mapping (2.1) reverses the orientation in the phase plane. Hence, it cannot be the Poincaré map of a three-dimensional flow.

Currey [10] was presumably first to observe that there exists a range of parameters for which two chaotic attractors coexist, and Derrida *et al.* [11] showed how the period-doubling cascade in the Hénon map for certain parameter values follows the Feigenbaum scenario. At about the same time, Newhouse [12] found that there are regions in parameter space in which a dense set of stable periodic orbits (sinks) with extremely small basins of attraction may exist. A decade later, this result was corroborated by Robinson [13], and today understanding the so-called *Newhouse regions* represents one of the major challenges to the description of dynamical systems. For more recent discussions of this problem we can refer the reader to the publications by Kan *et al.* [14] and by Gonchenko *et al.* [15].

The variety of different motions generated by the two-dimensional Hénon map in dependence of the parameters [16–18] and the characteristics of the transition to chaos are typical to a wide class of dynamical systems. As shown, for instance, by Anishchenko [5] the map (2.1) can be used as a good approximation to a three-dimensional model of a radiophysical generator with inertial nonlinearity. In particular, the organization of bifurcations in the plane of the control parameters α and β was found to be qualitatively equivalent to the bifurcations that are realized in the Poincaré map of the full three-dimensional model. Similar forms of equivalence can be found when analyzing models of several classes of automatic control systems, including systems with pulse-width modulation and relay system with hysteresis [19–23]. Such systems will be discussed in considerable detail in the following chapters. Let also notice that Lozi [24] has proposed a piecewise linear version of the Hénon map and that Misiurewicz [25] has shown that this map has a proper chaotic attractor, i.e., that sinks embedded in an otherwise chaotic attractor do not occur.

The characteristic organization of the domains of stability, referred to as a *swallow tail* or a *crossroad area*, and denoted here as Π_k^{Swallow}, that one finds in the Hénon map can be observed for a wide variety of different nonlinear dynamical systems, including the sine-circle map [26, 27] and other bimodal one-dimensional maps [28, 29] as well as various time-continuous nonlinear dynamical systems [30–32]. Detailed analyses of the swallow tail structure have been performed by Kuznetsov *et al.* [33–35] and, from a more mathematical point of view, by Mira *et al.* [36] (see also references in these works). It was shown, for instance, by Kuznetsov *et al.* [34] that a quantitative universality in the transition to chaos in two-dimensional Hénon-like maps exists. It is worth

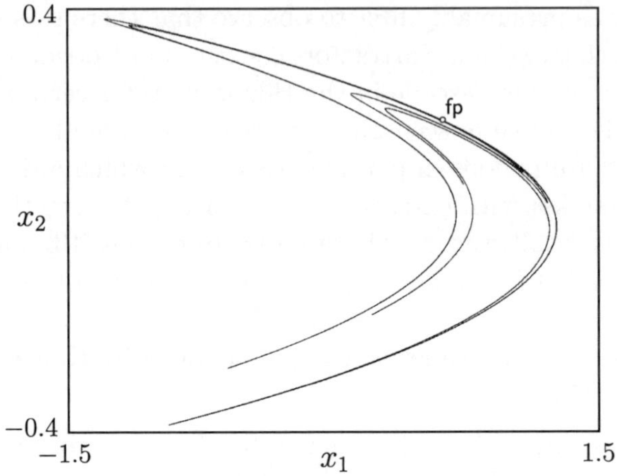

Figure 2.1. Hénon attractor as calculated from the two-dimensional mapping (2.1) for $\alpha = 1.4$ and $\beta = 0.3$. (An initial transient before the orbit settles in on the attractor has been deleted).

noticing, however, that the finer details of the structure, particularly around the so-called tricritical points, may differ between time-continuous systems and their bimodal map representations [37].

Let us finally mention that Hansen and Cvitanović [18] have described the bifurcation structure for a class of generic Hénon maps, i.e. maps with one folding per iteration, by constructing a series of n-unimodal approximations to the Hénon map and investigating the associated *symbolic dynamics.*

Figure 2.1 shows the results of plotting 10^5 successive points of the Hénon map. In this figure we have also indicated the position of the fixed point (denoted as fp)

$$x_{1c} = \frac{1}{2\alpha}\left[\beta - 1 + \sqrt{(1-\beta)^2 + 4\alpha}\right];$$
$$x_{2c} = \beta x_{1c}.$$

With $\alpha = 1.4$ and $\beta = 0.3$ this fixed point is a saddle point, and the points of the chaotic attractor fall along its infinitely folded unstable manifold.

The purpose of the present chapter is to use the Hénon map to introduce some of the basic concepts and methods that we are going to apply in the following chapters. We first determine the periodic orbits and examine their stability as a function of the parameters of the Hénon map. In this way we can construct a two-dimensional bifurcation diagram that partitions the parameter space into regions of existence for the different solutions [9]. We shall also discuss the characteristic folded structure of the one-dimensional bifurcation diagram and illustrate the structure of the basins of attraction under conditions when the map displays coexisting solutions. Our approach will be numerical and engineering oriented, leaving the discussion of some of the more theoretical problems to the final section. Nonetheless, the presentation will clearly illustrate how complicated the dynamics can be, even for relatively simple systems.

2.2 Division of the Parameter Plane

Let us begin with some preliminary notes.

Let X_{ci}, $i = \overline{1,m}$ be a periodic motion (an m-cycle) of the dynamic system (2.1). This implies that all X_{ci}, $i = \overline{1,m}$ satisfy the following equation

$$X_c - F^{(m)}(X_c) = 0, \; F^{(m)} \equiv \underbrace{F \circ F \circ \cdots \circ F}_{m\,times}. \tag{2.2}$$

The local stability of the m-cycle is determined by the eigenvalues ρ_1 and ρ_2 of the basic matrix Φ_m. These eigenvalues (or *Floquet multipliers*) are the roots of the equation

$$\det(\Phi_m - \rho E) = 0,$$

with

$$\Phi_i = \frac{\partial F(X_{c(i-1)})}{\partial X_{c(i-1)}} \Phi_{i-1}, \; \Phi_0 = E, \; i = \overline{1,m};$$

and, as evaluated for the Hénon map,

$$\frac{\partial F(X_{c(i-1)})}{\partial X_{c(i-1)}} = \begin{pmatrix} -2\alpha x_{1c(i-1)} & 1 \\ \beta & 0 \end{pmatrix}.$$

Here, E denotes the unit matrix, and $x_{1c(i-1)}$, $i = \overline{1,m}$ represent the values assumed by x_1 along the m-cycle. The m-cycle is locally stable, if both the multipliers ρ_1 and ρ_2 are located within the unit circle in the complex plane.

Let $\Pi = \{(\alpha, \beta) : \mu < \alpha < \nu; \; \vartheta < \beta < \gamma\}$ be the set of parameters to be considered for the dynamic system (2.1). A point in the plane $P = (\alpha, \beta) \in \Pi$ thus corresponds to a specific choice of parameter values. Let $X_{ci}(P)$, $i = \overline{1,m}$ denote a locally stable m-cycle corresponding to the point P. We would like to determine the set of periodic solutions that exist in Π and, in order to delineate the regions of existence for each of these solutions, to examine the bifurcations through which they lose their stability.

Let $\Pi_{k,j}$ be a simply connected set of parameters $\Pi_{k,j} \subset \Pi$, $k \geq 1$, $j = 1, 2, \ldots, s$ such, that for every $P \in \Pi_{k,j}$ there exists a stable k-cycle $X_{ci}(P)$, $i = \overline{1, k}$, which varies continuously with the parameters in $\Pi_{k,j}$. If several motions exist for a fixed P, different $\Pi_{k,j}$ correspond to different motions. The k-cycle for $\Pi_{k,j}$ is minimum, that is, there is no m-cycles with $m < k$ in $\Pi_{k,j}$. The index j is introduced to distinguish sets, having the same k. The value of s may be finite or infinite.

The boundaries of the set $\Pi_{k,j}$ may be exogenous, that is, there exists a stable m-cycle for any points $P \in \Gamma_{k,j}$, delineated by a restriction to the range of parameter variation. Alternatively, the boundaries $\Gamma_{k,j}$ may be formed by a collection of bifurcation points or by a collection of *points of accumulation*, corresponding to a threshold to chaos.

The point $P_* \in \Pi$ is a bifurcation point for X_{ci}, if the equation

$$\chi(P_*, \rho) = \det(\Phi_m - \rho E) = \rho^2 + \xi_1 \rho + \xi_2 = 0$$

has a root that falls on the unit circle in the complex plane. Here, $\xi_1 = \mathrm{tr}(\Phi_m)$ and $\xi_2 = \det(\Phi_m)$ denote the trace and determinant of the basic matrix, respectively.

The m-cycle loses its stability when the (numerically) larger multiplier crosses the unit circle. In the bifurcation point the multiplier becomes equal to $+1$ or to -1. This transition is referred to as *local bifurcation*. The corresponding bifurcation curves are determined by the equations

$$\chi(P, 1) = 1 + \xi_1 + \xi_2 = 0 \tag{2.3}$$

and

$$\chi(P, -1) = 1 - \xi_1 + \xi_2 = 0. \tag{2.4}$$

In accordance with standard textbooks [38, 39] we denote these curves by N_+ and N_-, respectively. N_+ represents a curve of saddle-node bifurcations where a saddle cycle and a node meet and are annihilated. N_- represents a curve of period-doubling bifurcations where the m-cycle loses its stability while producing a new stable cycle of twice the period. For the dissipative system (2.1), the case where two complex conjugated eigenvalues cross the unit circle, corresponding to a Hopf bifurcation, cannot occur. This follows from the fact that the product of the two eigenvalues $\rho_1\rho_2 = |\rho_i|^2 = \det(\Phi_m) = (-\beta)^m$, $i = 1, 2$ cannot be equal to 1 because $|\beta| < 1$.

It is interesting to study the properties of the sets $\Pi_{k,j}$, particularly the possibility of non-empty intersections of $\Pi_{k,j}$, in order to understand the great variety of different motions that are generated by the dynamical system (2.1). We may recall from our discussion in Chapter 1 that the logistic map for $\lambda \in [1, 4]$ has a unique attracting state and never exhibits coexisting solutions. In the following we shall restrain our examination to the region $\Pi = \{(\alpha, \beta) : -1 < \alpha < 4; |\beta| < 1\}$.

Let us first consider the solutions to Eq. (2.2) and examine the local stability properties for some of the periodic cycles.

For $m = 1$, the equation for the Hénon map gives

$$\begin{aligned} \alpha x_{1c}^2 + (1-\beta)x_{1c} - 1 = 0; \\ x_{2c} = \beta x_{1c} \end{aligned} \tag{2.5}$$

and, as long as $(1-\beta)^2 + 4\alpha > 0$ and $\alpha \neq 0$, the roots of this equation determine the coordinates of two period-1 cycles (fixed points):

$$\begin{aligned} x_{1c} &= \frac{1}{2\alpha}\left[\beta - 1 \pm \sqrt{(1-\beta)^2 + 4\alpha}\right]; \\ x_{2c} &= \beta x_{1c}. \end{aligned}$$

The solution with the plus sign is born as a stable node and the other solution as a saddle point.

The domain of existence for the stable fixed point is determined by Eqs. (2.3) and (2.4). When inserting the coordinates of the fixed point into the expressions for ξ_1 and ξ_2, the bifurcation curves N_+ and N_- are obtained as

$$(1-\beta)^2 + 4\alpha = 0 \tag{2.6}$$

and

$$3(1-\beta)^2 - 4\alpha = 0, \tag{2.7}$$

respectively. Let us denote this *domain of stability* as $\hat{\Pi}_{1,1}$:

$$\hat{\Pi}_{1,1} = \left\{(\alpha, \beta) : -\frac{1}{4}(\beta - 1)^2 < \alpha < \frac{3}{4}(\beta - 1)^2, |\beta| < 1\right\}.$$

If the parameters $P \in \tilde{\Pi}_1$:

$$\tilde{\Pi}_1 = \left\{(\alpha, \beta) : -1 < \alpha < -\frac{1}{4}(\beta - 1)^2, |\beta| < 1\right\},$$

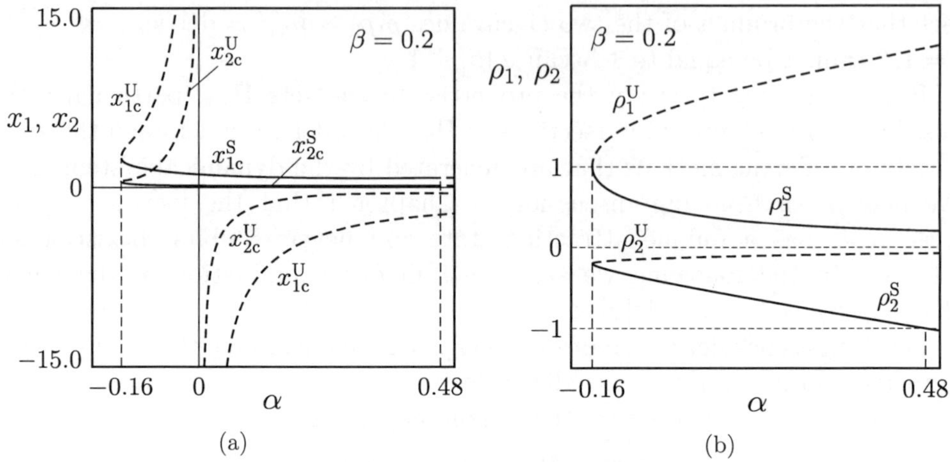

Figure 2.2. (a) Variation of the stable and unstable period-1 cycles with the nonlinearity parameter α. Note the different nature of these dependences through the point $\alpha = 0$. (b) Corresponding variation of the multipliers ρ_1 and ρ_2.

then Eq. (2.5) has no real roots, and along the bifurcation curve (2.6), $x_{1c} = (\beta - 1)/2\alpha$, $x_{2c} = \beta x_{1c}$ is a double root. Hence, when changing the parameters along some smooth curve (referred to as the *trajectory of deformation*) from $\tilde{\Pi}_1$ into $\hat{\Pi}_{1,1}$, two solutions arise abruptly in a saddle-node bifurcation, one corresponding to the stable fixed point and the other to the saddle point. In the reverse transition, the stable fixed point disappears as it is annihilated with the saddle point at the point of intersection between the bifurcation curve (2.6) and the trajectory of deformation. For $\beta = 0.2$, the saddle-node bifurcation takes place at $\alpha = -0.16$ (see, e.g., Fig. 2.2 and Fig. 2.3).

Let us use a moment to consider what happens to the two fixed points as α changes sign. For $\alpha = 0$, Eq. (2.2) only has one solution

$$x_{1c} = \frac{1}{1-\beta}, \; x_{2c} = \frac{\beta}{1-\beta}.$$

It is obvious, that

$$\lim_{\alpha \to \pm 0} x_{1c}^S = \frac{1}{1-\beta} \quad \text{and} \quad \lim_{\alpha \to \pm 0} x_{2c}^S = \frac{\beta}{1-\beta}.$$

We conclude that the solution to Eq. (2.5) that corresponds to the stable 1-cycle X_c^S depends smoothly on the parameters in $\hat{\Pi}_{1,1}$. The saddle point

X_c^U varies continuously in $\hat{\Pi}_{1,1}$ except for the points in which $\alpha = 0$. Here, the solution X_c^U displays an infinite discontinuity associated with a vertical asymptote.

The different nature of the parameter dependences for the solutions X_c^S and X_c^U is illustrated in Fig. 2.2(a). The corresponding variations of the multipliers of the stable and unstable 1-cycles are given in Fig. 2.2(b). As long as both of its multipliers fall between -1 and 1, the fixed point is stable. When the trajectory of deformation and the bifurcation curve (2.7) intersect, the fixed point loses its stability, and a stable 2-cycle arises softly as the result of a period-doubling bifurcation. For $\beta = 0.2$ (see Fig. 2.2(b) and Fig. 2.3), the period-doubling bifurcation occurs at $\alpha = 0.48$. The fixed point continues to exist as a saddle point over the whole range of parameters beyond the bifurcation curve (2.7).

For $m = 2$ equation (2.2) takes the form

$$\alpha^3 x_{1c}^4 - 2\alpha^2 x_{1c}^2 + (1-\beta)^3 x_{1c} - (1-\beta)^2 + \alpha = 0;$$
$$x_{2c} = \frac{\beta}{1-\beta}(1 - \alpha x_{1c}^2).$$

This equation has four real roots in the domain

$$\alpha > \frac{3}{4}(\beta - 1)^2.$$

Two of the roots correspond to unstable fixed points, and it is not difficult to show that the 2-cycle satisfies the equation

$$\begin{gathered} \alpha^2 x_{1c}^2 - \alpha(1-\beta)x_{1c} + (1-\beta)^2 - \alpha = 0; \\ x_{2c} = \frac{\beta}{1-\beta}(1 - \alpha x_{1c}^2). \end{gathered} \tag{2.8}$$

The set $\hat{\Pi}_{2,1}$, in which the stable 2-cycle is defined, is limited by the bifurcation curve (2.7) to the left and by a curve, defined by the equation

$$5\beta^2 - 6\beta - 4\alpha + 5 = 0 \tag{2.9}$$

on the other side. Hence,

$$\hat{\Pi}_{2,1} = \left\{ (\alpha, \beta) : \frac{3}{4}(\beta - 1)^2 \le \alpha \le \frac{1}{4}(5\beta^2 - 6\beta + 5), |\beta| < 1 \right\}.$$

When passing through the boundary (2.9) into the domain $\hat{\Pi}_{2,1}$ a stable 4-cycle arises softly, and the 2-cycle continues to exist, but becomes a saddle. For $\beta = 0.2$ (see Fig. 2.3), the second period-doubling bifurcation occurs at $\alpha = 1.0$.

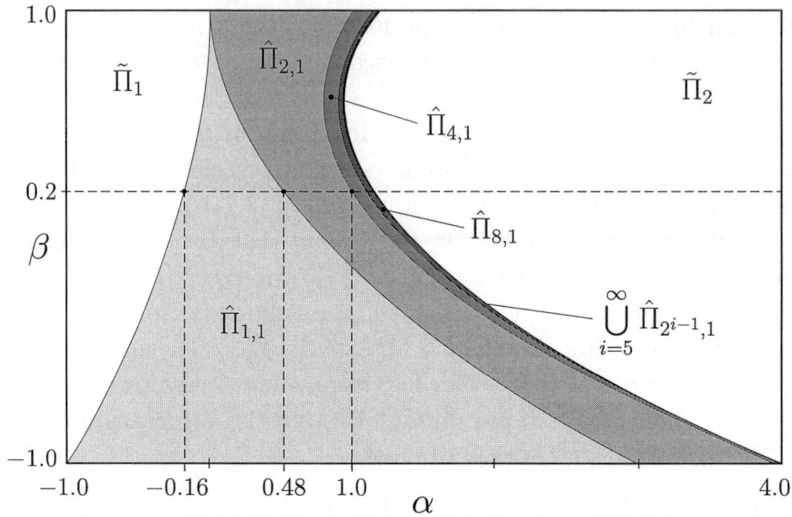

Figure 2.3. Collection of the sets $\hat{\Pi}_{2^{i-1},1}$, $i = 1, 2, \ldots$ that form $\Pi_{1,1}$. $\tilde{\Pi}_1$ and $\tilde{\Pi}_2$ denote domains of Π where the trajectories diverge to infinity for all initial conditions.

For $m = 2^{i-1}$, $i = 3, 4, \ldots$ it becomes increasingly difficult (if not impossible) to obtain the bifurcation curves in explicit form. Hence, the sets $\hat{\Pi}_{2^{i-1},1}$, $i = 3, 4, \ldots$ were obtained numerically. The collection of $\hat{\Pi}_{2^{i-1},1}$, $i = 1, 2, \ldots$ which forms $\Pi_{1,1}$, is shown in Fig. 2.3:

$$\Pi_{1,1} = \bigcup_{i=1}^{\infty} \hat{\Pi}_{2^{i-1},1}.$$

The set $\Pi_{1,1}$ consists of domains of existence of locally stable 2^{i-1}-cycles ($i = 1, 2, \ldots$). The boundaries, separating these domains in $\Pi_{1,1}$ correspond to period-doubling bifurcation curves. The set $\Pi_{1,1}$ is limited by the bifurcation curve (2.6) from the left and by a curve formed by the collection of accumulation points from the right.

Now consider

$$\Pi_{k,j} = \bigcup_{i=1}^{\infty} \hat{\Pi}_{2^{i-1}k,j}$$

for $k \neq 1$. From equation (2.2) with $m = 3$ we find

$$\alpha^6 x_{1c}^6 - \alpha^5(1-\beta)x_{1c}^5 + \alpha^4\big((1-\beta)^2 - 3\alpha\big)x_{1c}^4$$

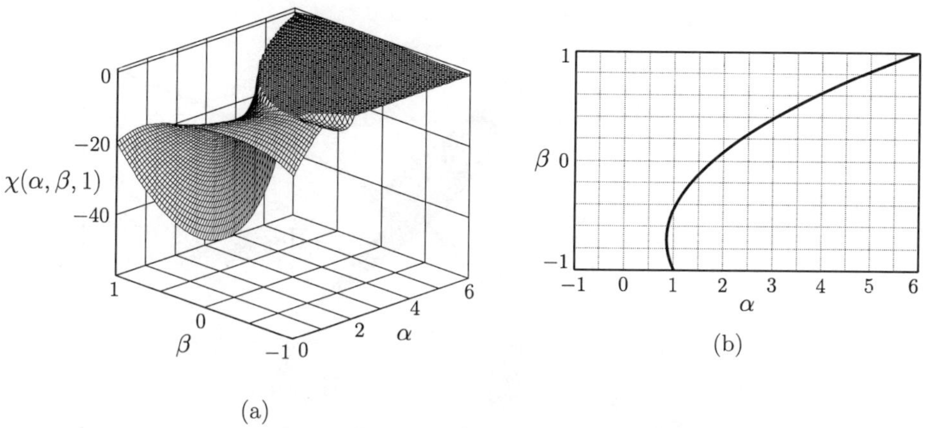

Figure 2.4. (a) 2D surface on which the bifurcation curve N_+ for the 3-cycle lies. (b) Curve of saddle-node bifurcation for the 3-cycle.

$$
\begin{aligned}
&-\alpha^3(1-\beta)(1+\beta^2-2\alpha)x_{1c}^3\\
&+\alpha^2\big(\beta^4-\beta^2+1-(3\alpha+2\beta)(1-\alpha+\beta^2)\big)x_{1c}^2\\
&+\alpha(\beta-1)\big((1+\beta+\beta^2)^2+\alpha^2-2\alpha\big)x_{1c}\\
&+1+\beta+\beta^2+\beta^4+\beta^5+\beta^6\\
&-\alpha(1-2\alpha+\alpha^2-4\beta-5\beta^2-4\beta^3+\beta^4-2\alpha\beta^2)=0.
\end{aligned}
\tag{2.10}
$$

The roots of this equation can only be found numerically. Let us try to find an equation for the bifurcation curve N_+ for the 3-cycle. Considering that Eq. (2.10), with parameter values located on the curve N_+, only has double roots, we obtain

$$x_{1c}^3-\frac{1-\beta}{2\alpha}x_{1c}^2+\sigma x_{1c}+\vartheta=0.$$

Here,

$$\sigma=\frac{1}{2\sqrt{\vartheta}}(2\alpha-\alpha^2-1-\beta-\beta^2+\beta^3+\beta^4+\beta^5+\alpha^2\beta-2\alpha\beta^3)$$

and

$$\vartheta=1+\beta+\beta^2+\beta^4+\beta^5+\beta^6-\alpha(1-2\alpha+\alpha^2-4\beta-5\beta^2-4\beta^3+\beta^4-2\alpha\beta^2).$$

Hence, the desired bifurcation curve will lie on the two-dimensional surface (Fig. 2.4(a))

$$\chi(\alpha,\beta,1)=-8\sqrt{\mu(\alpha,\beta)}+(1-\beta)(1+\beta)^2,$$

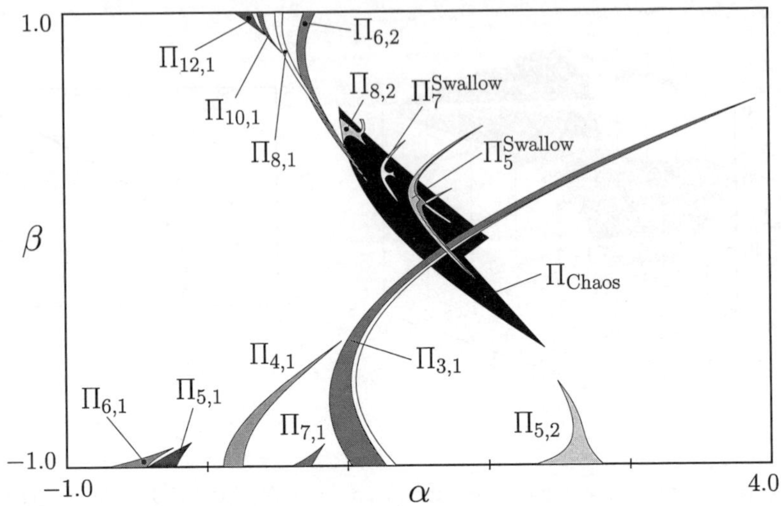

Figure 2.5. Collection of the sets $\Pi_{k,1}$, $k = 3, 5, 6, 7, 8, 10, 12$ and $\Pi_{k,2}$, $k = 5, 6, 8$. The black region denoted Π_{Chaos} is the domain of chaotic dynamics.

with

$$\mu(\alpha, \beta) = 1 + \beta + \beta^2 + \beta^4 + \beta^5 + \beta^6 \\ - \alpha(1 + \alpha^2 - 4\beta - 5\beta^2 - 4\beta^3 + \beta^4) + 2\alpha^2(1 + \beta^2).$$

We can find an equation for the bifurcation curve, by setting $\chi(\alpha, \beta, 1)$ equal to zero:

$$-8\sqrt{\mu(\alpha, \beta)} + (1 - \beta)(1 + \beta)^2 = 0.$$

The form of this curve is presented in Fig. 2.4(b). The remaining domains $\hat{\Pi}_{2^{i-1}\cdot 3,1}$, $i = 2, 3, \ldots$, as well as $\hat{\Pi}_{2^{i-1},1}$, $i = 3, 4, \ldots$, were obtained numerically.

Figure 2.5 shows the collection

$$\Pi_{3,1} = \bigcup_{i=1}^{\infty} \hat{\Pi}_{2^{i-1}\cdot 3,1};$$

$$\Pi_{k,1} = \bigcup_{i=1}^{\infty} \hat{\Pi}_{2^{i-1}k,1}, \; k = 5, 6, 7, 8, 10, 12,$$

and

$$\Pi_{k,2} = \bigcup_{i=1}^{\infty} \hat{\Pi}_{2^{i-1}k,2}, \; k = 5, 6 \text{ and } 8.$$

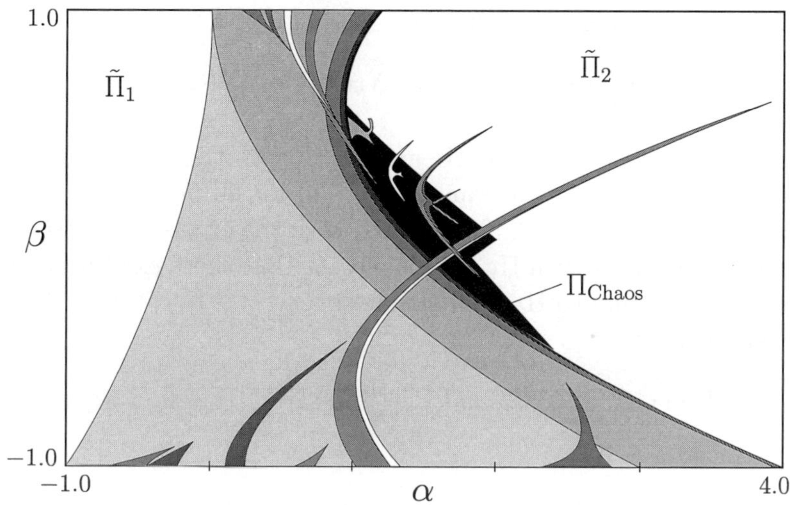

Figure 2.6. Resulting picture for the division of the parameter plane for the Hénon map into domains of various oscillatory modes.

Obviously, Fig. 2.5 represents only those sets that are relatively large in Π.

Figure 2.6 shows the resulting picture for the division of the parameter plane $\Pi = \{(\alpha, \beta) : -1 < \alpha < 4; |\beta| < 1\}$ of the Hénon map into domains of different oscillatory modes. The domains of chaotic dynamics adjoin to those parts of the boundaries of $\Pi_{k,j}$ that are formed by sets of accumulation points. The diagram depicts only one of the chaotic domains. This is the domain that occupies the largest region in the parameter plane. In Figs. 2.5 and 2.6 this domain is shown by a dark shading and denoted Π_{Chaos}. The other chaotic domains are too narrow to be visible in the diagram. There is a great number (presumably a dense set) of windows with regular dynamics within Π_{Chaos}. They are delineated to the left by curves of saddle-node bifurcation of m-cycles. The inner structure of these domains may be similar to that for $\Pi_{k,j}$ or it can be different. Domains, whose properties are different from the properties of $\Pi_{k,j}$, are denoted as Π_k^{Swallow} (see Fig. 2.5). As previously noted, the structure of Π_k^{Swallow} and the bifurcations associated with such domains were described in detail by Mira *et al.* [36, 40] and by Kuznetsov *et al.* [33–35]. We shall review some of these results in the following section.

For any point $P \in \Pi$, with the exception of $P = (0, 0)$, there is such a (simply or not simply connected) domain D_* in the phase space of the dynamic system (2.1) that if $X_0 \in D_*$, then

$$\lim_{k\to\infty} ||X_k(X_0)|| = \infty, \; X_k(X_0) = F^{(k)}(X_0). \tag{2.11}$$

In Fig. 2.6 (cf. also Fig. 2.3) $\tilde{\Pi}_1$ and $\tilde{\Pi}_2$ denote those domains of Π (not hatched), where Eq. (2.11) with any X_0 is valid for all points $P \in \tilde{\Pi}_1 \cup \tilde{\Pi}_2$.

As seen in Figs. 2.5 and 2.6, all sets $\Pi_{k,j}$, with the exception of Π_k^{Swallow}, have non-empty intersections with $\Pi_{1,1}$. This means, that there exist different stable m-cycles, corresponding to $\Pi_{i,l}$ and $\Pi_{q,p}$ for every $P \in \Pi_{i,l} \cap \Pi_{q,p}$.

It is easy to see, that

$$\Pi_{3,1} \cap \Pi_5^{\text{Swallow}} \neq \emptyset \text{ and } \Pi_i^{\text{Swallow}} \cap \Pi_j^{\text{Swallow}} \neq \emptyset, \; i \neq j.$$

2.3 Further Analysis of the Hénon Map

With the above introductory discussion let us now elaborate on the bifurcation scenarios observed when moving along the parameters in the domain

$$\left(\bigcup_{k=1}^{\infty} \bigcup_{j=1}^{s} \Pi_{k,j} \right) \cup \Pi_{\text{Chaos}}.$$

To be specific we shall consider the sections $\beta = 0.3$, $\beta = 0.68$, and $\beta = 0.5$ when changing the parameter α over the range

$$\Omega_\alpha = \left\{ \alpha : -\frac{1}{4}(\beta - 1)^2 < \alpha < 4, \; \beta = \text{const}, \; |\beta| < 1 \right\}.$$

The bifurcation diagrams calculated for these values of β and for the indicated range of α are depicted in Figs. 2.7(a), 2.8(a), and 2.9. These figures were obtained as so-called brute force bifurcation diagrams where the active parameter is slowly scanned while for each value of this parameter the values attained by the stable (periodic or chaotic) orbit are plotted on the vertical axis. To observe several coexisting solutions the diagram has been scanned from both sides, or several different initial conditions have been considered. In a brute force bifurcation diagram one cannot follow the unstable solutions. Hence, one cannot detect from where a stable periodic solution originates when it suddenly arises in a saddle-node bifurcation.

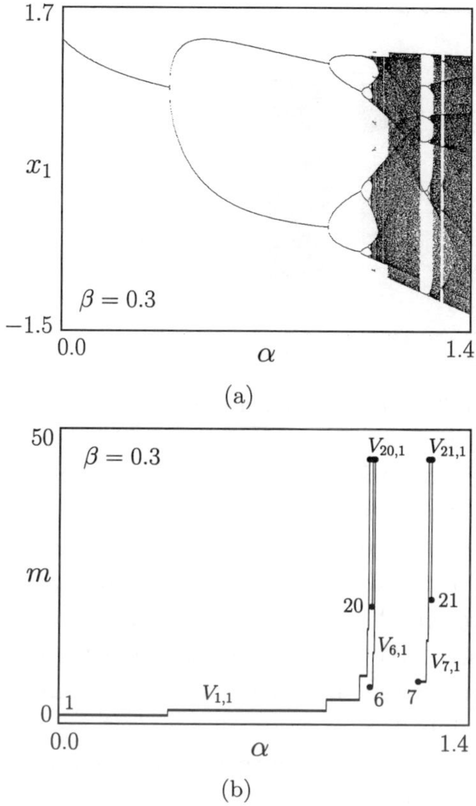

Figure 2.7. (a) Bifurcation diagram and (b) branching pattern for $0 \leq \alpha \leq 1.4$ and $\beta = 0.3$. The branching pattern specifies the periods for the stable cycles as a function of the bifurcation parameter.

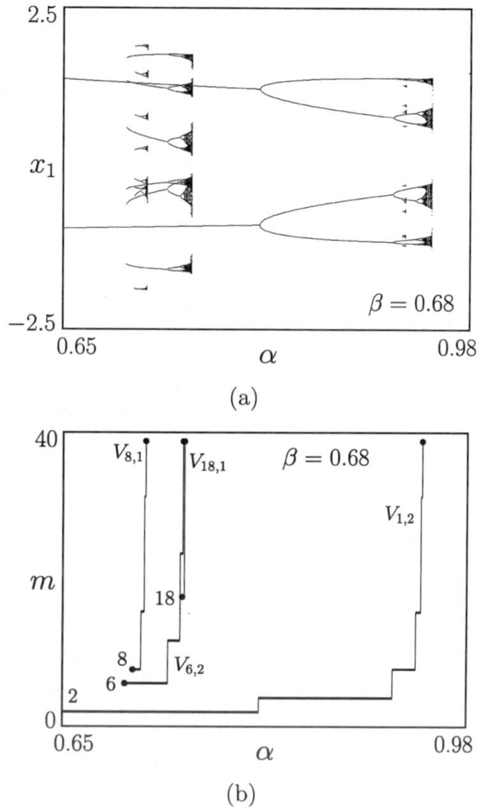

Figure 2.8. (a) Bifurcation diagram and (b) branching pattern for $0.65 \leq \alpha \leq 0.98$ and $\beta = 0.68$. Note the coexistence of several stable cycles and chaotic attractors.

It is apparent from the bifurcation diagram that the transition to chaos typically takes place via a period-doubling cascade as α is increased. Each such cascade leads to the appearance of aperiodic motion at some critical value of α. As we know from the logistic map and from many other systems displaying chaotic dynamics [5, 7], the ranges of α in which chaotic oscillations are observed are interrupted by small intervals (windows) where stable cycles exist.

In the interval

$$-\frac{1}{4}(\beta - 1)^2 < \alpha < 4, \ |\beta| < 1$$

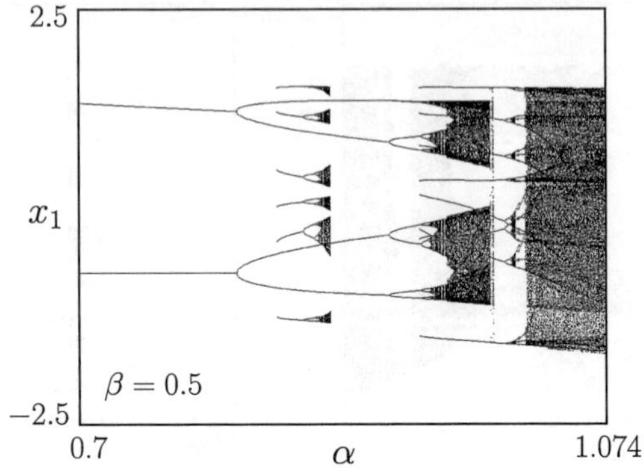

Figure 2.9. Bifurcation diagram for $0.7 \leq \alpha \leq 1.074$ and $\beta = 0.5$.

a great variety of new stable motions arise in saddle-node bifurcations (including 3-, 6-, 8-, and 18-cycles) with subsequent period-doubling bifurcations. In Fig. 2.9, for instance, we observe the appearance of a stable period-6 solution for $\alpha \approx 0.84$ and of a stable period-8 solution for $\alpha \approx 0.94$. These bifurcation diagrams are similar to the diagrams presented, e.g., by Kuznetsov *et al.* [33] for a system of two coupled logistic maps. We conclude that the pattern of cycle branching is a good deal more complicated than observed for the simple logistic map in Chapter 1 (see, e.g., Fig. 1.16).

Let us follow the evolution of various stable cycles with the variation of α. The results of numerical calculations for $\beta = 0.3$ and $\beta = 0.68$ are presented in Tables 2.1 and 2.2, respectively. In these tables the second column specifies the periodicity of a cycle, the forth and the fifth columns delineate the interval of α, in which a locally stable m-cycle exists. The last column contains the width of this range.

It is convenient to represent the data of the tables in the form of a diagram, called a *branching pattern* in some of our previous publications [19, 20]. In such diagrams (Figs. 2.7(b) and 2.8(b)) values of the active parameter are plotted along the horizontal axis and values of the period m for the stable cycles are plotted along the vertical axis. For this purpose, we use the same designations as Baushev *et al.* [19]. $V_{k,j}$ specify the various branches, with the index k indicating from which k-cycle the branch starts. The maximum value j indicates the number of observed branches, having the same k. All branches start with k-cycles that arise in saddle-node bifurcations. Some branches are determined on sets that were not shown in Fig. 2.6, as their sizes in the parameter plane

$V_{k,j}$	m	i	$\alpha_{2^{i-1}k,j}$	$\alpha_{2^{i}k,j}$	$\Delta\alpha_{2^{i-1}k,j}$
$V_{1,1}$	1	1	−0.1225	0.3675	0.49
	2	2	0.3675	0.9125	0.545
	4	3	0.9125	1.0255884050738	0.1133554050738
	8	4	1.0255884050738	1.0511256620352	0.0252702569614
	16	5	1.0511256620352	1.0565637581588	0.0054380961236
	32	6	1.0565637581588	1.0577304656609	0.0011667075021
	64	7	1.0577304656609	1.0579808932	0.0002255343391
	128	8	1.0579808932	1.0580344521	0.0000535589
	256	9	1.0580344521	1.0580459231	0.000011471
$V_{6,1}$	6	1	1.062372	1.0710703629071	0.0086983629071
	12	2	1.0710703629071	1.0750124047736	0.0039420418665
	24	3	1.0750124047736	1.0759204758	0.000908071087
	48	4	1.0759204758	1.0761186644	0.0001981886
	96	5	1.0761186644	1.0761612671	0.0000426027
$V_{7,1}$	7	1	1.2266174	1.2541834642429	0.0275660642429
	14	2	1.2541834642429	1.2600151726701	0.0058317084272
	28	3	1.2600151726701	1.2614289068416	0.0014137341715
	56	4	1.2614289068416	1.2617	0.0002710931584
$V_{9,1}$	9	1	1.293952248	1.2940570232	0.0001047748
	18	2	1.2940570232	1.2941079422	0.000050919
	36	3	1.2941079422	1.2941199612	0.000012019
	72	4	1.2941199612	1.2941226013	0.0000026401
$V_{12,1}$	12	1	1.0720080196	1.0728902059	0.0008821863
	24	2	1.0728902059	1.0733516722	0.0004614663
	48	3	1.0733516722	1.0734589837	0.0001073115
	96	4	1.0734589837	41.0734824547	0.000023471
$V_{20,1}$	20	1	1.0675791454	1.0677430954543	0.0001790954543
	40	2	1.0677430954543	1.0678313511	0.000088255646
	80	3	1.0678313511	1.0678520309	0.0000206798
$V_{20,2}$	20	1	1.10	1.1012085844	0.0012085844
	40	2	1.1012085844	1.1012928253	0.0000842409
	80	3	1.1012928253	1.1013113061	0.0000184808
$V_{21,1}$	21	1	1.26887	1.2692033433938	0.0003333433938

Table 2.1. Parameter-α ranges in which various periodic orbits can be observed for the Hénon map (2.1) with $\beta = 0.3$. $V_{k,j}$ specify the various branches, with k indicating the periodicity of the cycle from which the branch starts. m denotes the periodicity of the cycle.

are too small.

The range of existence of each branch $V_{k,j}$ is presented in the form of intervals:

$$\alpha_{k,j} < \alpha_{2k,j} < \cdots < \alpha_{2^ik,j} < \cdots < \alpha_{k,j*}.$$

Here the values $\alpha_{2^{i-1}k,j}$, $i = 2, 3, \ldots$ correspond to the loss of stability. of the

$V_{k,j}$	m	i	$\alpha_{2^{i-1}k,j}$	$\alpha_{2^ik,j}$	$\Delta\alpha_{2^{i-1}k,j}$
$V_{1,2}$	2	2	0.0768	0.808	0.7312
	4	3	0.808	0.9169381427467	0.1089381427467
	8	4	0.9169381427467	0.9361479503607	0.019209807614
	16	5	0.9361479503607	0.9400330556502	0.0038851052895
	32	6	0.9400330556502	0.9408680756308	0.0008350199806
	64	7	0.9408680756308	0.9410471449543	0.0001790693235
$V_{6,2}$	6	1	0.70142261	0.7343659105724	0.0329433005724
	12	2	0.7343659105724	0.7445017603884	0.010135849816
	24	3	0.7445017603884	0.7467731395103	0.0022713791219
	48	4	0.7467731395103	0.7472652867434	0.0004921472331
$V_{8,1}$	8	1	0.70791248	0.713479366256	0.005566886256
	16	2	0.713479366256	0.7158180118385	0.0023386455825
	32	3	0.7158180118385	0.7163564950326	0.0005384831941
	64	4	0.7163564950326	0.7164739420036	0.000117446971
$V_{18,1}$	18	1	0.74696954	0.7476245768696	0.0006550368696
	36	2	0.7476245768696	0.7479207404767	0.0002961636071

Table 2.2. Parameter-α ranges in which various periodic orbits can be observed for $\beta = 0.68$.

$2^{i-1}k$-cycle and the soft occurrence of a 2^ik-cycle. These values were calculated numerically by solving Eqs. (2.3) and (2.4) together with Eq. (2.2). When i increases, the intervals $\Delta\alpha_{2^{i-1}k,j} = \alpha_{2^ik,j} - \alpha_{2^{i-1}k,j}$ of existence for the $2^{i-1}k$-cycle become narrower (see Tables 2.1 and 2.2), so that each branch has its limiting value $\alpha_{k,j*}$, which corresponds to the transition to aperiodic motion. Following common practice we shall refer to these parameter values as *accumulation points*. By calculating the ratio of two subsequent values of $\Delta\alpha_{2^{i-1}k,j}$ for a particular branch $V_{k,j}$ one can see how the accumulation process asymptotically is controlled by the Feigenbaum constant $\delta_F = 4.6692016\ldots$. In Figs. 2.7(b) and 2.8(b), the accumulation points are shown as vertical lines (asymptotes). The beginning of the windows of periodicity in the bifurcation diagrams coincides with the beginning of the branches. Some branches $V_{k,j}$ $(k \neq 1)$ intersect with $V_{1,1}$. In those ranges of α where $V_{1,1} \cap V_{k,j} \neq \emptyset$, $k \neq 1$, a stable 2^{i-1}-cycle coexists with stable $2^{d-1}k$-cycles, $i = 1, 2, \ldots$; $d = 1, 2, \ldots$. This implies, that the phase plane (x_1, x_2) of the dynamic system (2.1) comprises the basins of attraction for different cycles. In this case, depending on the initial conditions either the 2^{i-1}-cycle or the $2^{d-1}k$-cycle are approached. The character of the division of phase plane for system (2.1) into basins of attraction for different cycles is shown in Figs. 2.10–2.12.

Figures 2.10(a) and (b) display the basins of attraction for the 1- and 3-cycles

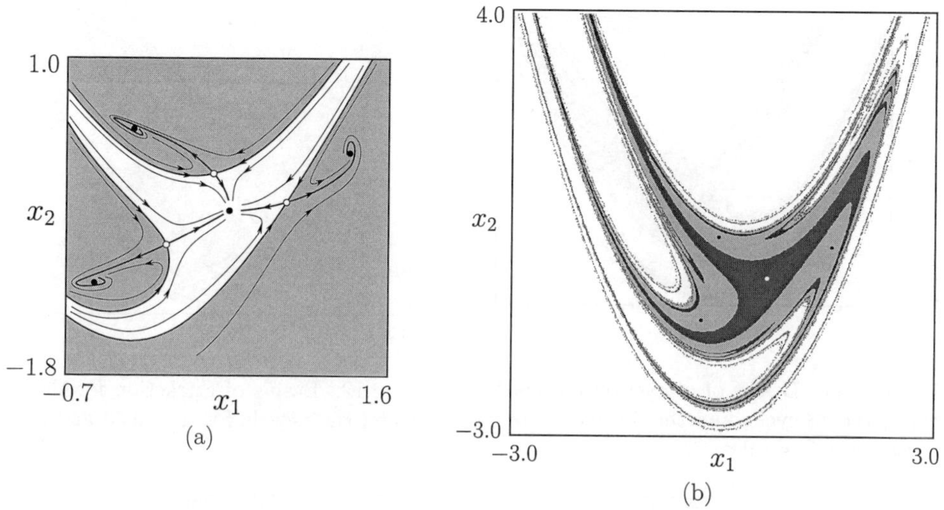

Figure 2.10. (a) Phase diagram and (b) basins of attraction for the coexisting 1- and 3-cycles at $\alpha = 0.949$ and $\beta = -0.744$. The phase diagram illustrates the stable and unstable manifolds of the various cycles. The position of the period-3 saddle cycle at the basin boundary is also indicated in this diagram.

that coexist for $\alpha = 0.949$ and $\beta = -0.744$. The stable period-3 cycle has been formed together with a period-3 saddle cycle in a subharmonic saddle-node bifurcation. The stable manifold of the saddle cycle, i.e., the set of points that asymptotically converge to the period-3 saddle cycle, forms the basin boundary. This boundary is clearly indicated in the phase portrait of Fig. 2.10(a). Also shown in this diagram are a number of typical trajectories (thin curves). It is obvious that the stable period-3 cycle has been converted into a focus, i.e., the trajectories approach the stationary solution in a spiralling manner. In Fig. 2.10(b) the locations of the two coexisting stable cycles are marked by dots.

Figure 2.11 shows the basins of attraction for the coexisting 1-, 18-, and 3-cycles with $P = (1.0295, -0.8)$, and Fig. 2.12 illustrates the basins of attraction for the coexisting 2- and 8-cycles ($P = (0.53, 0.88)$). In both cases it appears that the basins have a fractal boundary [41, 42]. As mentioned in Chapter 1, *fractal basin boundaries* arise in global bifurcations when the stable manifold of a saddle cycle under variation of the parameters starts to intersect the unstable manifold to the same (or to another) cycle. This gives rise to a so-called *tangle* between the two manifolds. Since a point of intersection of the

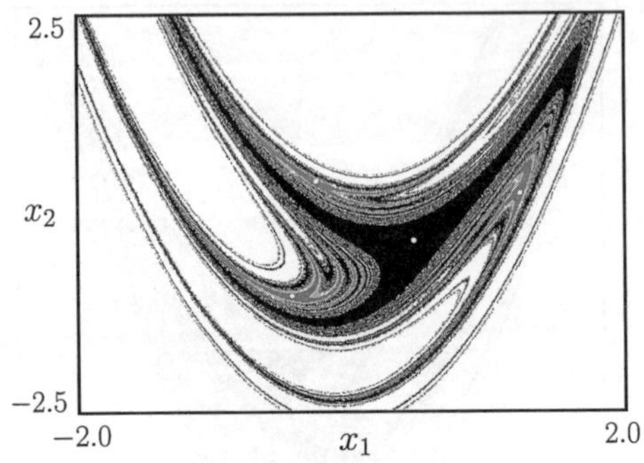

Figure 2.11. Basins of attraction for the 1-cycle, the 18-cycle, and the 3-cycle for $\alpha = 1.0295$ and $\beta = -0.8$.

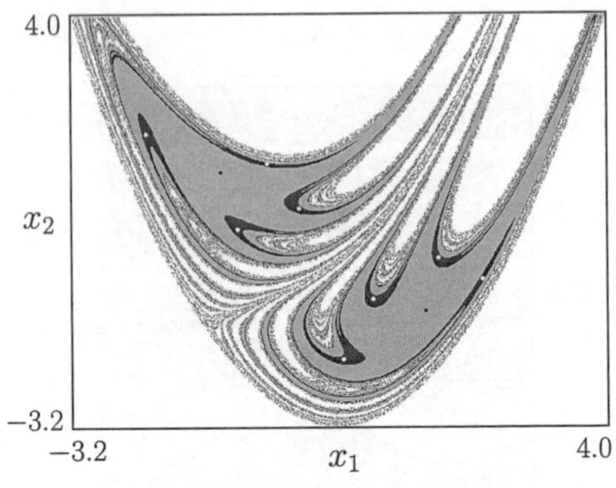

Figure 2.12. Basins of attraction for the 2-cycle and the 8-cycle for $\alpha = 0.53$ and $\beta = 0.88$.

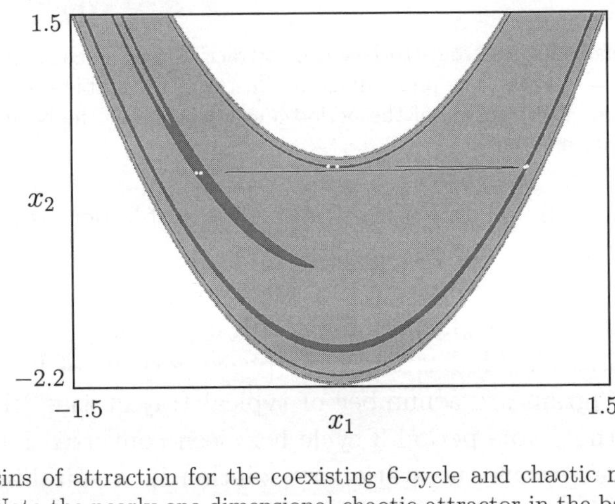

Figure 2.13. Basins of attraction for the coexisting 6-cycle and chaotic motion for $\alpha = 1.7$ and $\beta = -0.03$. Note the nearly one-dimensional chaotic attractor in the basin with light gray shading.

unstable manifold with the stable manifold must be repeated in all subsequent iterations along the stable manifold and since, likewise, a point of intersection must be repeated in all backwards iterations along the unstable manifold, the two manifolds develop a *horseshoe structure* [43] with infinitely many foldings and intersections. As the stable manifold forms the boundary of a basin, it is clear that this basin must develop an infinitely folded (fractal) structure.

Some of the first examples of fractal basin boundaries were obtained by

Ueda [44] in his studies of the forced Duffing oscillator. But this type of structure has also been observed for the forced pendulum (Fig. 1.2) [31], in a model of interacting business cycles [30], and in many other nonlinear dynamical systems. For a more detailed discussion of this phenomenon we refer to the book by Ott on Chaos in Dynamical Systems [45]. As discussed in Chapter 1, the presence of fractal basin boundaries gives rise to a so-called *final state sensitivity* [41]. If the initial conditions near a basin boundary are specified with an uncertainty ε, then in the limit $\varepsilon \to 0$ a fraction $f(\varepsilon) \approx \varepsilon^{\nu}$ of trajectories will give uncertain outcomes. For a regular basin boundary, $\nu = 1$. For a fractal basin boundary, however, $\nu < 1$, and the uncertainty in the final outcome diminishes more slowly with ε that does the uncertainty in the initial conditions. Trajectories starting from initial conditions in Figs. 2.10–2.12 that are left blank diverge to infinity.

By comparing the branching patterns with the bifurcation diagrams one can see that there are intervals within the limits of $V_{1,1}$ where periodic motions coexist with chaotic oscillations. Here, the system may choose a periodic motion or chaotic oscillations, depending on the initial conditions. As an example Fig. 2.13 exhibits the results of plotting the basins of attraction for the coexisting 6-cycle and chaotic motion ($P = (1.7, -0.03)$).

To complement the above results, Fig. 2.14 presents a chart of dynamical regimes with their associated basins of attraction for different attracting states for the parameter values $P \in \Pi_{i,l} \cup \Pi_{q,p}$. The numerical data are given in Table 2.3. In the first column the table lists the number of the inset that contains the basins of attraction which are indicated in Fig. 2.14. The second and the third columns specify the parameter values at which the basins have been calculated. Periods of coexisting stable cycles are indicated in the fourth column of the table (∞ means aperiodic oscillations), and the last four columns contain the boundary values of the variables x_1 and x_2.

Now let us consider the difference between the properties of the set Π_k^{Swallow} and the set $\Pi_{k,j}$. Similarly to the approach used by Anishchenko [5] to analyze a model of a radiophysical generator with inertial nonlinearity, we introduce a three-dimensional space $\big(\alpha, \beta, x_{\rm c}(P)\big)$. Here, $x_{\rm c}(P)$ denotes one of the coordinates of the fixed point $X_{\rm c}(P)$ of the mapping $X_{\rm c} = F^{(m)}(X_{\rm c})$. As illustrated in Fig. 2.15 we can now depict $X_{\rm c}$ as a function of α and β. The locus corresponding to stable and unstable m-cycles forms a two-dimensional surface $\Theta_{m,j}$ in $(\alpha, \beta, x_{\rm c}(P))$. The first index in the designation of $\Theta_{m,j}$ denotes the value of

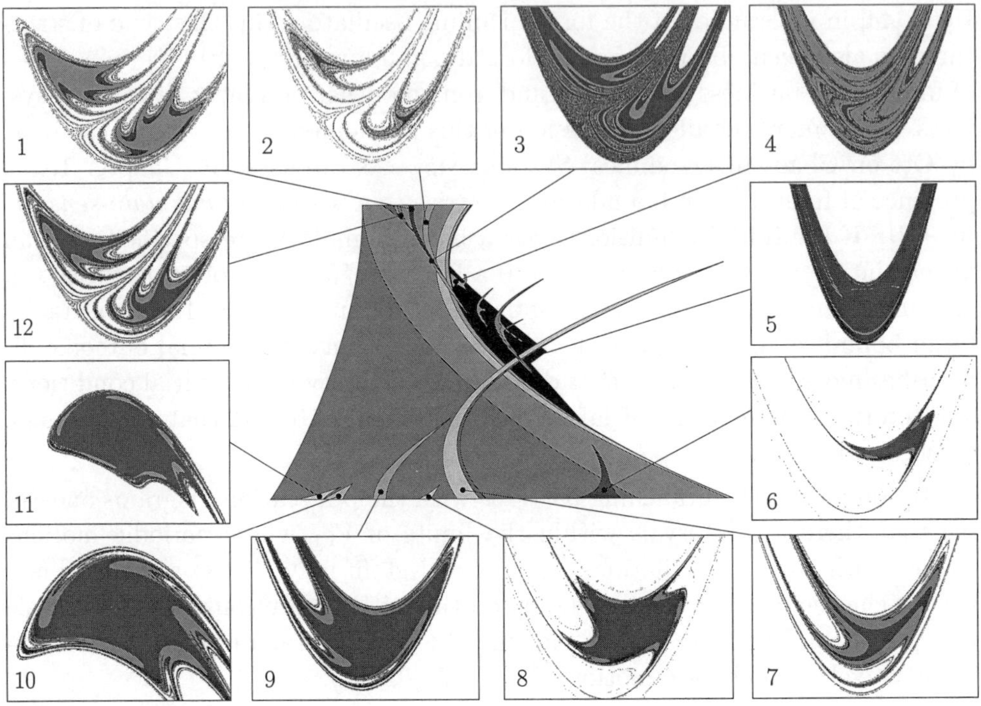

Figure 2.14. Chart of dynamical regimes and basins of attraction for different values of the parameters α and β. The values of the parameters for which the basins were calculated are given in Table 2.3.

m, for which the surface was built, and the second index is introduced to distinguish different surfaces $\Theta_{m,j}$, having the same m. The plot of such a surface for $m = 5$ and $m = 10$ is schematically shown in Fig. 2.15(a). Figure 2.15(b) shows a projection of the surface $\Theta_{m,j}$ onto the plane of control parameters.

In order to make the illustration clear, the bifurcation curves in Fig. 2.15 are not drawn to scale. The domains of existence for the stable 5-cycle and for the two stable 10-cycles are denoted as $\hat{\Pi}_{5,1}^{\text{Swallow}}$ and by $\hat{\Pi}_{10,1}^{\text{Swallow}}$ and $\hat{\Pi}_{10,2}^{\text{Swallow}}$, respectively. The set $\hat{\Pi}_{5,1}^{\text{Swallow}}$ is delineated by the period-doubling bifurcation curves $\Gamma_{1,5}^{-}$ and $\hat{\Gamma}_{1,5}^{-}$ and by the curves $\Gamma_{0,5}^{+}$, $\Gamma_{1,5}^{+}$, and $\hat{\Gamma}_{1,5}^{+}$, $\Gamma_{0,5}^{+}$ being the saddle-node bifurcation curve at which the period-5 cycle is first created. $\Gamma_{1,5}^{+}$ and $\hat{\Gamma}_{1,5}^{+}$ are the fold lines of the surface $\Theta_{5,1}$, corresponding the merging of the stable 5-cycle with an unstable 5-cycle (the curve N_{+}). The curves $\Gamma_{1,5}^{+}$ and $\hat{\Gamma}_{1,5}^{+}$ are

No.	α	β	m	$x_{1\min}$	$x_{1\max}$	$x_{2\min}$	$x_{2\max}$
1	0.387	0.977	2, 10	−4.0	4.0	−3.5	5.0
2	0.719	0.914	2, 6	−3.0	3.0	−3.0	3.0
3	0.719	0.672	2, 6, 8	−3.0	3.0	−3.0	3.0
4	0.973	0.492	8, ∞	−3.0	3.0	−3.0	3.0
5	1.559	−0.086	3, ∞	−3.0	3.0	−2.2	4.0
6	2.613	−0.945	1, 5	−1.0	1.0	−1.5	1.0
7	0.934	−0.805	1, 3	−3.0	3.0	−3.0	4.0
8	0.66	−0.984	1, 6	−3.0	3.0	−3.0	3.0
9	0.250	−0.820	1, 4	−9.5	9.0	−8.0	12.0
10	−0.297	−0.969	1, 5	−3.5	5.0	−6.0	3.5
11	−0.531	−0.984	1, 6	−2.0	3.5	−4.0	2.0
12	0.504	−0.938	2, 8	−3.5	3.5	−3.2	4.0

Table 2.3.

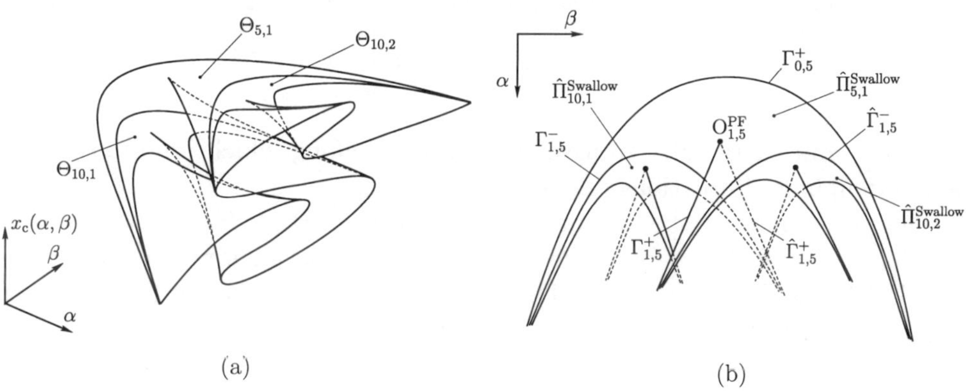

Figure 2.15. (a) 2D surface illustrating the peculiar organization of the domains of existence for the 5-cycle $\hat{\Pi}_{5,1}^{\text{Swallow}}$ and the 10-cycles $\hat{\Pi}_{10,1}^{\text{Swallow}}$ and $\hat{\Pi}_{10,2}^{\text{Swallow}}$ associated with the swallow tail structure. (b) Bifurcation curves bounding the domains $\hat{\Pi}_{5,1}^{\text{Swallow}}$, $\hat{\Pi}_{10,1}^{\text{Swallow}}$ and $\hat{\Pi}_{10,2}^{\text{Swallow}}$. Note that these plots are not drawn to scale. Similar figures may be found in the works by Mira *et al.* [36] and by Carcasses and Mira [40].

supported by the *codimension-2 bifurcation* point $O_{1,5}^{PF}$. This is a pitchfork bifurcation point, and it is the meeting point of the two fold lines in the projection of $\Theta_{5,1}$. These fold lines produce a region in parameter space where two stable and two unstable period-5 cycles coexist. One of the unstable 5-cycles is annihilated with the stable 5-cycle on the boundary $\Gamma_{0,5}^{+}$. The picture below the curves $\Gamma_{1,5}^{-}$ and $\hat{\Gamma}_{1,5}^{-}$ repeats the structure described for $\hat{\Pi}_{5,1}^{\text{Swallow}}$ (Fig. 2.15 and Fig. 2.16). The bifurcation curves $\Gamma_{j,2^{i-1}\cdot 5}^{-}$ and $\hat{\Gamma}_{j,2^{i-1}\cdot 5}^{-}$, $j = 1, \ldots, 2^{i-1}$; $i = 1, 2, \ldots$ ac-

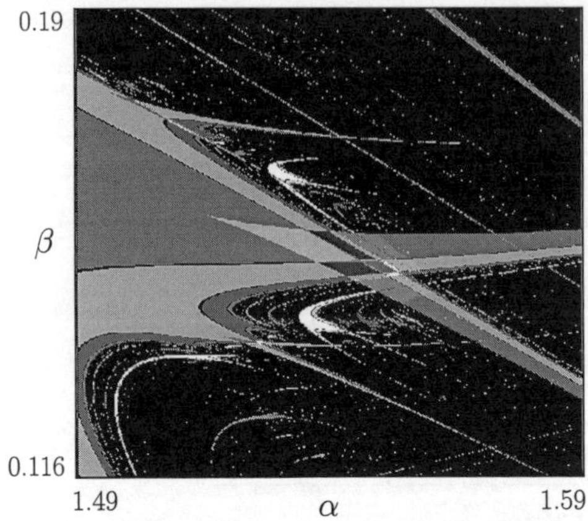

Figure 2.16. Results of numerical calculations of the domains Π_k^{Swallow}, $k = 5, 6, 7, 8$ etc.

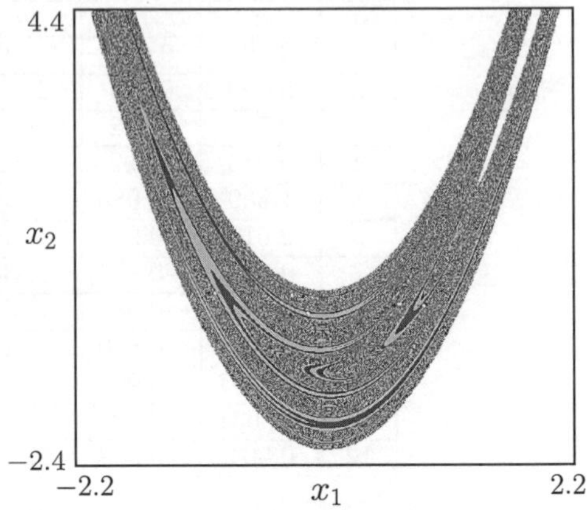

Figure 2.17. Basins of attraction for two coexisting 5-cycles in the overlapping region of the swallow tail Π_5^{Swallow}. $\alpha = 1.531$ and $\beta = 0.153$.

cumulate, and there exist transversal directions along which infinite series of period-doubling bifurcations take place.

The projections of the two-dimensional surfaces $\Theta_{m,j}$, ($m = 2^{i-1} \cdot 5$, $j = 1, \ldots, 2^{i-1}$, $i = 2, 3, \ldots$) onto the plane of control parameters have the same peculiarities as $\Theta_{5,1}$ (Fig. 2.16). The domain Π_5^{Swallow} consists of the collection of the sets $\Pi_{2^{i-1}\cdot 5,j}^{\text{Swallow}}$, $j = 1, \ldots, 2^{i-1}$, $i = 1, 2, \ldots$:

$$\Pi_5^{\text{Swallow}} = \bigcup_{i=1}^{\infty} \left(\bigcup_{j=1}^{2^{i-1}} \Pi_{2^{i-1}\cdot 5,j}^{\text{Swallow}} \right).$$

Figure 2.16 shows the collection of some Π_k^{Swallow} that begin with different stable k-cycles

$$\Pi_k^{\text{Swallow}} = \bigcup_{i=1}^{\infty} \left(\bigcup_{j=1}^{2^{i-1}} \Pi_{2^{i-1}k,j}^{\text{Swallow}} \right), \quad k = 5, 6, 7, 8$$

and Fig. 2.17 shows the basins of attraction for two coexisting 5-cycles in the overlap region within the swallow tail Π_5^{Swallow}. This figure provides a new example of fractal basin boundaries.

Let us denote the part of Π_k^{Swallow} where a particular stable motion exists and is uninterrupted along the parameters as $\Pi_{k,\text{conv}}^{\text{Swallow}}$. The properties of $\Pi_{k,\text{conv}}^{\text{Swallow}}$

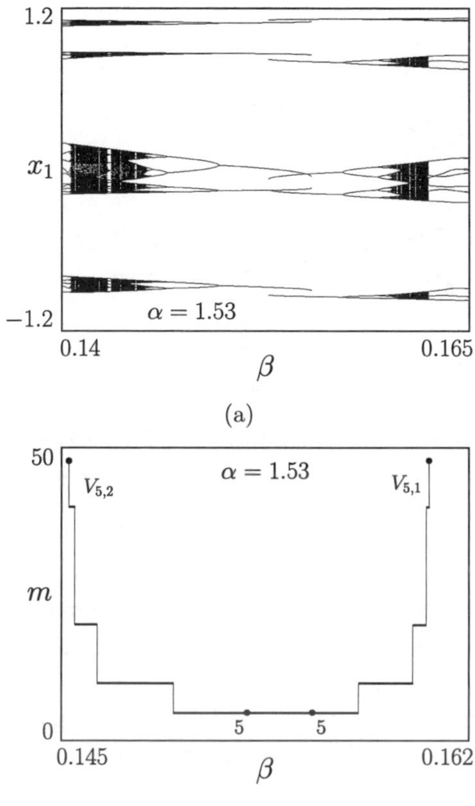

Figure 2.18. (a) Bifurcation diagram for $\alpha = 1.53$ and $0.14 \leq \beta \leq 0.165$ and (b) branching pattern for $\alpha = 1.53$ and $0.145 \leq \beta \leq 0.162$. This scan intersects that region of Π_5^{Swallow} where two stable 5-cycles coexist.

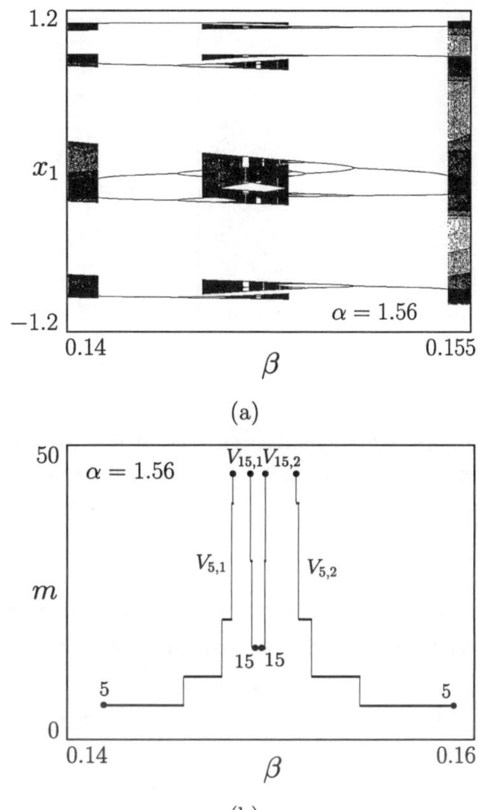

Figure 2.19. (a) Bifurcation diagram and (b) corresponding branching pattern for $\alpha = 1.56$ and $0.14 \leq \beta \leq 0.16$. This scan intersects that region of the domain Π_5^{Swallow} where two chaotic motions can coexist. We can also observe the coexistence of two different periodic cycles and of periodic cycles with chaotic motions.

are similar to the above adduced properties of $\Pi_{k,j}$. There is non-uniqueness of motions in the domain $\Pi_k^{\text{Swallow}} \setminus \Pi_{k,\text{conv}}^{\text{Swallow}}$, which lies between the bifurcation curves $\Gamma^+_{j,2^{i-1}\cdot 5}$ and $\hat{\Gamma}^+_{j,2^{i-1}\cdot 5}$, $j = 1, \ldots, 2^{i-1}$, $i = 1, 2, \ldots$ (Figs. 2.15 and 2.16).

When the boundaries of $\Pi_k^{\text{Swallow}} \setminus \Pi_{k,\text{conv}}^{\text{Swallow}}$ are intersected in the points of the curves $\Gamma^+_{j,2^{i-1}\cdot 5}$ and $\hat{\Gamma}^+_{j,2^{i-1}\cdot 5}$, one can observe hard transitions from one mode to

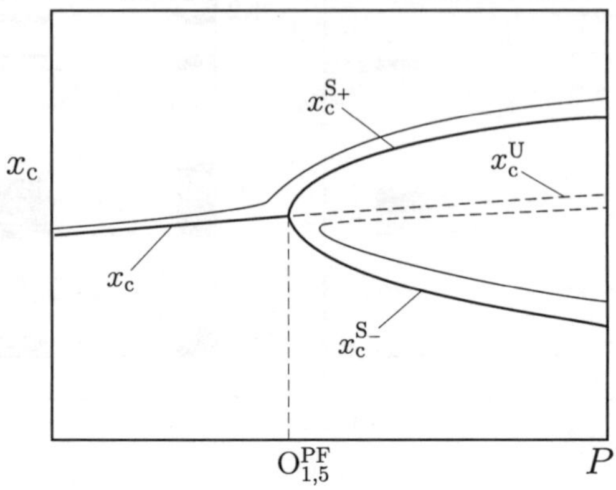

Figure 2.20. Qualitative plot of the variation of the stable and unstable 5-cycles with the parameters when the trajectory of deformation passes through (heavy curve) and close to (fine curves) the bifurcation point of codimension-2 $O_{1,5}^{PF}$.

another with hysteresis as it is typical for such transitions. Figures 2.18 and 2.19 illustrate only two of the possible variations in the character of the dynamics with change of the parameters. In other sections, the branching patterns may be even more complicated.

Let us finally consider the case where the trajectory of deformation passes through $O_{j,2^{i-1}\cdot 5}^{PF}$, $j = 1, \ldots, 2^{i-1}$, $i = 1, 2, \ldots$ The dependence of the solutions $x_c(P)$ of Eq. (2.2) on the parameters for this case is qualitatively shown in Fig. 2.20, where $x_c^{S-}(P)$ and $x_c^{S+}(P)$ correspond to the stable 5-cycles, and $x_c^{U}(P)$ corresponds to the unstable 5-cycle. As P approaches the point $O_{1,5}^{PF}$, the solutions $x_c^{S-}(P)$, $x_c^{S+}(P)$ and $x_c^{U}(P)$ come closer to meet in $O_{1,5}^{PF}$ and form a single stable 5-cycle that varies smoothly with P. If the trajectory of deformation is shifted such that it no longer passes through $O_{1,5}^{PF}$, $x_c(P)$ falls into two isolated solutions as the symmetric pitchfork bifurcation is replaced by a saddle-node bifurcation [7].

2.4 Piecewise-Smooth Two-Dimensional Maps

The purpose of this section is to present a few examples of the bifurcation phenomena that one can observe in piecewise-smooth dynamical systems. The idea is to contrast these phenomena with the phenomena we have just discussed for the smooth Hénon map.

In accordance with our introductory comments in Chapter 1, one can distinguish between two different types of bifurcations for piecewise-smooth dynamical systems. The first type is similar to the bifurcations that we know from smooth systems. These are the *local bifurcations* (such as the period-doubling, saddle-node, and Hopf bifurcations) that correspond to loss of stability for the periodic solution as one (or two) of its eigenvalues passes out through the unit circle and the *global bifurcations* associated with the formation of a homoclinic (or heteroclinic) connection or with the destruction of a chaotic attractor in a boundary crisis. Bifurcations of the second type cannot occur in smooth systems. They are connected with situations where the trajectory becomes tangent to one of the so-called sewing surfaces. As discussed in Chapter 1, a sewing surface is a surface that divides the phase space of the dynamical system into domains such that the motion in each domain is described by equations with continuous and sufficiently smooth right-hand sides, but the equations differ from domain to domain. Such violations of the topological structure are connected with *border-collision bifurcations*, or *C-bifurcations*.

The simplest bifurcation pattern corresponding to a border-collision bifurcation is the continuous transition from one type of solution to a solution of a different type. More complicated bifurcational phenomena are also possible, for example, period-tripling, merging and disappearance of two solutions of different types, and direct transitions to chaos. We have discussed some of these phenomena in connection with the tent maps considered in Chapter 1. However, the interplay between the ordinary bifurcation phenomena that one can observe in smooth systems and the border-collision bifurcations is even more complicated for piecewise-smooth systems.

Let us start the discussion by considering the simple two-dimensional piecewise-smooth map [20]:

$$\begin{cases} w_{1k} = \mathrm{e}^{a\lambda_1} w_{1(k-1)} + \mathrm{e}^{a\lambda_1(1-z_k)} - \mathrm{e}^{a\lambda_1}; \\ w_{2k} = \mathrm{e}^{a\lambda_2} w_{2(k-1)} + \mathrm{e}^{a\lambda_2(1-z_k)} - \mathrm{e}^{a\lambda_2}. \end{cases} \tag{2.12}$$

Here z_k is calculated by the algorithm:

$$z_k = \begin{cases} 0, & \varphi(0) \le 0; \\ z^*, & \varphi(0) > 0 \text{ and } \varphi(1) < 0; \\ 1, & \varphi(0) > 0 \text{ and } \varphi(1) \ge 0, \end{cases}$$

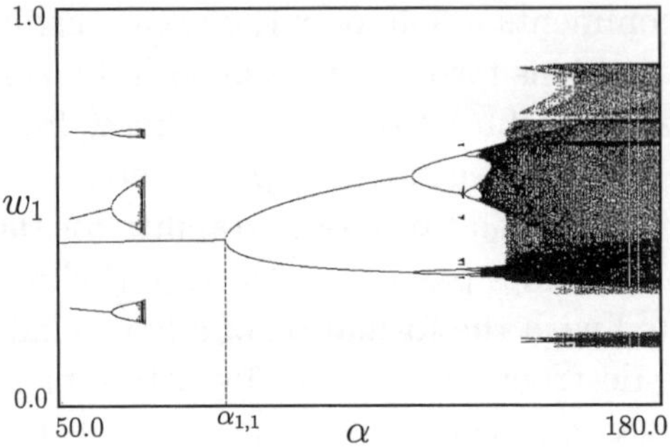

Figure 2.21. Bifurcation diagram for the piecewise-smooth map (2.12). Note how a four-band chaotic attractor arises abruptly for $\alpha \approx 150$. The appearance of the period-3 solution and the transition to chaos from this solution also exhibit a number of peculiarities.

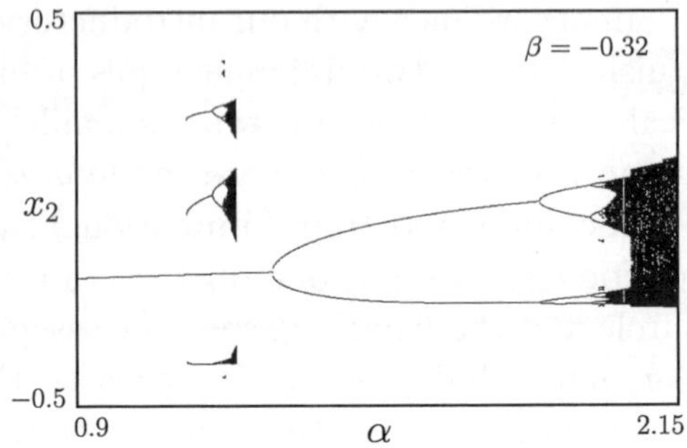

Figure 2.22. Bifurcation diagram for the Hénon map. At a first glance this diagram appears similar to the diagram for the piecewise-smooth map in Fig. 2.21. However, as described in the text, there are a number of essential differences.

where z^* is the smallest root of the equation $\varphi(z) = 0$,

$$\varphi(z) = \vartheta\left(1 - w_{2(k-1)}\right) e^{a\lambda_2 z} - \left(1 - w_{1(k-1)}\right) e^{a\lambda_1 z} - \frac{Q}{\alpha} z + q - \frac{\lambda_1}{\lambda_2} + 1.$$

The parameter values for this model are indicated in Chapter 5.

The map (2.12) was introduced by Baushev and Zhusubaliyev [20] as a stroboscopic map for a two-dimensional set of nonautonomous differential equations with the constant matrix A describing the behavior of a DC/DC converter with pulse-width modulation. We will not repeat the description of how the map is obtained because it is not essential for the purpose of the present analysis. The general method is given in Chapter 5. Let us note, however, that a is the clock period and λ_1 and λ_2 are the eigenvalues of the matrix A. Q is a normalized reference signal and q is a normalized amplitude of the sawtooth voltage of the pulse-width modulator. The bifurcation parameter α is an amplification constant for the error signal amplifier. A detailed analysis of bifurcations and transitions to chaos for the map (2.12) may be found in our original publication [20].

Let us start with the analysis of local bifurcations. Figure 2.21 reproduces the bifurcation diagram obtained for $50.0 \leq \alpha \leq 180.0$. Numerical results of calculations of the branching pattern are presented in Table 2.4 [20].

The range $0 < \alpha < \alpha_{1,1}$ ($\alpha_{1,1} \approx 86.58$) is the domain of existence for the stable fixed point. At the point $\alpha = \alpha_{1,1}$ the maximum (in absolute value) mul-

$V_{k,j}$	m	i	$\alpha_{2^{i-1}k,j}$	$\alpha_{2^{i}k,j}$	$\Delta\alpha_{2^{i-1}k,j}$
$V_{1,1}$	1	1	0	86.58	86.58
	2	2	86.58	126.84	40.26
	4	3	126.84	137.96	11.12
	8	4	137.96	140.41	2.45
	16	5	140.41	140.94	0.53
	32	6	140.94	141.06	0.12
$V_{3,1}$	3	1	52.67	61.50	8.83
	6	2	61.50	68.0	6.5
	12	3	68.0	68.0153	0.0153
$V_{5,1}$	5	1	144.60	144.85	0.125
	10	2	144.85	144.92	0.07
	20	3	144.92	144.95	0.03
$V_{6,1}$	6	1	136.526	137.10	0.574
	12	2	137.10	137.30	0.2
$V_{7,1}$	7	1	144.26	144.30	0.04
	14	2	144.30	144.325	0.025
$V_{9,1}$	9	1	66.56	66.78	0.22
$V_{15,1}$	15	1	68.68	68.698	0.018
$V_{21,1}$	21	1	68.393	68.40	0.027

Table 2.4. Survey of the regions of existence for periodic solutions corresponding to different branches for the piecewise-smooth two-dimensional map (2.12).

tiplier of the fixed point crosses the unit circle through -1 and a period-2 cycle softly arises. With further increase of α, a period-doubling sequence leading to chaotic dynamics occurs similar to that of the Hénon map (see Table 2.4). One can compare the bifurcation diagram in Fig. 2.21 with the diagram obtained for the Hénon map (see Fig. 2.22, where $0.9 < \alpha < 2.15$, $\beta = -0.32$). In both cases the chaotic domain corresponding to the main branch (i.e., the branch that develops from the 1-cycle) is interrupted by a large number of windows with regular dynamics starting from saddle-node bifurcations. Increasing the control parameter α within one of these windows leads to an infinite period-doubling sequence ending with a transition to chaotic dynamics. The scaling constant $\delta_{\mathrm{F}} = 4.6692016\ldots$ still describes the rate at which $\Delta\alpha_{2^{i-1}k,j}$ decreases from period-doubling to period-doubling.

Let us now consider some of the signatures of border-collision bifurcations. Figure 2.23 shows a magnified part of the bifurcation diagram corresponding to a range of α where border-collision bifurcations occur. The results of a *continuation analysis* are given in Fig. 2.24. Continuation is a numerical method that allows us to follow stable as well as unstable periodic solutions under

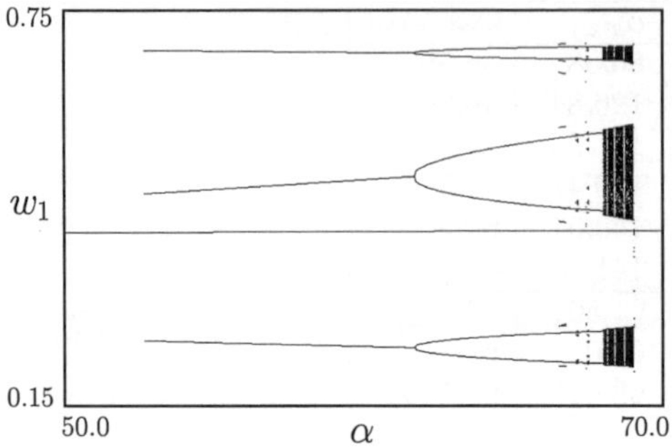

Figure 2.23. Magnified part of the diagram presented in Fig. 2.21 showing the branch beginning from the 3-cycle.

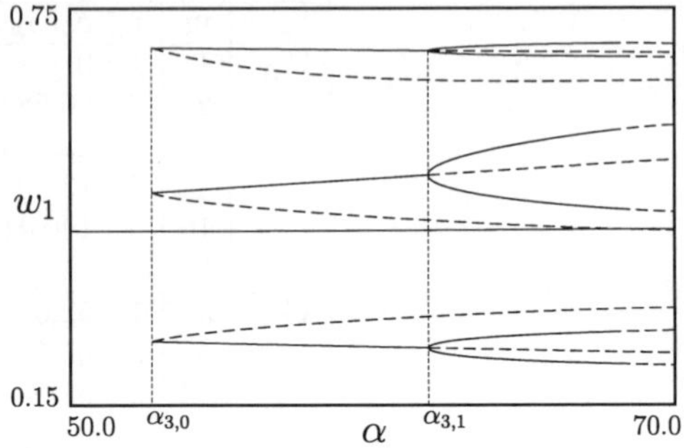

Figure 2.24. Results of a continuation analysis.

variation of a parameter. The solid lines in this figure represent the evolution of stable cycles, and dashed lines the behavior of unstable cycles.

The period-3 cycle appears without warning at the point $\alpha = \alpha_{3,0} \approx 52.67$. This bifurcation is similar in several respects to a saddle-node bifurcation for a smooth system. However, under variation of α we notice how the stable and unstable cycles move away from each other in a nearly linear manner, rather than in the parabolic way characteristic of a smooth system. Moreover, inspection of Fig. 2.25 shows how the eigenvalues jump abruptly. When $\alpha = \alpha_{3,1} \approx 61.5$ (see Table 2.4) the 3-cycle loses stability as the numerically larger multiplier $\rho_3^{(3)\mathrm{S}}$ of the stable cycle leaves the unit circle through -1, and a 6-cycle softly arises via a local period-doubling bifurcation. With further increase of α, the period of the 6-cycle is again doubled via a local period-doubling bifurcation, but the 12-cycle loses its stability with a sudden transition to chaos.

The described scenario for the transition to chaos is typical for many piecewise-smooth systems [23]. The transition to chaos involves a finite sequence of local period-doubling and border-collision bifurcations unlike the dynamics of the Hénon map where the transition from the 3-cycle to chaos occurs via an infinite period-doubling sequence (see Fig. 2.26).

We may recall from our discussion of the skew tent map in Chapter 1, that this map only exhibits a single period-doubling bifurcation before the transition to chaos. The reason for this is that at the moment when the period-2 cycle becomes unstable as its eigenvalue passes through -1, the period-4, period-8, etc., cycles also lose their stability. For a piecewise smooth map, as illustrated

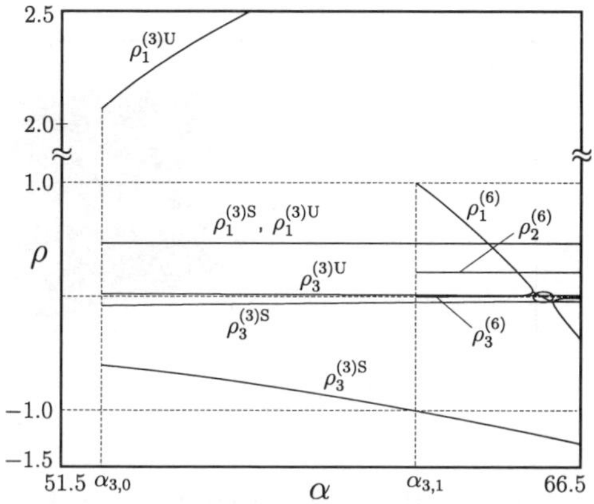

Figure 2.25. The multiplier dependences of stable and unstable cycles on the parameter. $\rho_i^{(3)S}$ and $\rho_i^{(3)U}$, $i = 1, 2, 3$ denote the multipliers of the stable and the unstable 3-cycle, respectively.

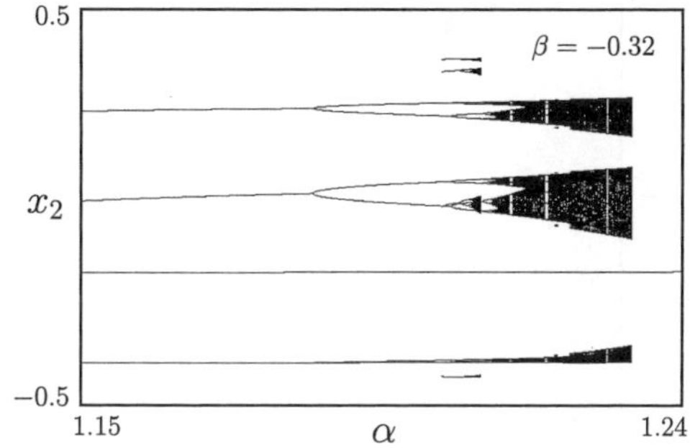

Figure 2.26. Magnified part of the diagram presented in Fig. 2.22 showing the branch developing from the 3-cycle in the Hénon map.

by the above example, one can observe finite sequences of period-doubling bifurcations before the transition to chaos occurs. We will show in Chapter 5 that a finite sequence of local period-doubling bifurcations are always followed by a border-collision bifurcation that involves a change in the solution type and then by a border-collision transition that leads either to the soft appearance of a cycle of chaotic intervals or to a sudden transition to chaos.

At the point $\alpha = \alpha_{3,0}$, the stable 3-cycle merges with the unstable 3-cycle and disappears via a border-collision bifurcation. The trajectory of the stable 3-cycle is constructed from four parts, while for the unstable 3-cycle it is sewed up from six parts. In the bifurcation point, one of the sections of the phase trajectory of the 3-cycle becomes tangent to the border of the domain of smoothness. This leads to the disappearance of two sections of the trajectory of the unstable cycle. As mentioned above, this bifurcation is similar to a saddle-node bifurcation. However, there are certain differences. The bifurcation can involve two unstable cycles as well as a pair of stable and unstable cycles. In any case one of the involved cycles changes its type in the bifurcation point and merges with another cycle. The multipliers of the first cycle show finite discontinuities at the bifurcation point (see Fig. 2.25). Unlike saddle-node bifurcations, border-collision bifurcations are not associated with the continuous crossing of one of the multipliers through $+1$.

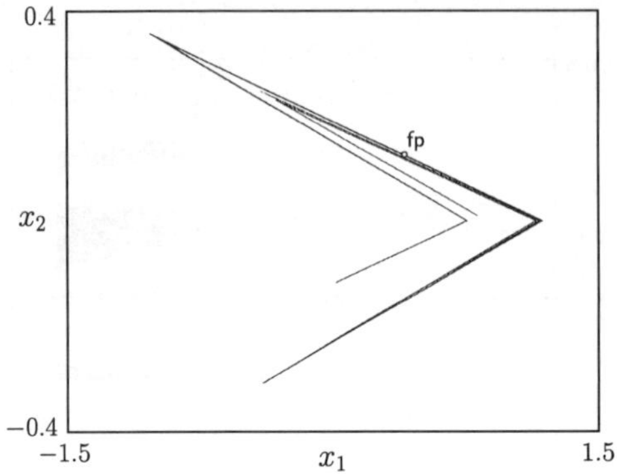

Figure 2.27. Quasihyperbolic Lozi attractor for $\alpha = 1.7$ and $\beta = 0.3$. The unstable fixed point fp is also indicated.

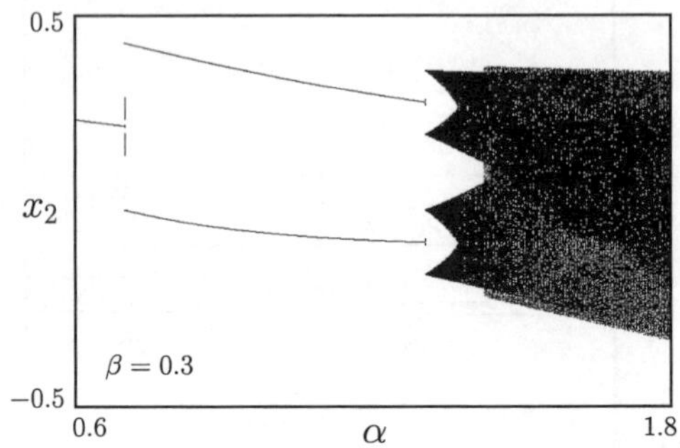

Figure 2.28. Bifurcation diagram for the Lozi map. Note the peculiar form of the bifurcations that take place at $\alpha = 0.7$ and $\alpha = 1.3$. These transitions do not show hysteretic phenomena.

Let us consider one more example of a two-dimensional map, namely the well-known piecewise-linear map referred to as the Lozi map [24]:

$$\begin{cases} x_k = 1 - \alpha|x_{k-1}| + y_{k-1}; \\ y_k = \beta x_{k-1}. \end{cases} \qquad k = 1, 2, \ldots \tag{2.13}$$

The curve $x = 0$ divides the state space of the map (2.13) into two regions. In region $x < 0$ the map has one functional form and in region $x > 0$ the map has a different form. The map (2.13) is differentiable on each side of the border $x = 0$. At $x = 0$ the map has two sets of one-sided partial derivatives, depending on whether the border is approached from the positive or negative side. For the Lozi map, Misiurewicz [25] has proved the existence of a chaotic attractor. This attractor (see Fig. 2.27), known as the quasihyperbolic Lozi attractor, is well studied in some range of parameter values, both from a topological and from a metrical point of view [24, 25, 48–50].

The routes to chaos for the Lozi map are similar to the bifurcation scenarios we discussed in Chapter 1 for the skew tent map. Figure 2.28 shows that the map does not follow the usual period-doubling route to chaos. At $\alpha = 0.7$, the fixed point loses its stability as one of its multipliers passes through -1 and the solution jumps abruptly to a stable period-2 cycle of finite amplitude. This transition is not hysteretic, however. The 2-cycle remains stable until, at $\alpha = 1.3$, the system abruptly jumps into a four-band chaotic attractor. With further increase of α, the four-band attractor undergoes a homoclinic bifurcation in which the four bands merge into two and, finally, into a single band.

2.5 Some Remaining Problems

In this chapter we have considered a number of simple two-dimensional maps in order to illustrate both some of the similarities and some of the differences that exist between smooth and piecewise-smooth systems. The domains of stability for the periodic and chaotic oscillations in the plane of parameters were delineated using analytical as well as numerical techniques. Two types of domains for the stable cycles were determined: $\Pi_{k,j}$ and Π_k^{Swallow}, $k = 1, 2, \ldots$, and the structure of these sets as well as their general properties were studied. It was shown that the motions observed on the sets $\Pi_{k,j}$ depend smoothly on the parameters. When moving continuously within the boundaries of $\Pi_{k,j}$, the transitions from some stable cycles to other cycles occur softly through a sequence of direct (or reverse) period-doubling bifurcations. In Π_k^{Swallow}, on the other hand, hysteretic transitions are possible.

When intersecting the boundaries of $\Pi_{k,j}$, catastrophic transitions from one stable motion to another or to chaos are possible. Such transitions are not hysteretic. The sets $\Pi_{k,j}$ intersect non-emptily. Moreover, some of the regions $\Pi_{k,j}$ have intersections with domains where chaotic oscillations occur and that is the basis for a great variety of different bifurcations.

We realize that our approach has been application-oriented and numerical, and that we have left a number of more theoretical problems unmentioned. The properties of the swallow-tail structure, for instance, is substantially more complicated than described above, and many questions in relation to the so-called Newhouse regions remain essentially unanswered.

The *swallow-tail structure* already arises in bimodal one-dimensional maps. In the analysis of such maps, the concepts of *superstable* and *doubly superstable* orbits play a significant role. A *bimodal map* is a map with two extrema, a maximum and a minimum. The sine-circle map

$$\vartheta_k = \vartheta_{k-1} + \Omega - \frac{K}{2\pi} \sin 2\pi\vartheta_{k-1} \mod 1, \tag{2.14}$$

for instance, displays bimodal behavior for $K > 1$.

As mentioned in Chapter 1, the sine-circle map [51] is the basic model to describe entrainment (or frequency locking) in systems of two interacting oscil-

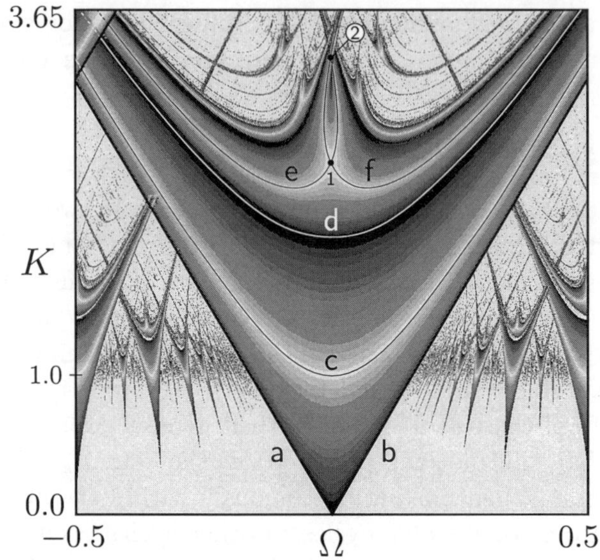

Figure 2.29. Map of dynamical regimes for the sine-circle map. The levels of gray indicate the absolute value of the cycle multiplier, with white corresponding to zero and black to one. The curves a and b are the saddle-node curves delineating the 1 : 1 resonance tongue, c is the curve of the superstability for the fixed point, d is the first period-doubling curve, and e and f are the curves of superstability for the period-2 solutions. 1 is the point of double superstability of the period-2 solution and 2 is the point, where two different mutually symmetric superstable period-2 solutions exist.

lators. It may be used, for instance, to explain the response of chicken heart cells to external periodic stimuli [52] or to analyze the synchronization of various business cycles in the economy [53]. The variable ϑ_k represents the (normalized) phase of one of the oscillators as measured stroboscopically with the period of the other oscillator. Ω is the phase advance per period in the absence of coupling, and K is a measure of the coupling strength.

As long as $K < 1$, the sine-circle map is a diffeomorphism, i.e. differentiable and invertible, and periodic and quasiperiodic motions are the only types of behavior that can occur. 1 : 1 frequency locking, for instance, is observed in the region of parameter space where the fixed point ϑ^* is stable, i.e. for $|\Omega| < K/2\pi$ (between the curves a and b in Fig. 2.29), and similar resonance (or Arnol'd) tongues describe the regions where other rational relations between the periods of the two oscillators can be observed.

At $K = 1$ a critical value is reached at which the sine-circle map develops an inflection point with zero slope at $\vartheta_k = 0$. At this point the map becomes bimodal. The resonance tongues start to overlap, allowing for the coexistence of two or more synchronized states, and quasiperiodic behavior ceases to exist. At the same time, folding starts to occur, and with further increase of K a number of bifurcations take place within the individual tongue, leading, as illustrated in Fig. 2.29, to an example of the swallow tail structure [26, 27].

For the one-dimensional map $\vartheta_k = f(\vartheta_{k-1})$ a periodic orbit ϑ_{ci}, $i = \overline{1, m}$ is said to be *superstable* if it passes through an extremum point for the map. In this case

$$\frac{\mathrm{d}}{\mathrm{d}\vartheta_{ci}} f^{(m)}(\vartheta_{ci}) = 0$$

and the transient approach to the orbit is faster than exponential. The superstable period-1 solution (corresponding to the state of $1:1$ synchronization) is determined by the condition that the slope of the map vanishes at the fixed point, i.e.

$$f'(\vartheta^*) = 1 - K \cos 2\pi\vartheta^* = 0$$

or by substitution into the expression (2.14) with $\vartheta_k = \vartheta_{k-1} = \vartheta^*$

$$\Omega = \frac{K}{2\pi} \sin 2\pi\vartheta^* = \pm\frac{K}{2\pi}\sqrt{1 - \frac{1}{K^2}}\,. \tag{2.15}$$

Curves in parameter space where superstable periodic orbits exist are referred to as *bones* [26, 27] and, as we shall see, the bones together form the *skeleton of the swallow tail* structure for the bimodal map. We immediately note from the above result (2.15) that the superstable period-1 solution (see the curve c in Fig. 2.29) continues to exist as K goes to infinity and that it follows a curve that asymptotically approaches the edges $\Omega = \pm K/2\pi$ of the synchronization tongue.

For $\Omega = 0$, the period-1 orbit is superstable for $K = 1$. As the coupling parameter is increased, a period-doubling bifurcation takes place when the slope of the map $f(\vartheta_k)$ becomes -1 at the fixed point. This is the case for

$$\Omega = \pm\frac{K}{2\pi}\sqrt{1 - \left(\frac{2}{K}\right)^2}\,. \tag{2.16}$$

The curve (2.16) (curve d in Fig. 2.29) also has the edges of the $1:1$ synchronization tongue as asymptotes. This provides an explanation to the characteristic narrow wings of the swallow tail structure (compare, e.g., with Fig. 2.6). For $\Omega = 0$, the first period-doubling bifurcation takes place at $K = 2$.

Since the map (2.14) is bimodal for $K > 1$, we can now ask if it is possible for the period-2 orbit to pass through both of the extremum points. In this case, the orbit is said to be *doubly superstable.* By virtue of the symmetry of the map this can only occur for $\Omega = 0$, and with this provision the conditions that the period-2 orbit passes both extrema yield [30]

$$\arcsin\sqrt{1 - \frac{1}{K^2}} = \frac{K}{2}\sqrt{1 - \frac{1}{K^2}}$$

or $K \approx 2.5366$ (see the point 1 in Fig. 2.29).

From the point where the doubly superstable period-2 orbit exists one can follow two curves in parameter space where the period-2 solution passes either one or the other extremum of the map (curves e and f in Fig. 2.29). Hence, we have two curves (bones) with superstable period-2 orbits. It turns out that these curves intersect at another point on the vertical axis, namely for $K = \sqrt{1 + \pi^2} \approx 3.2969$ (the point 2 in Fig. 2.29). At this point, the sine-circle map displaces two mutually symmetric superstable period-2 solutions, one passing through the minimum of the map and one passing through the maximum.

Formation of these two mutually symmetric period-2 orbits occurs in a pitchfork bifurcation where the period-2 solution passes two symmetrically placed points on the sine-circle map where the slope is 1. This condition gives $K = \pi \approx 3.1415$. The pitchfork bifurcation forms the cusp point for two saddle-node bifurcation curves that branch to both sides in parameter space and give rise to a folding of the period-2 solution. This is the starting point for the characteristic folded bifurcation structure that we meet in the swallow tails (as illustrated, e.g., in Fig. 2.15).

As K is further increased, each period-2 solution undergoes a second period-doubling, leading to two (partly overlapping) regions of parameter space where stable period-4 solutions can be observed. In each of these regions one can locate a point where a doubly superstable period-4 solution exists, and from each of these points one can follow two curves in parameter space along which the period-4 orbit is superstable. Each such pair of curves again displays a second point of intersection where two coexisting superstable period-4 solutions are observed. Hence, in each region with period-4 dynamics one can again locate a point where a pitchfork bifurcation occurs, and the symmetric period-4 solution splits into two mutually symmetric period-4 solutions. From each of these pitchfork bifurcation points, curves of saddle-node bifurcations (in which the period-4 solutions are folded) branch out in parameter space. In this way, by continuation of the above description, we can construct a picture of the swallow tail as an infinite set of consistently organized period-doubling and saddle-node bifurcations. The periodic windows in the Hénon map will generate a swallow tail of a similar type. As discussed in the following paragraphs, however, the finer details may differ between the swallow tails in one-dimensional maps and those of the two-dimensional Hénon map [34].

The classical Feigenbaum theory [54, 55] is essentially a one-parameter analysis. We have previously mentioned how Feigenbaum's renormalization group analysis leads to the existence of universal scaling behaviors in phase space as well as in parameter space. The Feigenbaum constant $\delta_{\mathrm{F}} \approx 4.6692\ldots$ determines the rate of convergence for the period-doubling cascade in parameter space, and the constant $\alpha_{\mathrm{F}} \approx -2.5029\ldots$ controls the self-similar rescaling in phase space. We have also indicated how the scaling constant δ_{F} applies to the infinite period-doubling transition to chaos observed for many parameter scans in the Hénon map.

When the analysis is extended to multiparameter systems, the parameter space may contain a set of smooth codimension-1 surfaces at which the period-doubling bifurcations occur, and these surfaces may converge to a critical set, which is also a codimension-1 surface, at which the transition to chaos takes place. An example of this type of behavior is shown in Fig. 4.23 for a relay control system with hysteresis. Here, the main parameters are the input voltage and the load resistance. A typical signature is the observation of symmetrical (complete or incomplete) opposing cascades of forward and reverse period-doubling bifurcations as a parameter is scanned. However, more complicated situations can also arise, and the swallow tail structure is an example of such a situation.

For two-parameter systems, besides *critical curves* in the parameter plane with Feigenbaum type behavior, the threshold to chaos may contain a *Cantor set* of critical points at which the scaling behavior is different and, as the number of parameters increases, new types of critical behavior can be expected to arise.

As a simple illustration of this phenomenon we may consider a one-dimensional map with a single extremum. If the map has three free parameters, a *codimension-2 curve* may exist in parameter space along which the second and third derivatives of the map vanish. If the transition to chaos occurs along this curve, the scaling behavior will be characterized by a different set of Feigenbaum constants, namely by the values of α_F and δ_F that characterize a *quartic* (rather than a quadratic) *extremum*.

Similarly for the sine-circle map and for other bimodal one-dimensional maps. The condition that a quadratic extremum is mapped into another quadratic extremum specifies a single relation between the parameters (e.g., Ω and K for the sine-circle map). This defines a curve in the two-parameter plane along which the twice iterated map $f(f(\vartheta))$ has a quartic maximum, and where the transition to chaos follows a different scaling behavior. The limit point for the period-doubling cascade found under these conditions is said to be tricritical [56]. *Tricriticality* was examined for several one-dimensional maps already in the late 80's [57, 58]. Typically, one observes that there is a gap in the Feigenbaum critical curve with a tricritical point at each end.

Kuznetsov *et al.* [34] have performed a detailed study of the scaling behavior for Hénon-type maps, and Kuznetsov *et al.* [35] have examined the scaling behavior for the two-dimensional Ikeda map that arises as an approximate return map for a laser excited optical ring cavity with absorber [30]. They conclude that tricritical behavior in two-dimensional maps typically occurs as a three-parameter phenomenon as opposed to the two-parameter tricriticality observed in one-dimensional maps. Moreover, the tricritical behavior observed, for instance, in the Ikeda map is pseudo-tricritical. The scaling behavior is found to follow tricritical behavior for a number of iterations of the two-dimensional map, but then a different type of scaling behavior takes over. As one would expect, this type of *cross-over dynamics* again displays certain universal traits.

Finally, it is worth mentioning that Kuznetsov *et al.* [37] have performed a comparative analysis of a kicked oscillator system, its two-dimensional Poincaré map, and the approximate one-dimensional return map (assuming strong dissipation). Each of these systems display a swallow tail structure. However, the details in the structure differs from the time-continuous model to its one-dimensional map representation. From our application-oriented point of view the existence of a Cantor set of points with special scaling properties may not seem so important. However, the conclusion we have to draw is that seemingly well-argumented approximations in the model formulation can change the type of criticality that occurs. Depending on the number of free parameters, very specific situations can arise. Considering the paramount significance of universality in the field of nonlinear dynamics, understanding of the different forms of critical behavior that one can encounter is, of course, essential from a more theoretical point of view.

However, our problems are not over yet. In Chapter 1 we defined chaotic dynamics in terms of three basic criteria: (a) the system displays sensitivity to the initial conditions (at least one positive Lyapunov exponent), (b) there is a dense set of unstable (i.e. saddle) periodic orbits of different periodicities, and (c) there are trajectories that come everywhere (ergodicity). It is relatively easy to show that these conditions apply to the tent map, and by appealing to the concept of *topological conjugacy* [45] one can also show, for instance, that the logistic map $x_k = \lambda x_{k-1}(1 - x_{k-1})$ is chaotic for $\lambda = 4$.

For two-dimensional maps (and for three-dimensional flows), the situation is much more complicated, and the problem of explaining the bifurcation structure in detail remains largely unsolved. The difficulties are associated with the problem of establishing the necessary and sufficient conditions for the *structural stability* of solutions to such systems. In 1979, Newhouse [59] presented an important theorem stating that in any neighborhood of a C^n-smooth multidimensional and strongly nonlinear dynamical system ($n \geq 2$) there are regions in the space of dynamical systems where homoclinic tangencies are dense. Hence, as a parameter is changed, new *homoclinic tangencies* continue to occur. Such regions in parameter space are now referred to as Newhouse regions.

Following Shilnikov [60] we may distinguish between three types of strange attractors:

The first type, denoted *hyperbolic attractors*, are the limit sets for systems that satisfy conditions similar to those of *Smale's Axiom A attractors* [43, 61]. The periodic orbits and the homoclinic trajectories are dense, but they all belong to the same type of saddle topology, i.e., the dimensions of the stable and unstable manifolds are the same for all orbits. Moreover, the manifolds intersect transversally, and tangencies between the manifolds do not arise. Unfortunately, very few (if any) practical systems satisfy these requirements. In particular, neither the Hénon map nor, for that matter, the Rössler oscillator satisfy the conditions for a hyperbolic attractor.

The second type of strange attractors, denoted *quasihyperbolic attractors*, are not structurally stable although they have structurally stable homoclinic and heteroclinic orbits. One can also show, that stable periodic orbits embedded in the chaotic set do not arise under small parameter variations. Like hyperbolic attractors, quasi-hyperbolic attractors are ergodic, and statistical measures (such as the invariant density [45]) can be applied to calculate various mean values (e.g., the Lyapunov exponents). The Lorenz attractor and the Lozi attractor discussed in Sec. 2.4 are both examples of quasihyperbolic attractors.

Finally, the third type of strange attractors, called *quasiattractors*, are limit sets in which orbits of different topological type are embedded, including, for instance, nodes (stable cycles) and saddle cycles with one and two unstable directions. Moreover, tangencies occur between the stable and unstable manifolds, and new tangencies continue to occur (*persistent homoclinic tangencies*) as a parameter is changed. Actually, as shown by Gonchenko *et al.* [62] systems with infinitely many homoclinic orbits of any order of tangency and with infinitely many arbitrarily degenerate periodic orbits are dense in the Newhouse regions. The mathematical difficulties associated with such regions are truly astonishing [63, 64].

For models with an equilibrium of saddle-focus type with a one-dimensional unstable manifold and a homoclinic loop, Ovsyannikov and Shilnikov [65, 66] have found that systems with a countable set of stable solutions are everywhere dense. Similarly, for the Hénon family

$$f_\lambda(x, y) = (\lambda - x^2 + by, x)$$

with $b = 0.3$ and $\lambda \in [1.270, 1.420]$, Kan *et al.* [14] have found that there exist three intervals (comprising 93% of the considered range for λ) where the two-dimensional map displays homoclinic tangencies for a dense set of parameter values. Hence in these intervals, which extends from the end of a large period-7 window at $\lambda \approx 1.270$ to the boundary crisis at $\lambda \approx 1.420$ where the attractor disappears, the system is not structurally stable.

Kan *et al.* [14] substantiate their findings by a combination of theoretical arguments and numerical calculations, including an evaluation of the intersection between the stable and the unstable manifolds. This involves an estimate of the *thickness of a Cantor set*. In this connection it is important to notice that the presence of Newhouse regions cannot be detected in our usual one- and two-dimensional bifurcation diagrams. Windows of stable orbits, for instance, that appear to represent structurally stable systems may, in reality, not be so. A large number of homoclinic tangencies can occur in regions of phase space away from the considered periodic cycle.

Let us conclude this discussion by emphasizing that although the results we are going to present in the following chapters may appear complicated, there are a number of theoretical problems associated with chaos in multidimensional systems that are even more complicated, and to which we do not know the solutions. Nonetheless, for most practical purposes the analysis that we present will be sufficient.

Bibliography

[1] P. S. Landa, Nonlinear Oscillations and Waves in Dynamical Systems (Kluwer Academic Publ., Dordrecht, 1996).

[2] E. Mosekilde, Yu. Maistrenko, and D. Postnov, Chaotic Synchronization — Applications to Living Systems (World Scientific, Singapore, 2002).

[3] M. Hénon, *A Two-Dimensional Mapping with a Strange Attractor*, Comm. Math. Phys. **6**, 69–77 (1976).

[4] M. Hénon and Y. Pomeau, *Two Strange Attractors with a Simple Structure*, in Turbulence and Navier-Stokes Equation, Lecture Notes in Mathematics, vol. 668 (Springer, Berlin, 1975).

[5] V. S. Anishchenko, Dynamical Chaos — Models and Experiments. Appearance Routes and Structure of Chaos in Simple Dynamical Systems (World Scientific, Singapore, 1995).

[6] P. Bergé, Y. Pomeau, and Ch. Vidal, Order within Chaos — Towards a Deterministic Approach to Turbulence (John Wiley and Sons, New York, 1984).

[7] J. Guckenheimer and P. J. Holmes, Nonlinear Oscillations, Dynamical Systems, and Bifurcations of Vector Fields (Springer–Verlag, Berlin, 1983).

[8] H. G. Schuster, Deterministic Chaos — An Introduction (VCH Verlagsgesellschaft, Weinheim, 1989).

[9] Zh. T. Zhusubaliyev, V. N. Rudakov, E. A. Soukhoterin, and E. Mosekilde, *Bifurcation Analysis of the Hénon Map*, Discrete Dynamics in Nature and Society **4**, 203–221 (2000).

[10] J. Currey, *On the Hénon Transformation*, Comm. Math. Phys. **68**, 129–140 (1979).

[11] B. Derrida, A. Gervois, and Y. Pomeau, *Universal Metric Properties of Bifurcations and Endomorphisms*, J. Phys. A **12**, 269–296 (1979).

[12] S. Newhouse, *Diffeomorphisms with Infinitely Many Sinks*, Topology **13**, 9–18 (1974).

[13] C. Robinson, *Bifurcation to Infinitely Many Sinks*, Commun. Math. Phys. **90**, 433–459 (1983).

[14] I. Kan, H. Kocak, and J. A. Yorke, *Persistent Homoclinic Tangencies in the Hénon Family*, Physica D **83**, 313–325 (1995).

[15] S. V. Gonchenko, L. P. Shilnikov, and D. V. Turaev, *Dynamical Phenomena in Systems with Structurally Unstable Poincaré Homoclinic Orbits*, Chaos **6**, 15–31 (1996).

[16] C. Simo, *On the Hénon-Pomeau Attractor*, J. Stat. Phys. **21**, 465–494 (1979).

[17] P. J. Holmes and D. C. Whitley, *Bifurcations of One- and Two-Dimensional Maps*, Phil. Trans. R. Soc. Lond. A **311**, 43–102 (1984).

[18] K. Hansen and P. Cvitanović, *Bifurcation Structures in Maps of Hénon Type*, Nonlinearity **11**, 1233–1261 (1998).

[19] V. S. Baushev, Zh. T. Zhusubaliyev, and S. G. Michal'chenko, *Stochastic Features in the Dynamic Characteristics of a Pulse-Width Controlled Voltage Stabilizer*, Electrical Technology (1), 137–150 (1996), Elsevier Science.

[20] V. S. Baushev and Zh. T. Zhusubaliyev, *Indeterminable States of a Voltage Regulator with Pulse-Width Control*, Electrical Technology (3), 85–98 (1992), Elsevier Science.

[21] Zh. T. Zhusubaliyev, *Bifurcations and Chaotic Motions in the Dynamics of Relay Automatic Control Systems*, in Proc. of the 3rd International Conf. "Recognition-97", Kursk, 25–30 (1997), in Russian.

[22] Zh. T. Zhusubaliyev, *Investigation of Chaotic Regimes of a Voltage Converter with Pulse-Width Modulation*, Elektrichestvo **6**, 40–46 (1997), in Russian.

[23] Zh. T. Zhusubaliyev, E. A. Soukhoterin, and E. Mosekilde, *Border-Collision Bifurcations and Chaotic Oscillations in a Piecewise-Smooth Dynamical System*, Int. J. Bifurcation and Chaos **11**, 2977–3001 (2001).

[24] R. Lozi, *Un Attracteur Étrange(?) du Type Attracteur de Hénon*, J. Phys. **39**, 9–10 (1978).

[25] M. Misiurewicz, *Strange Attractors for the Lozi Mappings*, in Nonlinear Dynamics, edited by R. G. H. Helleman (New York Academy of Sciences, New York, Annals **357**, 1980), pp. 348–358.

[26] L. Glass and R. Perez, *Fine Structure of Phase Locking*, Phys. Rev. Lett. **48**, 1772–1775 (1982).

[27] M. Schell, S. Fraser, and R. Kapral, *Subharmonic Bifurcations in the Sine Map: An Infinite Hierarchy of Cusp Bistabilities*, Phys. Rev. A **28**, 373–378 (1983).

[28] R. S. MacKay and C. Tresser, *Some Flesh on the Skeleton: The Bifurcation Structure of Bimodal Maps*, Physica D **27**, 412–422 (1987).

[29] J. Gallas, *Dissecting Shrimps: Results for Some One-Dimensional Physical Models*, Physica A **202**, 196–223 (1994).

[30] E. Mosekilde, Topics in Nonlinear Dynamics. Applications to Physics, Biology and Economic Systems (World Scientific, Singapore, 1996).

[31] F. C. Moon, Chaotic Vibrations. An Introduction for Applied Scientists and Engineers (Wiley, New York, 1987).

[32] A. P. Kuznetsov, S. P. Kuznetsov, I. R. Sataev, and L. O. Chua, *Multi-Parameter Criticality in Chua's Circuit at Period-Doubling Transitions to Chaos*, Int. J. Bifurcation and Chaos **6**, 119–148 (1996).

[33] A. P. Kuznetsov, S. P. Kuznetsov, and I. R. Sataev, *Variety of Types of Critical Behavior and Multistability in Period-Doubling Systems with Unidirectional Coupling near the Onset of Chaos*, Int. J. Bifurcation and Chaos **3**, 139–152 (1993).

[34] A. P. Kuznetsov, S. P. Kuznetsov, and I. R. Sataev, *From Bimodal One-Dimensional Maps to Hénon-like Two-Dimensional Maps: Does Quantitative Universality Survive?*, Physics Letters A **184**, 413–421 (1994).

[35] A. P. Kuznetsov, S. P. Kuznetsov, E. Mosekilde, and L. V. Turukina, *Two-Parameter Analysis of the Scaling Behavior at the Onset of Chaos: Tricritical and Pseudo-Tricritical Points*, Physica A **300**, 367–385 (2001).

[36] C. Mira, J.-P. Carcasses, G. Millérioux, and L. Gardini, *Plane Foliation of Two-Dimensional Noninvertible Maps*, Int. J. Bifurcation and Chaos **6**, 1439–1462 (1996).

[37] A. P. Kuznetsov, L. V. Turukina, and E. Mosekilde, *Dynamical Systems of Different Classes as Models of the Kicked Nonlinear Oscillator*, Int. J. Bifurcation and Chaos **11**, 1065–1077 (2001).

[38] N. V. Butenin, Yu. I. Neimark, and N. A. Fufayev, Introduction to the Theory of Nonlinear Oscillations (Nauka Publ., Moscow, 1987), in Russian.

[39] Yu. I. Neimark and P. S. Landa, Stochastic and Chaotic Oscillations (Kluwer Academic Publ., Dordrecht, 1992).

[40] J.-P. Carcasses and C. Mira, *"Crossroad Area–Spring Area" Transition (I) Parameter Plane Representation*, Int. J. Bifurcation and Chaos **1**, 183–196 (1991).

[41] C. Grebogi, S. W. Donald, E. Ott, and J. A. Yorke, *Final State Sensitivity: An Obstruction to Predictability*, Phys. Lett. A **99**, 415–418 (1983).

[42] C. Grebogi, E. Ott, and J. A. Yorke, *Fractal Basin Boundaries, Long-Lived Chaotic Transients, and Unstable-Unstable Pair Bifurcations*, Phys. Rev. Lett. **50**, 935–938 (1983).

[43] S. Smale, *Differentiable Dynamical Systems*, Bull. Amer. Math. Soc. **73**, 747–817 (1967).

[44] Y. Ueda, *Survey of Regular and Chaotic Phenomena in the Forced Duffing Oscillator*, Chaos, Solitons and Fractals **1**, 199–231 (1991).

[45] E. Ott, Chaos in Dynamical Systems (Cambridge University Press, England, 1993).

[46] V. I. Arnol'd, Catastrophe Theory (Springer–Verlag, 1992).

[47] R. Gilmore, Catastrophe Theory for Scientists and Engineers (Dover, New York, 1993).

[48] P. Collet and Y. Levy, *Ergodic Properties of the Lozi Mappings*, Commun. Math. Phys. **93**, 461–481 (1984).

[49] V. S. Anishchenko and G. I. Strelkova, *Irregular Attractors*, Discrete Dynamics in Nature and Society **2**, 53–72 (1998).

[50] V. S. Anishchenko, T. E. Vadivasova, G. I. Strelkova, and A. S. Kopeikin, *Chaotic Attractors of Two-dimensional Invertible Maps*, Discrete Dynamics in Nature and Society **2**, 249–256 (1998).

[51] V. I. Arnol'd, *Small Denominators, I: Mappings of the Circumference into Itself*, AMS Trans. Series 2 **46**, 213–284 (1965).

[52] M. R. Guevara, L. Glass, and A. Shrier, *Phase Locking, Period-Doubling Bifurcations, and Irregular Dynamics in Periodically Stimulated Cardiac Cells*, Science **214**, 1350 (1981).

[53] Chr. Haxholdt, Chr. Kampmann, E. Mosekilde, and J. D. Sterman, *Entrainment in a Disaggregated Economic Long Wave Model*, System Dynamics Review **1**, 178–198 (1995).

[54] M. J. Feigenbaum, *Quantitative Universality for a Class of Nonlinear Transformations*, J. Stat. Phys. **19**, 25–52 (1978).

[55] M. J. Feigenbaum, *Universal Behavior in Nonlinear Systems*, Physica D **7**, 16–39 (1983).

[56] S. J. Chang, M. Wortis, and J. A. Wright, *Iterative Properties of a One-Dimensional Quartic Map. Critical Lines and Tricritical Behavior*, Phys. Rev. A **24**, 2669–2684 (1981).

[57] B. P. Bezruchko, Yu. V. Gulyaev, S. P. Kuznetsov, and E. P. Seleznev, *New Type of Critical Behavior in Coupled Systems of the Transition to Chaos*, Sov. Phys. Dokl. **31**, 258–260 (1986).

[58] R. S. MacKay and J. B. J. Zeijts, *Period-Doubling for Bimodal Map: A Horseshoe for a Renormalization Operator*, Nonlinearity **1**, 253–277 (1988).

[59] S. E. Newhouse, *The Abundance of Wild Hyperbolic Sets and Non-Smooth Stable Sets for Diffeomorphisms*, Publ. Math. IHES **50**, 101–151 (1979).

[60] L. P. Shilnikov, *Chua's Circuit: Rigorous Results and Future Problems*, Int. J. Bifurcation and Chaos **4**, 489–519 (1994).

[61] S. Smale, *Diffeomorphisms with Many Periodic Points*, in Differential Combinational Topology, edited by S. S. Cairns (Princeton University Press, 1963), pp. 63–80.

[62] S. V. Gonchenko, D. V. Turaev, and L. P. Shilnikov, *On Models with a Non-Rough Homoclinic Poincaré Curve*, Physica D **62**, 1–14 (1993).

[63] S. V. Gonchenko, L. P. Shilnikov, O. V. Sten'kin, and D. V. Turaev, *Bifurcations of System with Structurally Unstable Homoclinic Orbits and Moduli of Ω Equivalence*, Comp. Math. Appl. **34**, 111–142 (1997).

[64] S. V. Gonchenko, L. P. Shilnikov, and D. V. Turaev, *Quasiattractors and Homoclinic Tangencies*, Comp. Math. Appl. **34**, 195–227 (1997).

[65] I. M. Ovsyannikov and L. P. Shilnikov, *On Systems with Saddle-Focus Homoclinic Curve*, Math. USSR Sborn. **58**, 557–574 (1987).

[66] I. M. Ovsyannikov and L. P. Shilnikov, *Systems with a Homoclinic Curve of Multidimensional Saddle-Focus Type and Spiral Chaos*, Math. USSR Sborn. **73**, 415–443 (1992).

Chapter 3

RELAY CONTROL SYSTEMS

3.1 Bifurcations and Chaos in Relay Systems

Relay control systems [1–3] (also referred to as relay feedback systems [4–6]) constitute an important class of non-linear regulation systems and are widely used in many fields of engineering and applied science [3, 4, 7–12]. Examples are power converters, AC and DC electric drives [13–19], adaptive control systems [20–23] and a variety of other mechanical and electromechanical systems [1, 2]. Johansson *et al.* [24–26] provide examples of applications in variable-structure systems [8], supervisory switch-control systems [27], hybrid systems [28, 29], and delta-sigma modulators for signal processing [30, 31].

Typically, systems of this type are free running which means that the oscillations are determined by the properties of the system itself [1] and not sustained by an external periodic action.

The advantages of using relay control systems rather than systems with continuous control are that the relay element serves as an amplifier that converts the continuous control signal into a well-defined switching sequence. Relay systems have a broad variety of applications in engineering due to the large power gains of the relay element and because of their simplicity. Oscillatory relay systems exploit self-oscillations as a main operating mode. Such systems include, for example, systems for automatic regulation of the voltage (current) for power converters and control systems for AC and DC electric drives with semiconductor converters. Relay systems can provide arbitrarily small ripple on the controlled voltages or currents, and they can significantly increase the control accuracy and the speed of operation.

The oscillatory properties of relay systems have been studied in numerous works. However, most authors pay attention only to periodic motions (see, e.g., [1, 13, 15, 20, 32–34]). On the other hand, both experiments and numerical simulations have revealed that more complicated motions occur, including periodic solutions with chattering and sliding [25, 26] as well as quasiperiodic and chaotic oscillations (see, for example, [22, 23, 35]). *Chattering* denotes an extremely rapid series of switchings and *sliding* refers to a nearly infinite number of relay switchings [8, 36]. The existence of periodic solutions with sliding was reported by Neimark [37] and Fedosenko [38] and independently by Johansson *et al.* [25]. In a series of papers, di Bernardo *et al.* [29, 39–41] studied bifurcation scenarios for the appearance of asymmetric periodic solutions and of sliding modes in unforced relay feedback systems with a symmetric relay characteristics. They showed the existence of specific classes of local bifurcations that the authors refer to as sliding and multisliding bifurcations. These bifurcations result in the appearance of orbits having one or more sliding sections. Sequences of multisliding bifurcations constitute a sliding-adding scenario, with subsequent orbits in a sequence having an increasing number of sliding sections. In one of the papers, Kowalczyk and di Bernardo [40] demonstrate that asymmetric chaotic attractors can occur even in systems with symmetric relay characteristics (see also [29]).

Research on relay systems with chaotic behavior is scarce and work on specific bifurcation scenarios is nearly non-existing. For particular systems, some results can be found in a series of paper by Krutova [42, 43], Postnikov [21–23], Zhusubaliyev [35], and Zhusubaliyev *et al.* [44]. Among the first works to deal with aperiodic and chaotic motions, according to Neimark and Landa [45], are those of Alekseev [46, 47]. Alekseev [47] used the point mapping method to investigate the dynamics of a relay temperature regulator. He found a parameter domain in which complex periodic oscillations existed and, when the parameters were changed, underwent an infinite number of period-doubling bifurcations resulting in the emergence of aperiodic motion. Neimark and Landa [45] note that the range of parameters within which one can observe chaotic behavior is interrupted by gaps where periodic motions exist, with the periodic domains being everywhere dense, i.e., in any arbitrarily small neighborhood of a set of parameters that produce chaotic oscillations one can find stable limit cycles. As discussed in Chapters 1 and 2, this structure is typical of nonlinear dynamical systems, and it arises already for systems whose motion is described by a one-dimensional return map with a smooth extremum (see, e.g., [45, 48, 49]).

Cook [50], Amrani and Atherthon [51], and Genesio and Tesi [52] have shown that quasiperiodic and chaotic motions can occur even in low-order relay systems with hysteresis, and Holzhüter and Klinker [53] have investigated some aspects of the Ott-Grebogi-Yorke (OGY) method of controlling chaos for a relay system with hysteresis. Krutova [42, 43] investigated a relay control system aided process of stabilizing a space vehicle about the set position for its orientation. The first paper [42] demonstrated that the control action can excite elastic vibrations that lead to chaotic dynamics.

Postnikov [21–23] studied the dynamics of a nuclear reactor with a relay control system. His research showed that if the control object is oscillatorily unstable, relay systems with hysteresis can display not only periodic, but also chaotic motions. In one of the papers [23] one can find the domains of periodicity and chaos determined in the parameter plane for systems of second and third order with an oscillatorily unstable linear part. This work also discusses the problem of how to choose a feedback control law which, for a system of arbitrary order, will produce chaotic motions with the same mean switching frequency of the relay element as a second or third order system.

One would expect that the appearance of chaos in relay systems with hysteresis and oscillatorily unstable linear parts was preceded by complex bifurcational transitions similar to those that have been described for other self-oscillatory systems with chaotic dynamics [45, 48, 49, 54–57]. However, the papers by Postnikov [21–23] do not contain detailed results on bifurcation mechanisms and the transition to chaos. Apparently it was assumed in these works that the transition to chaos followed one of the typical scenarios known for smooth systems [45, 48, 49, 54–57]. It is obvious from our discussion in Chapters 1 and 2, however, that the transitions to chaos in relay systems with hysteresis and deadzones may display several peculiarities connected with the occurrence of *border-collision bifurcations*. However, as we shall show, the investigation of border-collision bifurcations in relay systems with hysteresis involves specific mathematical difficulties. For particular applications one can use the results described in publications by Zhusubaliyev *et al.* [58], Feigin [59], di Bernardo *et al.* [60], and Banerjee *et al.* [61].

As piecewise-smooth multi-dimensional systems, relay systems can display situations where several locally stable limit cycles with different dynamic characteristics coexist for a wide range of parameters [44, 62]. These cycles typically arise in hard transitions, for example through saddle-node bifurcations, and with changing parameters they can undergo either a finite or an infinite sequence of period-doubling bifurcations, resulting in the transition to chaos. Moreover, the transition to chaos can take place through the appearance and subsequent destruction of a quasiperiodic motion on a two-dimensional torus. As a result we have parameter domains wherein, alongside with locally stable limit cycles, there are co-existing modes of chaotic or quasiperiodic oscillations. Under such conditions the action of external noise, even of low intensity, can induce a sudden transition from one dynamic state to another state with different characteristics and, in particular, from regular to chaotic dynamics [44].

The main purpose of the present chapter is to describe a set of methods that can be used for the analysis of nonlinear dynamic phenomena in relay control systems. The chapter is organized as follows. In Sec. 3.2 we introduce a system of autonomous differential equations with discontinuous right-hand sides that can be serve as a model of relay systems. We first consider a piecewise-linear system of autonomous differential equations. Investigation of this model can be reduced to the study of the properties of a multi-dimensional piecewise-smooth mapping defined in the space of the same dimension as the phase space of the original system of differential equations. In Sec. 3.3 we describe the method for numerical determining the periodic solutions. The problem of determining limit cycles is reduced to solving a system of transcendental equations with respect to the switching instants of the relay element. Such an approach allows us to obtain not only the stable but also the unstable periodic orbits.

Section 3.4 is devoted to an analysis of the local stability of periodic solutions for sets of differential equations with discontinuous right-hand sides. First we consider dynamical systems with external periodic action and then we generalize the obtained results to autonomous systems with discontinuous right-hand sides.

Investigation of local stability follows the same procedure as for smooth dynamical systems. However, the presence of discontinuities results in some peculiarities in the analysis, namely, the solution for the system of linearized equations with respect to small disturbances displays a jump at the points of discontinuity. The value of this jump is determined by the *updating matrix*.

In order to obtain the updating matrix we use an analytical approach based on the application of generalized functions. In many cases, the construction of the updating matrix is uncertain because of the non-uniqueness of its definition. We establish a condition that ensures uniqueness for this matrix. In the deterministic description of die tossing [63], an example of a piecewise-smooth mechanical system mentioned in Chapter 1, the so-called *collision map* plays the role of the updating matrix.

Section 3.5 contains two practically important examples of relay control systems. First, we consider a relay system with hysteresis, which is represented by a system of piecewise-smooth autonomous differential equations with variable matrix. In this connection, we describe the technique for analyzing the stability of limit cycles. Another example is represented by a relay system without hysteresis, the behavior of which was studied, for instance, by di Bernardo *et al.* [29, 39–41]. The mathematical model of such a system is a set of piecewise-smooth autonomous differential equations with a constant matrix. For this model we present both the techniques for obtaining the Poincaré map and the methods and algorithms for determining the limit cycles and analyzing their stability.

Section 3.6 establishes the equations of motion for the DC/DC converter with relay control used as a basic model in our investigation of nonlinear phenomena in piecewise-smooth autonomous dynamical systems. A detailed bifurcation analysis of this model will be presented in Chapter 4.

Finally, in Sec. 3.7 we consider a practical implementation of relay systems designed to control high-power electric drives with DC motors. As a particular example, we describe the system that has been used to automatize the traction and braking modes of the electric motors on the first USSR high-speed train ER200. Servicing the line between Moscow and Saint-Petersburg, this train was set in operation on March 1, 1984. The section provides a relatively detailed account of the technical aspects of this control system. The section also describes the formulation of a mathematical model for a DC electric drive with relay control. The system considered in Sec. 3.7 differs from that discussed in Sec. 3.6 by the presence of a time deadzone for the relay element. Analysis of the complex oscillatory behavior of this system will also be given in Chapter 4.

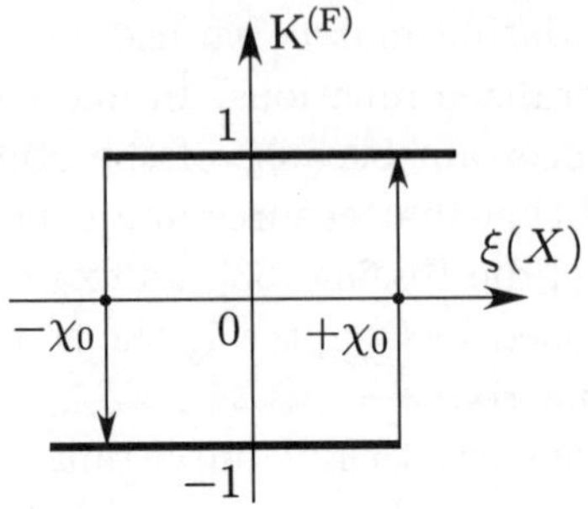

Figure 3.1. Switching characteristics of the relay element.

3.2 Modeling Relay Systems with Hysteresis

Relay control systems may be described by systems of autonomous differential equations of the form

$$\frac{\mathrm{d}X}{\mathrm{d}t} = G(X). \tag{3.1}$$

Here $X = (x_1, x_2, \dots, x_n)^{\mathrm{T}}$ and $G = (g_1, g_2, \dots, g_n)^{\mathrm{T}}$ are column matrices (vectors); $(\cdot)^{\mathrm{T}}$ denotes the operation of transposition; and n is the order of the system. To represent the switching conditions, the right-hand side of Eq. (3.1) has simple (finite) discontinuities on certain surfaces (hypersurfaces) which are determined by the equations

$$\zeta_j(X) = 0,\ j = 1, 2, \dots$$

Let the right-hand side of system (3.1) has the form

$$G(X) = A(\mathrm{K}^{(\mathrm{F})})X + B(\mathrm{K}^{(\mathrm{F})}) \tag{3.2}$$

with $A(\mathrm{K}^{(\mathrm{F})})$ being a piecewise constant $n \times n$-matrix and $B(\mathrm{K}^{(\mathrm{F})})$ an n-dimensional vector for which

$$A(\mathrm{K}^{(\mathrm{F})}) = \begin{cases} A_1, & \mathrm{K}^{(\mathrm{F})} = 1; \\ A_2, & \mathrm{K}^{(\mathrm{F})} = -1; \end{cases} \qquad B(\mathrm{K}^{(\mathrm{F})}) = \begin{cases} B_1, & \mathrm{K}^{(\mathrm{F})} = 1; \\ B_2, & \mathrm{K}^{(\mathrm{F})} = -1. \end{cases}$$

Here, A_1 and A_2 are stable $n \times n$-matrices and B_1 and B_2 are n-dimensional vectors. $\mathrm{K}^{(\mathrm{F})} = \mathrm{K}^{(\mathrm{F})}(\xi)$, where $\xi = \xi(X)$ and $\mathrm{K}^{(\mathrm{F})}(\xi)$ are the input and output signals to the relay element, respectively.

Let us assume that the relay element has positive hysteresis with the following *switching algorithm* (see Fig. 3.1):

$$\begin{aligned}&\mathrm{K}^{(\mathrm{F})} = -1 \to \mathrm{K}^{(\mathrm{F})} = 1,\ \xi(X) \geq \chi_0,\\&\mathrm{K}^{(\mathrm{F})} = 1 \to \mathrm{K}^{(\mathrm{F})} = -1,\ \xi(X) \leq -\chi_0,\\&\xi(X) = UX - U_{\mathrm{ref}}.\end{aligned} \tag{3.3}$$

Here U_{ref} is the reference signal; $+\chi_0$ and $-\chi_0$ are the threshold values of the relay element; and U is an n-dimensional row matrix that determines the control law.

The system (3.1) with the right-hand side in the form (3.2) has a definite form for each state of the relay element:

$$\frac{\mathrm{d}X}{\mathrm{d}t} = A_1X + B_1 \qquad \text{for } \mathrm{K}^{(\mathrm{F})} = 1 \tag{3.4}$$

and

$$\frac{\mathrm{d}X}{\mathrm{d}t} = A_2X + B_2 \qquad \text{for } \mathrm{K}^{(\mathrm{F})} = -1. \tag{3.5}$$

Between two subsequent switchings the phase trajectories of the dynamical system (3.1) can be determined by solving either the system (3.4) or the system (3.5). In phase space a single stable fixed point (equilibrium point) corresponds to each of these systems. This gives rise to a solution of the Cauchy problem in which the actual state of the system tends towards the present equilibrium point. Changing the state of the relay element reverses the direction of motion because the position of the attracting point is switched.

It is not difficult to show that the condition of existence for oscillatory modes of the dynamical system (3.1) is expressed by the inequalities

$$-UA_1^{-1}B_1 + \chi_0 \leq U_{\mathrm{ref}} \leq -UA_2^{-1}B_2 - \chi_0. \tag{3.6}$$

Assume that the condition (3.6) is satisfied. Let t_k, $k = 1, 2, \ldots$ be the switching instants of the relay element, corresponding to the crossing of the error signal $\xi(X)$ over the threshold values $+\chi_0$ (in the positive) and $-\chi_0$ (in the negative direction), that is, when $\xi(X(t_k)) \pm \chi_0 = 0$. Finally, let $\mathrm{K}_{k-1}^{(\mathrm{F})}$ and $\mathrm{K}_k^{(\mathrm{F})}$ be the output signal values ($\mathrm{K}^{(\mathrm{F})} = -1$ or $\mathrm{K}^{(\mathrm{F})} = 1$) of the relay element in the ranges $t_{k-1} < t < t_k$ and $t_k < t < t_{k+1}$, respectively.

Applying the above designations together with conditions (3.3), the equation of motion may be written as [35]

$$\frac{\mathrm{d}X}{\mathrm{d}t} = A_kX + B_k \tag{3.7}$$

with

$$
\begin{aligned}
A_k &= A(\mathrm{K}_k^{(\mathrm{F})}),\ A_k \equiv A_{k+2};\\
B_k &= B(\mathrm{K}_k^{(\mathrm{F})}),\ B_k \equiv B_{k+2};\\
\mathrm{K}_k^{(\mathrm{F})} &= \mathrm{K}_{k-1}^{(\mathrm{F})}\,\mathrm{sign}\left\{\chi_0 + (-1)^{N_{k-1}}\xi(X)\right\},\ \mathrm{K}_0^{(\mathrm{F})} = -1,\\
N_k &= N_{k-1} + \frac{1-\mathrm{sign}\{\mathrm{K}_k^{(\mathrm{F})}\mathrm{K}_{k-1}^{(\mathrm{F})}\}}{2},\ N_0 = 0,\ k = 1, 2, \ldots
\end{aligned}
\tag{3.8}
$$

With this formulation it is not difficult to pass from (3.7) to the corresponding mapping:

In the range $t_{k-1} < t < t_k$ the system (3.7) assumes the form

$$\frac{\mathrm{d}X}{\mathrm{d}t} = A_k X + B_k.$$

The solution to this system is

$$X(t) = \mathrm{e}^{A_k(t-t_{k-1})}X_{k-1} - (E - \mathrm{e}^{A_k(t-t_{k-1})})A_k^{-1}B_k$$

with $X_{k-1} \equiv X(t_{k-1})$.

For $t = \theta$, $\theta > t_{k-1}$ the solution has the form

$$X(\theta) = \mathrm{e}^{A_k(\theta-t_{k-1})}X_{k-1} - (E - \mathrm{e}^{A_k(\theta-t_{k-1})})A_k^{-1}B_k.$$

Hence

$$X_k = X(t_k) = C_k X_{k-1} + V_k \tag{3.9}$$

with

$$C_k = \mathrm{e}^{A_k\tau_k} \quad \text{and} \quad V_k = -(E - C_k)A_k^{-1}B_k.$$

Here, $\tau_k = t_k - t_{k-1}$ is the smallest root of the equation

$$\varphi_k(\tau, X_{k-1}) = (-1)^{k-1}\chi_0 + U_{\mathrm{ref}} - UX(\tau, X_{k-1}) = 0,\ k = 1, 2, \ldots \tag{3.10}$$

The recurrent relations (3.9) in combination with Eq. (3.10) determines a shift map in a space of the same dimension as the original phase space [64].

The algorithm for solving the Cauchy problem by means of Eqs. (3.9) and (3.10) proceeds as follows: For a given X_{k-1} one should find τ_k numerically by solving Eq. (3.10), and then calculate X_k from Eq. (3.9). Obtaining the explicit dependence between X_{k-1} and X_k in the form (3.9) is connected with finding the first integrals of the differential equations (3.1). In the general

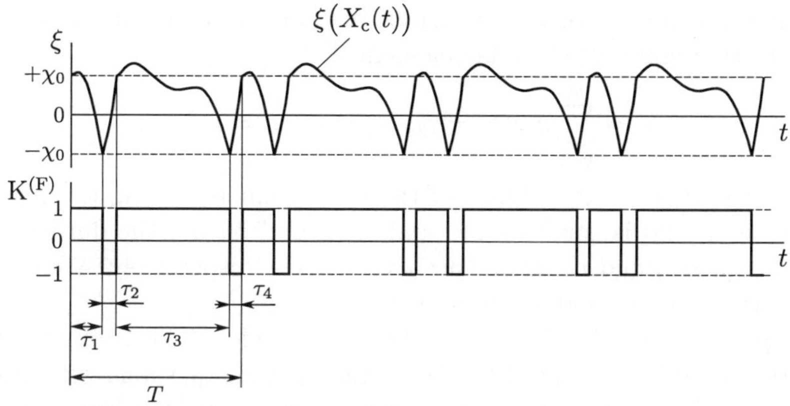

Figure 3.2. Response of the relay element in the presence of the periodic solution $X_c(t)$. χ_0 and $-\chi_0$ denote the threshold values for the switching process.

case this problem is unsolvable. Thus, when Eqs. (3.1) cannot be analytically integrated, the mapping

$$\begin{cases} X_k = F(\tau_k, X_{k-1}); \\ \varphi_k(\tau_k, X_{k-1}) = 0, \end{cases} \qquad k = 1, 2, \ldots$$

is obtained by solving numerically the original system (3.1).

3.3 Algorithms for Determining Limit Cycles

Let us consider the periodic motion $X_c(t)$ of the autonomous system (3.7) with period T

$$X_c(t + T) \equiv X_c(t). \tag{3.11}$$

According to (3.11), when t increases, the function $\xi(X_c(t))$ will also vary periodically:

$$\xi(X_c(t + T)) \equiv \xi(X_c(t)),$$

passing across the threshold values $+\chi_0$ and $-\chi_0$ at the instants $t = t_k$, that is,

$$(-1)^{k-1}\chi_0 + \xi(X_c(t_k)) = 0, \; k = 1, 2, \ldots$$

In this case $K_k^{(F)}\big(\xi(X_c(t))\big)$ is a periodic sequence of rectangular pulses (see Fig. 3.2). The period T of the periodic motion $X_c(t)$ is determined as a sum of the time intervals τ_k for the motion of the point in phase space between the subsequent switchings of the relay element

$$T = \sum_{k=1}^{m} a_k, \; a_k = \tau_{2k-1} + \tau_{2k}, \; m = 1, 2, \ldots$$

Here the number of switchings of the relay element during the period T is even-numbered and is equal to $2m$; and τ_{2k-1} and τ_{2k} are the durations of the positive (negative) and negative (positive) pulses, respectively. We denote the motion with such a period as an m-cycle.

Let X_{ck}, $k = \overline{0, p-1}$ $(p = 2m)$ be an m-cycle. The problem of determining the m-cycle is reduced to the system of the equations with respect to $\tau_1, \tau_2, \ldots, \tau_p$. By analogy with the approach used by Baushev and Zhusubaliyev [65], applying the recurrence equation (3.9), we obtain the sequence

$$\begin{aligned}
&X_1 = C_1 X_0 + V_1;\\
&X_2 = C_2 X_1 + V_2 = C_2 C_1 X_0 + C_2 V_1 + V_2;\\
&\quad \ldots\\
&X_k = C_k C_{k-1} \ldots C_1 X_0 + C_k C_{k-1} \ldots C_2 V_1 + C_k C_{k-1} \ldots C_3 V_2 + \ldots\\
&\qquad + C_k C_{k-1} V_{k-2} + C_k V_{k-1} + V_k;\\
&X_k = \prod_{i=1}^{k} C_{k+1-i} X_0 + \sum_{i=1}^{k-1} \left(\prod_{j=1}^{k-i} C_{k+1-j} \right) V_i + V_k.
\end{aligned}$$

From the periodicity condition $X_{cp} = X_{c0}$ one obtains

$$\left[E - \prod_{i=1}^{p} C_{p+1-i} \right] X_0 = \sum_{i=1}^{p-1} \left(\prod_{j=1}^{p-i} C_{p+1-j} \right) V_i + V_p,$$

and, hence,

$$\begin{aligned}
X_{ck} = &\prod_{i=1}^{k} C_{k+1-i} \left[E - \prod_{i=1}^{p} C_{p+1-i} \right]^{-1} \left\{ \sum_{i=1}^{p-1} \left(\prod_{j=1}^{p-i} C_{p+1-j} \right) V_i + V_p \right\}\\
&+ \sum_{i=1}^{k-1} \left(\prod_{j=1}^{k-i} C_{k+1-j} \right) V_i + V_k.
\end{aligned}$$

After substituting this expression into (3.10) we obtain a system of transcendental equations with respect to $\tau_1, \tau_2, \ldots, \tau_p$ [35, 44]:

$$\varphi_k(\tau_1, \tau_2, \ldots, \tau_p) = 0, \; k = 1, 2, \ldots, p,$$

where

$$\varphi_k(\tau_1, \tau_2, \ldots, \tau_p) = (-1)^{k-1}\chi_0 + \vartheta$$
$$- U\left\{\prod_{i=1}^{k} C_{k+1-i}\left[E - \prod_{i=1}^{p} C_{p+1-i}\right]^{-1}\left\{\sum_{i=1}^{p-1}\left(\prod_{j=1}^{p-i} C_{p+1-j}\right)V_i + V_p\right\}\right.$$
$$\left. + \sum_{i=1}^{k-1}\left(\prod_{j=1}^{k-i} C_{k+1-j}\right)V_i + V_k\right\}, \quad k = 1, 2, \ldots, p. \quad (3.12)$$

This approach allows us to obtain not only the stable but also the unstable periodic motions. This is extremely important when determining the domains of existence and stability for the periodic motions and calculating the bifurcation parameters at the transition point.

The above procedure is not new, but the development of concrete algorithms for numerically solving (3.12) is connected with difficulties of a principal nature. Because of their complex and transcendental form, Eqs. (3.12) can be solved only numerically by iteration methods. There is a number of rather comprehensive reviews of the most effective methods in the special literature (see, e.g. [66]). But general ways of numerically solving a system of nonlinear equations do not exist. In this work we solve the system (3.12) using Newton's method [66].

The difficulties connected with solving the system (3.12) increase sharply as the dimension of the system grows when the value of m rises. On the other hand, the described algorithm is extremely selective. Because it prespecifies the type of motion sought, it allows us to obtain only one particular periodic motion, and it discriminates against other possible motions. However, the existence of the specified solution for the selected parameters is unknown a priori. Moreover, the solution can be found only with a proper choice of the initial conditions, and the problem becomes even more complicated if the dynamical system is piecewise non-integrable.

3.4 Local Stability of Periodic Solutions

Our investigation of the local stability of periodic motions for dynamical systems with discontinuous right-hand parts is now reduced to considering a set of linearized equations with discontinuous solutions. The magnitude of the discontinuity is determined by the updating matrix M_k. It is a complicated problem to obtain this matrix at the discontinuities. In some cases one can apply the results derived by Aizerman and Gantmakher [67] and Filippov [36] who used a geometrically oriented approach to construct M_k. Most often, however, as shown by Baushev *et al.* [68] (see also [69]), the construction of the updating matrix in this way is incorrect, because the matrix contains undefined values. The existence of problems of this type are also noted in the works by Aizerman and Gantmakher [67] and by Filippov [36], but these authors did not pay particular attention to them.

In this section we describe a general approach for analyzing the local stability of dynamical systems with discontinuous right-hand parts. Unlike Aizerman and Gantmakher [67] and Filippov [36] we use an analytical approach involving generalized functions to obtain the updating matrix at the points of discontinuity [44, 68, 69]. The main results of our analysis are based on a theory by Baushev *et al.* [68] (see also [69]).

3.4.1 Nonautonomous dynamical systems

First, let us consider dynamical systems with external periodic action

$$\frac{\mathrm{d}X}{\mathrm{d}t} = G(t, X);\ X = (x_1, x_2, \ldots, x_n)^{\mathrm{T}};\ G = (g_1, g_2, \ldots, g_n)^{\mathrm{T}}. \tag{3.13}$$

As before, X and G are n-dimensional column matrices. The right-hand side of Eq. (3.13) is assumed to be time-periodic with the period a:

$$G(t + a, X) \equiv G(t, X)$$

and to have simple (finite) discontinuities on a certain hypersurface determined by the equation

$$\zeta(t, X) = 0,\ \zeta(t + a, X) \equiv \zeta(t, X).$$

We consider only one discontinuity surface, because it is not difficult to generalize the results to the case of several surfaces.

Let $X_{\mathrm{c}}(t)$ be a periodic solution of the system (3.13) with the period T:

$$X_{\mathrm{c}}(t + T) \equiv X_{\mathrm{c}}(t),\ T = ma,\ m = 1, 2, \ldots$$

The general approach to investigating the local stability of periodic motions of nonlinear dynamical systems is well-known and consists of the following steps:

1. Determine the periodic motion $X_c(t)$.

2. Linearize the equations of motion near $X_c(t)$.

3. Determine the basic matrix $F(T)$ by solving the matrix equation

$$\frac{\mathrm{d}F(t)}{\mathrm{d}t} = \frac{\partial G(t, X_c)}{\partial X} F(t), \; F(0) = E, \tag{3.14}$$

   where $\partial G(t, X_c)/\partial X$ is the Jacobian matrix.

4. Calculate the eigenvalues of the basic matrix $F(T)$ (multipliers) ρ_i, $i = \overline{1, n}$, i.e. the roots of secular equation

$$\det\bigl(F(T) - \rho E\bigr) = 0 \tag{3.15}$$

   and examine how they are situated within the complex plane with respect to the unit circle $|\rho| = 1$.

Let t_k $(0 < t_k < T)$ be the discontinuity points of the right-hand side of (3.1) on the periodic solution, i.e. $\zeta(t_k, X_c(t_k)) = 0$, $k = \overline{1, p}$. Here p is the number of discontinuities encountered in the period T. Since the right-hand side of (3.1) is discontinuous, the solution to (3.14) is also discontinuous and, at the points of discontinuity,

$$F_k^+ = M_k F_k^-, \; F_k^\pm = \lim_{t \to t_k \pm 0} F(t).$$

Thus, we need the updating matrix M_k to determine the basic matrix $F(T)$.

For definiteness, we assume, in accordance with our previous work [68], that the function $\zeta(X_c(t_k))$ changes sign from negative to positive at the point $t = t_k$ as t increases. On the periodic solution in the interval $t_{k-1} < t < t_{k+1}$ the right-hand side of (3.1) can be written as

$$G(t, X) = G_-(t, X) + \Delta G(t, X)\eta(\zeta)\big|_{X = X_c}. \tag{3.16}$$

Here $\Delta G(t, X) = G_+(t, X) - G_-(t, X)$ with $G_-(t, X)$ and $G_+(t, X)$ being sufficiently smooth vector functions, defined in the intervals $t_{k-1} < t < t_k$ and $t_k < t < t_{k+1}$, respectively. $\eta(\cdot)$ is the Heaviside step function.

Considering the perturbed solution as the sum of a periodic solution and a small disturbance ε:

$$X(t) = X_c(t) + \varepsilon(t),$$

from (3.13) (accounting for (3.16)) we find that

$$\frac{\mathrm{d}X_{\mathrm{c}}}{\mathrm{d}t}+\frac{\mathrm{d}\varepsilon}{\mathrm{d}t}=G_{-}(t,X_{\mathrm{c}}+\varepsilon)+\Delta G(t,X_{\mathrm{c}}+\varepsilon)\,\eta\big(\zeta(t,X_{\mathrm{c}}+\varepsilon)\big).$$

Expanding the right-hand side of this equation in series by the powers of ε and neglecting terms $\mathrm{O}(\varepsilon^2)$ we obtain [68]

$$\frac{\mathrm{d}\varepsilon}{\mathrm{d}t}=\left[\frac{\partial G_{-}(t,X_{\mathrm{c}})}{\partial X}+\frac{\partial\Delta G(t,X_{\mathrm{c}})}{\partial X}\eta\big((\zeta(t,X_{\mathrm{c}})\big)\right]\varepsilon \\ +\Delta G(t,X_{\mathrm{c}})\,\frac{\mathrm{d}\eta\big(\zeta(t,X_{\mathrm{c}})\big)}{\mathrm{d}\zeta}\left(\frac{\partial\zeta(t,X_{\mathrm{c}})}{\partial X}\right)^{\mathrm{T}}\varepsilon.$$

Considering the change of the sign of ζ at the point $t=t_k$ (see, e.g. [1]) we can write

$$\delta(\zeta)\frac{\mathrm{d}\zeta}{\mathrm{d}t}=\delta(t-t_k), \tag{3.17}$$

with $\delta(\cdot)$ being the Dirac function.

Since

$$\frac{\mathrm{d}\zeta}{\mathrm{d}t}=\frac{\partial\zeta}{\partial t}+\left(\frac{\partial\zeta}{\partial X}\right)^{\mathrm{T}}G(t,X),\ \left(\frac{\partial\zeta}{\partial X}\right)^{\mathrm{T}}=\left(\frac{\partial\zeta}{\partial x_1},\frac{\partial\zeta}{\partial x_2},\dots,\frac{\partial\zeta}{\partial x_n}\right),$$

by virtue of the expression (3.16) for $G(t,X)$

$$\left.\frac{\mathrm{d}\zeta}{\mathrm{d}t}\right|_{X=X_{\mathrm{c}}}=\left[\frac{\mathrm{d}\zeta(t,X_{\mathrm{c}})}{\mathrm{d}t}\right]_{-}+\left\{\left[\frac{\mathrm{d}\zeta(t,X_{\mathrm{c}})}{\mathrm{d}t}\right]_{+}-\left[\frac{\mathrm{d}\zeta(t,X_{\mathrm{c}})}{\mathrm{d}t}\right]_{-}\right\}\eta\big(\zeta(t,X_{\mathrm{c}})\big).$$

The signs "$-$" and "$+$" correspond to the domains $t_{k-1}<t<t_k$ and $t_k<t<t_{k+1}$, respectively. As a result of this derivation, a product of two generalized functions appears in the left-hand side of (3.17). This product is undefined, but there exist such sequences of smooth functions converging to $\delta(\zeta)$ and $\eta(\zeta)$ (see, e.g., [36]), that we can write

$$\delta(\zeta)\eta(\zeta)\equiv\vartheta_1\delta(\zeta),\ 0\le\vartheta_1\le1. \tag{3.18}$$

Applying these equations, Eq. (3.17) gives

$$\delta(\zeta)=\frac{\delta(t-t_k)}{\left(\frac{\mathrm{d}\zeta}{\mathrm{d}t}\right)^{-}+\vartheta_1\left[\left(\frac{\mathrm{d}\zeta}{\mathrm{d}t}\right)^{+}-\left(\frac{\mathrm{d}\zeta}{\mathrm{d}t}\right)^{-}\right]},$$

where

$$\left(\frac{\mathrm{d}\zeta}{\mathrm{d}t}\right)^{\pm} \equiv \lim_{t\to t_k\pm 0} \frac{\mathrm{d}\zeta}{\mathrm{d}t}.$$

Let us introduce the following designation:

$$\hat{A}_k(\vartheta_1) = \frac{1}{\left(\frac{\mathrm{d}\zeta}{\mathrm{d}t}\right)^{-} + \vartheta_1\left[\left(\frac{\mathrm{d}\zeta}{\mathrm{d}t}\right)^{+} - \left(\frac{\mathrm{d}\zeta}{\mathrm{d}t}\right)^{-}\right]}\Delta G_k \left(\frac{\mathrm{d}\zeta}{\mathrm{d}X}\right)^{\mathrm{T}}\Bigg|_{\substack{X=X_{\mathrm{c}}\\ t=t_k}}$$

with

$$\Delta G_k = G_k^+ - G_k^-,\ G_k^{\pm} = \lim_{t\to t_k\pm 0} G_{\pm}(t, X_{\mathrm{c}});$$
$$\hat{A}_k^- = \hat{A}_k(0);\ \hat{A}_k^+ = \hat{A}_k(1).$$

The system of linearized equations with respect to ε then takes the form [68, 69]

$$\frac{\mathrm{d}\varepsilon}{\mathrm{d}t} = \left[A(t) + \hat{A}_k(\vartheta_1)\,\delta(t-t_k)\right]\varepsilon, \tag{3.19}$$
$$A(t) = \frac{\partial G_-(t, X_{\mathrm{c}})}{\partial X} + \frac{\partial \Delta G(t, X_{\mathrm{c}})}{\partial X}\eta\big(\zeta(t, X_{\mathrm{c}})\big),$$

and Eq. (3.14) becomes

$$\frac{\mathrm{d}F}{\mathrm{d}t} = \left[A(t) + \hat{A}_k(\vartheta_1)\,\delta(t-t_k)\right]F,\ F(0) = E. \tag{3.20}$$

In the domain $t_{k-1} < t < t_{k+1}$ we are to look for the solution of the equation (3.19) in the form

$$\varepsilon(t) = \varepsilon_-(t) + \big(\varepsilon_+(t) - \varepsilon_-(t)\big)\eta(t-t_k).$$

Substituting this solution into (3.20) and considering that

$$\eta(t-t_k)\,\delta(t-t_k) \equiv \vartheta_2\delta(t-t_k),\ 0 \le \vartheta_2 \le 1,$$

we find

$$\big[\varepsilon_k^+ - \varepsilon_k^- - \hat{A}_k(\vartheta_1)\varepsilon_k^- - \vartheta_2\hat{A}_k(\vartheta_1)(\varepsilon_k^+ - \varepsilon_k^-)\big]\delta(t-t_k) = 0,\ \varepsilon_k^{\pm} = \lim_{t\to t_k\pm 0}\varepsilon(t).$$

Setting the expression within the brackets equal to zero, we finally obtain

$$\varepsilon_k^+ = M_k\varepsilon_k^-,$$
$$M_k = \big[E - \vartheta_2\hat{A}_k(\vartheta_1)\big]^{-1}\big[E + (1-\vartheta_2)\hat{A}_k(\vartheta_1)\big].$$

Let us pause for a moment to discuss some of the mathematical ramifications of this result:

Remark 3.1. If

$$\left.\frac{\partial\zeta}{\partial t}\right|_{t=t_k} = h\,\delta(t-t_k),\ h = \text{const},$$

then the matrix $\hat{A}_k(\vartheta_1)\,\delta(t-t_k)$ degenerates into a constant matrix and therefore $F(t)$ is continuous at $t = t_k$. The proof is based on the fact that one of the terms in the denominator of the expression for $\hat{A}_k(\vartheta_1)$ has $\delta(t-t_k)$ as a co-factor.

The updating matrix M_k has the following properties [44, 68, 69].

Proposition 3.1. *The matrices* $\hat{A}_k^+$ *and* $\hat{A}_k^-$ *are related through*

$$\hat{A}_k^- - \hat{A}_k^+ = \hat{A}_k^- \hat{A}_k^+. \tag{3.21}$$

Proof. Let us calculate the matrices on the left- and right-hand sides of the equality (3.21).

$$\hat{A}_k^- - \hat{A}_k^+ = \frac{1}{\left(\frac{\mathrm{d}\zeta}{\mathrm{d}t}\right)^-}\Delta G_k\left(\frac{\mathrm{d}\zeta}{\mathrm{d}X}\right)^{\mathrm{T}} - \frac{1}{\left(\frac{\mathrm{d}\zeta}{\mathrm{d}t}\right)^+}\Delta G_k\left(\frac{\mathrm{d}\zeta}{\mathrm{d}X}\right)^{\mathrm{T}}$$

$$= \frac{\left(\frac{\mathrm{d}\zeta}{\mathrm{d}X}\right)^{\mathrm{T}} G_k^+ - \left(\frac{\mathrm{d}\zeta}{\mathrm{d}X}\right)^{\mathrm{T}} G_k^-}{\left(\frac{\mathrm{d}\zeta}{\mathrm{d}t}\right)^-\left(\frac{\mathrm{d}\zeta}{\mathrm{d}t}\right)^+}\Delta G_k\left(\frac{\mathrm{d}\zeta}{\mathrm{d}X}\right)^{\mathrm{T}} = \frac{\left(\frac{\mathrm{d}\zeta}{\mathrm{d}X}\right)^{\mathrm{T}}\Delta G_k}{\left(\frac{\mathrm{d}\zeta}{\mathrm{d}t}\right)^-\left(\frac{\mathrm{d}\zeta}{\mathrm{d}t}\right)^+}\Delta G_k\left(\frac{\mathrm{d}\zeta}{\mathrm{d}X}\right)^{\mathrm{T}}.$$

The right-hand side of the equality:

$$\hat{A}_k^- \hat{A}_k^+ = \frac{1}{\left(\frac{\mathrm{d}\zeta}{\mathrm{d}t}\right)^-\left(\frac{\mathrm{d}\zeta}{\mathrm{d}t}\right)^+}\Delta G_k\left(\frac{\mathrm{d}\zeta}{\mathrm{d}X}\right)^{\mathrm{T}}\Delta G_k\left(\frac{\mathrm{d}\zeta}{\mathrm{d}X}\right)^{\mathrm{T}}.$$

Using the associative property of the matrix product we obtain

$$\hat{A}_k^- \hat{A}_k^+ = \frac{1}{\left(\frac{\mathrm{d}\zeta}{\mathrm{d}t}\right)^-\left(\frac{\mathrm{d}\zeta}{\mathrm{d}t}\right)^+}\Delta G_k\left[\left(\frac{\mathrm{d}\zeta}{\mathrm{d}X}\right)^{\mathrm{T}}\Delta G_k\right]\left(\frac{\mathrm{d}\zeta}{\mathrm{d}X}\right)^{\mathrm{T}}.$$

The matrix product within the square brackets is a scalar which can be interchanged with the matrix ΔG_k:

$$\hat{A}_k^- \hat{A}_k^+ = \frac{\left(\frac{\mathrm{d}\zeta}{\mathrm{d}X}\right)^{\mathrm{T}}\Delta G_k}{\left(\frac{\mathrm{d}\zeta}{\mathrm{d}t}\right)^-\left(\frac{\mathrm{d}\zeta}{\mathrm{d}t}\right)^+}\Delta G_k\left(\frac{\mathrm{d}\zeta}{\mathrm{d}X}\right)^{\mathrm{T}}.$$

The proposition 3.1 is proved. □

Proposition 3.2. *If* $\vartheta_1 = \vartheta_2 \equiv \vartheta$, *then*

$$M_k = E + \hat{A}_k^- = \left(E - \hat{A}_k^+\right)^{-1}. \tag{3.22}$$

Proof. Let us use the equality

$$\hat{A}_k^- - \hat{A}_k(\vartheta) = \vartheta \hat{A}_k^- \hat{A}_k(\vartheta). \tag{3.23}$$

One can prove this equality similarly to Eq. (3.21). We can hereafter transform the matrix M_k according to

$$\begin{aligned} M_k &= \left(E - \vartheta \hat{A}_k(\vartheta)\right)^{-1}\left(E + (1-\vartheta)\hat{A}_k(\vartheta)\right) \\ &= \left(E - \vartheta \hat{A}_k(\vartheta)\right)^{-1}\left(E - \vartheta \hat{A}_k(\vartheta) + \hat{A}_k(\vartheta)\right) \\ &= E + \left(E - \vartheta \hat{A}_k(\vartheta)\right)^{-1}\hat{A}_k(\vartheta). \end{aligned}$$

Since from (3.23) it results that

$$\hat{A}_k^- = \left(E - \vartheta \hat{A}_k(\vartheta)\right)^{-1}\hat{A}_k(\vartheta),$$

we have

$$M_k = E + \hat{A}_k^-.$$

Finally, taking Eq. (3.21) into account, we obtain

$$(E + \hat{A}_k^-)(E - \hat{A}_k^+) = E + \hat{A}_k^- - \hat{A}_k^+ - \hat{A}_k^- \hat{A}_k^+ = E.$$

The proposition 3.2 is proved. □

Proposition 3.3. *If the condition*

$$\left(\frac{\mathrm{d}\zeta}{\mathrm{d}X}\right)^{\mathrm{T}} \Delta G_k \Bigg|_{\substack{X=X_c \\ t=t_k}} = 0 \tag{3.24}$$

is fulfilled, the updating matrix is determined by expression (3.22) for any values of ϑ_1, ϑ_2.

Proof. At first let us show that if the condition (3.24) is fulfilled then

$$\hat{A}_k^- \hat{A}_k(\vartheta) = 0.$$

Indeed, the product of these matrices contains the term

$$\Delta G_k \left(\frac{\mathrm{d}\zeta}{\mathrm{d}X}\right)^{\mathrm{T}} \Delta G_k \left(\frac{\mathrm{d}\zeta}{\mathrm{d}X}\right)^{\mathrm{T}}\Bigg|_{\substack{X=X_{\mathrm{c}}\\ t=t_k}} = \Delta G_k \left[\left(\frac{\mathrm{d}\zeta}{\mathrm{d}X}\right)^{\mathrm{T}} \Delta G_k\right] \left(\frac{\mathrm{d}\zeta}{\mathrm{d}X}\right)^{\mathrm{T}}\Bigg|_{\substack{X=X_{\mathrm{c}}\\ t=t_k}},$$

which vanishes under condition (3.24).

From (3.23) it follows that

$$\hat{A}_k^- = \hat{A}_k(\vartheta_1).$$

The series of subsequent transformations does not require any comments:

$$\begin{aligned}
\hat{A}_k^- - \vartheta_2 \hat{A}_k^- \hat{A}_k(\vartheta_1) &= \hat{A}_k(\vartheta_1);\\
\hat{A}_k^-(E - \vartheta_2 \hat{A}_k(\vartheta_1)) &= \hat{A}_k(\vartheta_1);\\
\hat{A}_k^- &= \big(E - \vartheta_2 \hat{A}_k(\vartheta_1)\big)^{-1} \hat{A}_k(\vartheta_1);\\
E + \hat{A}_k^- &= E + \big(E - \vartheta_2 \hat{A}_k(\vartheta_1)\big)^{-1} \hat{A}_k(\vartheta_1);\\
E + \hat{A}_k^- &= \big(E - \vartheta_2 \hat{A}_k(\vartheta_1)\big)^{-1} \big(E + (1-\vartheta_2)\hat{A}_k(\vartheta_1)\big);\\
E + \hat{A}_k^- &= M_k.
\end{aligned}$$

The proposition 3.3 is proved. □

There is a Floquet-Lyapunov representation for the fundamental matrix $F(t)$:

$$F(t) = Z(t)\mathrm{e}^{Dt}.$$

Here $Z(t+T) \equiv Z(t)$, and the matrix D is constant and given by the logarithm of the basic matrix:

$$D = \frac{1}{T} \ln F(T).$$

The eigenvalues $\lambda_1, \lambda_2, \ldots, \lambda_n$ of the matrix D are referred to as *characteristic exponents*. When all multipliers are different there is a connection:

$$\lambda_k = \frac{1}{T} \ln \rho_k, \ k = \overline{1, n}.$$

Let $\rho_{\mathrm{r}k}$, $\rho_{\mathrm{i}k}$, $\lambda_{\mathrm{r}k}$, $\lambda_{\mathrm{i}k}$ be the real and imaginary parts of multipliers and characteristic exponents, respectively. Then [69]:

$$\lambda_{\mathrm{r}k} = \frac{1}{T}\ln|\rho_k|;$$

$$\lambda_{\mathrm{i}k} = \frac{1}{T}\times\begin{cases}\arctan\left|\dfrac{\rho_{\mathrm{i}k}}{\rho_{\mathrm{r}k}}\right|, & \rho_{\mathrm{r}k}>0;\\ \pi-\arctan\left|\dfrac{\rho_{\mathrm{i}k}}{\rho_{\mathrm{r}k}}\right|, & \rho_{\mathrm{r}k}<0.\end{cases}$$

Asymptotic stability is realized when

$$|\rho_k| < 1 \text{ or } \lambda_{\mathrm{r}k} < 0 \text{ for all } k.$$

The matrix D can be obtained as an eigenmatrix of the boundary problem [69]

$$\frac{\mathrm{d}Z}{\mathrm{d}t} = \frac{\partial G(t, X_{\mathrm{c}})}{\partial X} Z(t) - Z(t)D,\ Z(0) = Z(T) = E.$$

It should be noted that $Z(t)$ has discontinuities at $t = t_k$ with the updating matrix M_k.

3.4.2 Autonomous dynamical systems

Let $X_{\mathrm{c}}(t)$ be a periodic solution of the autonomous system (3.1). Investigation of the local stability of periodic solutions is performed following the same steps as for the system with external periodic action (3.13).

For the autonomous system (3.1) the linearized equation with respect to ε [44]

$$\frac{\mathrm{d}\varepsilon}{\mathrm{d}t} = \left[A(t) + \hat{A}_k(\vartheta_1)\,\delta(t-t_k)\right]\varepsilon, \tag{3.25}$$

$$A(t) = \frac{\partial G_-(X_{\mathrm{c}})}{\partial X} + \frac{\partial \Delta G(X_{\mathrm{c}})}{\partial X}\eta(\zeta(X_c))$$

differs from Eq. (3.19) in only one respect. Namely, that the derivatives $(\mathrm{d}\zeta/\mathrm{d}t)^{\pm}$ in $\hat{A}_k(\vartheta_1)$ contain a single rather than two terms [44]:

$$\left(\frac{\mathrm{d}\zeta}{\mathrm{d}t}\right)^{\pm} = \left(\frac{\partial\zeta}{\partial t}\right)^{\mathrm{T}} G_k^{\pm}(X).$$

3.5 Application to Relay Systems

3.5.1 Relay system with hysteresis

As a first example, let us consider the mathematical model of a relay system with hysteresis represented by Eq. (3.7). The method for constructing the equations for limit cycles is described in Sec. 3.3, hence, we do not repeat it here but turn to the problem of estimating the basic matrix.

Let $t_1, t_2 \ldots, t_p$ be switching instants of the relay element during the period of the cyclic motion such, that:

$$\zeta\big(X_{\rm c}(t_k)\big) = (-1)^{k-1}\chi_0 + \xi\big((X_{\rm c}(t_k)\big) = 0, \ k = 1, 2, \ldots, p,$$

or

$$(-1)^{k-1}\chi_0 + UX_{\rm c}(t_k) - U_{\rm ref} = 0,$$

and let

$$\mathrm{K}_k^{(\mathrm{F})} = \begin{cases} -1, & t_{k-1} < t < t_k, \\ 1, & t_k < t < t_{k+1}. \end{cases}$$

In this way, the type of periodic solution, i.e. the sequence of changes of the state of the relay element during the period T, is specified.

In the domain $t_{k-1} < t < t_k$ the equation for $F(t)$ corresponding to Eq. (3.20) has the form

$$\frac{\mathrm{d}F}{\mathrm{d}t} = A_k F,$$

with $F(t_{k-1} + 0) = F_{k-1}$.

The solution to this system is

$$F(t) = \mathrm{e}^{A_k(t - t_{k-1})} F_{k-1},$$

and it follows from this expression that

$$F(t_k - 0) = \mathrm{e}^{A_k \tau_k} F_{k-1} = C_k F_{k-1}.$$

Taking into account that $F(t_k+0) = M_k F(t_k-0)$, we can obtain the recurrent formula for calculating the basic matrix $F(T) = F_p$:

$$F_k = M_k C_k F_{k-1}, \ F_k = F(t_k + 0), \ F_0 = E.$$

Here the matrix $\hat{A}_k(\vartheta_1)$ in the expression for M_k is determined by

$$\hat{A}_k(\vartheta_1) = \frac{1}{U(G_k^- + \vartheta_1 \Delta G_k)} \Delta G_k U,$$

with

$$G_k^+ = A_{k+1}X_{ck} + B_{k+1},\ G_k^- = A_kX_{ck} + B_k,$$

$$X_{\mathrm{c}k} = \left[\prod_{i=1}^{k} C_{k+1-i}\right] X_{\mathrm{c}0} + \sum_{i=1}^{k-1}\left[\prod_{j=1}^{k-i} C_{k+1-j}\right] V_i + V_k,$$

$$X_{\mathrm{c}0} = \left[E - \prod_{i=1}^{p} C_{p+1-i}\right]^{-1}\left[\sum_{i=1}^{p-1}\left(\prod_{j=1}^{p-i} C_{p+1-j}\right) V_i + V_p\right].$$

Hence,

$$F(T) = \prod_{k=1}^{p} M_{p+1-k}C_{p+1-k}.$$

It is not so hard to find an expression for the basic matrix for autonomous systems that are synthesized from nonintegrable partial systems. The principal difficulty in this case is to develop a concrete algorithm that allows us to obtain the periodic motions rather than to construct the updating matrix.

3.5.2 Relay system without hysteresis

Equations of motion.

As another example let us consider a relay system without hysteresis whose equation of motion has the form

$$\frac{\mathrm{d}X}{\mathrm{d}t} = G(X), \tag{3.26}$$

with

$$X = \begin{pmatrix} x_1 \\ x_2 \\ \vdots \\ x_n \end{pmatrix};\ G(X) = \begin{pmatrix} g_1 \\ g_2 \\ \vdots \\ g_n \end{pmatrix};$$

$$G(X) = AX + B\mathrm{K}^{(\mathrm{F})}(\xi);$$

$$\mathrm{K}^{(\mathrm{F})}(\xi) = \operatorname{sign}\{\xi\},\ \xi(X) = -UX.$$

Here, A is a constant $n \times n$ matrix, B is an n-dimensional vector, and U is an n-dimensional row matrix.

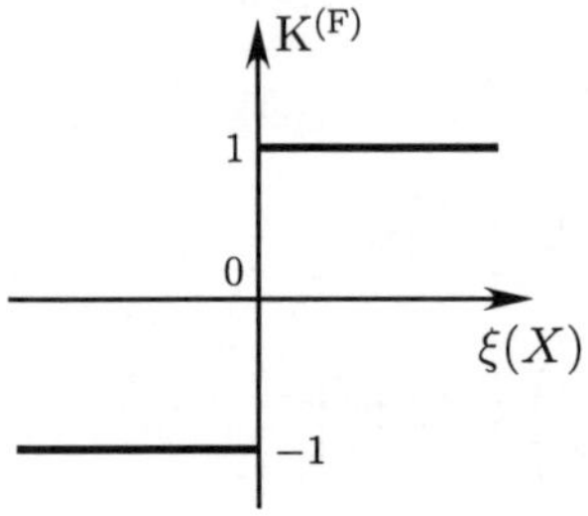

Figure 3.3. Switching characteristics of the relay element without hysteresis.

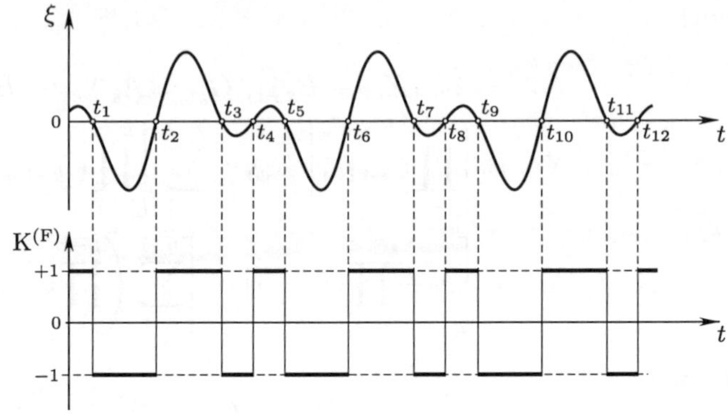

Figure 3.4. Time diagram illustrating the principle for producing the control pulses.

The characteristics for the relay element is shown in Fig. 3.3. Figure 3.4 displays time diagrams illustrating the principle of producing the control pulses.

Johansson *et al.* [24–26] consider a model of relay system without hysteresis presented in so-called *observer canonical form*:

$$A = \begin{pmatrix} -a_1 & 1 & 0 & \dots & 0 \\ -a_2 & 0 & 1 & \dots & 0 \\ \dots & \dots & \dots & \dots & \dots \\ -a_{n-1} & 0 & 0 & \dots & 1 \\ -a_n & 0 & 0 & \dots & 0 \end{pmatrix}; \; B = \begin{pmatrix} b_1 \\ b_2 \\ \vdots \\ b_{n-1} \\ b_n \end{pmatrix}; \; U = \begin{pmatrix} 1 \\ 0 \\ \vdots \\ 0 \\ 0 \end{pmatrix}^{\mathrm{T}}.$$

As a specific example of such a system we may consider a third-order relay system described by the matrices

$$A = \begin{pmatrix} -(2\mu\omega + \nu) & 1 & 0 \\ -(2\mu\omega\nu + \omega^2) & 0 & 1 \\ -\nu\omega^2 & 0 & 0 \end{pmatrix}; \; B = \begin{pmatrix} \Omega \\ 2\Omega\sigma\beta \\ \Omega\beta^2 \end{pmatrix}; \; U = \begin{pmatrix} 1 \\ 0 \\ 0 \end{pmatrix}^{\mathrm{T}}.$$

The dynamics of this system was studied in detail by Johansson *et al.* [24–26] (see also [39]).

However, contrary to the rather specific approach described by Johansson *et al.* [24–26], we treat the system (3.26) as belonging to a general class of a piecewise-linear autonomous differential equations.

Poincaré map.

In the domain $t_{k-1} < t < t_k$ the system (3.26) has the form

$$\frac{\mathrm{d}X}{\mathrm{d}t} = AX + B_k,\ X(t_{k-1}) = X_{k-1},$$
$$B_k = \begin{cases} B, & \xi > 0; \\ -B, & \xi < 0 \end{cases}$$

with the solution

$$X(t) = \mathrm{e}^{A(t-t_{k-1})}\left(X_{k-1} + \int_{t_{k-1}}^{t} e^{-A(\theta-t_{k-1})} B_k\,\mathrm{d}\theta\right)$$
$$= \mathrm{e}^{A(t-t_k)} X_{k-1} - \left(E - \mathrm{e}^{A(t-t_{k-1})}\right) A^{-1} B_k.$$

Hence, the Poincaré map for the system (3.26) is given by

$$X_k = \mathrm{e}^{A\tau_k} X_{k-1} + V_k, \tag{3.27}$$
$$V_k = -(E - \mathrm{e}^{A\tau_k}) A^{-1} B_k.$$

Here, τ_k is the smallest positive root of the equation

$$\varphi(\tau, X_{k-1}) = 0, \tag{3.28}$$

where

$$\varphi(\tau, X_{k-1}) = U\left(\mathrm{e}^{A\tau_k}(X_{k-1} + X_{sk}) - X_{sk}\right), X_{sk} = A^{-1} B_k.$$

Equations for limit cycles.

Let us use the method described in Sec. 3.3 to construct equations for the limit cycles.

From the recurrence expression (3.27) we can successively obtain (see also Sec. 3.3)

$$X_1 = \mathrm{e}^{A\tau_1} X_0 + V_1;$$
$$X_2 = \mathrm{e}^{A\tau_2} X_1 + V_2 = \mathrm{e}^{A(\tau_1+\tau_2)} X_0 + \mathrm{e}^{A\tau_2} V_1 + V_2;$$
$$X_3 = \mathrm{e}^{A\tau_3} X_2 + V_3 = \mathrm{e}^{A(\tau_1+\tau_2+\tau_3)} X_0 + \mathrm{e}^{A(\tau_2+\tau_3)} V_1 + \mathrm{e}^{A\tau_3} V_2 + V_3;$$
$$\dots$$
$$X_k = \mathrm{e}^{A(\tau_1+\tau_2+\dots+\tau_k)} X_0 + \mathrm{e}^{A(\tau_2+\tau_3+\dots+\tau_k)} V_1 + \mathrm{e}^{A(\tau_3+\tau_4+\dots+\tau_k)} V_2 + \dots + \mathrm{e}^{A\tau_k} V_{k-1} + V_k;$$
$$X_k = \mathrm{e}^{As_k} X_0 + \sum_{i=1}^{k} \mathrm{e}^{As_{k-i}} V_i, \tag{3.29}$$

where

$$s_k = \sum_{j=1}^{k} \tau_{k+1-j}.$$

Accounting for the periodicity condition $X_0 = X_p$ we obtain

$$(E - e^{AT})X_0 = \sum_{i=1}^{p} e^{As_{p-i}} V_i,$$

where $T = \sum_{i=1}^{p} \tau_i$ is the period of the solution.

Substituting $X_0 = \left(E - e^{AT}\right)^{-1} \sum_{i=1}^{p} e^{As_{p-i}} V_i$ into the Eq. (3.29) we may derive the expression

$$X_k = e^{As_k}(E - e^{AT})^{-1} \sum_{i=1}^{p} e^{As_{p-i}} V_i + \sum_{i=1}^{k} e^{As_{k-i}} V_i, \tag{3.30}$$

and substituting this expression into Eq. (3.28) we finally obtain a system of transcendental equations with respect to $\tau_1, \tau_2, \ldots, \tau_p$

$$\begin{cases} \varphi_1(\tau_1, \tau_2, \ldots \tau_p) = 0; \\ \varphi_2(\tau_1, \tau_2, \ldots \tau_p) = 0; \\ \ldots \\ \varphi_p(\tau_1, \tau_2, \ldots \tau_p) = 0 \end{cases} \tag{3.31}$$

or

$$\varphi_k(\tau_1, \tau_2, \ldots \tau_p) = 0, \; k = \overline{1, p}.$$

Here,

$$\varphi_k(\tau_1, \tau_2, \ldots \tau_p) = U\left(e^{As_k}(E - e^{AT})^{-1} \sum_{i=1}^{p} e^{As_{m-i}} V_i + \sum_{i=1}^{k} e^{As_{k-i}} V_i\right).$$

Local stability of limit cycles.

The linearized equation with respect to ε similar to Eq. (3.25) now takes the form

$$\frac{d\varepsilon}{dt} = \frac{\partial G(X_c)}{\partial X} \varepsilon, \; \varepsilon(0) = \varepsilon_0 \tag{3.32}$$

Here,

$$\frac{\partial G(X_c)}{\partial X} = A + \hat{A}_k(\vartheta_1)\delta(t - t_k),$$

with

$$\hat{A}_k(\vartheta_1) = \frac{1}{\left(\frac{\partial \xi}{\partial X}\right)^{\mathrm{T}} (G_k^- + \vartheta_1 \Delta G_k)} \Delta G_k \left(\frac{\partial \xi}{\partial X}\right)^{\mathrm{T}},$$

$$\left(\frac{\partial \xi}{\partial X}\right)^{\mathrm{T}} = U;$$

$$G_k^- = AX_{ck} + B_k;$$

$$\Delta G_k = B_{k+1} - B_k = -2B_k.$$

X_{ck} can be determined by substituting the solution $\tau_1,\ \tau_2,\ \ldots \tau_p$ of the system (3.31) into the expression (3.30).

Let us construct the expression for the basic matrix. Within the domain $t_{k-1} < t < t_k$ the system (3.32) has the form

$$\frac{\mathrm{d}\varepsilon}{\mathrm{d}t} = A\varepsilon,\ \varepsilon(t_{k-1} + 0) = \varepsilon_{k-1}.$$

with the solution

$$\varepsilon(t) = \mathrm{e}^{A(t - t_{k-1})}\varepsilon_{k-1}.$$

Hence,

$$\varepsilon(t_k - 0) = \mathrm{e}^{A\tau_k}\varepsilon_{k-1}.$$

At the discontinuity points

$$\varepsilon_k = M_k \mathrm{e}^{A\tau_k}\varepsilon_{k-1}, \qquad (3.33)$$
$$\varepsilon_k = \varepsilon(t_k + 0).$$

Using Eq. (3.33) we can hereafter obtain the succession

$$\varepsilon_1 = M_1 \mathrm{e}^{A\tau_1}\varepsilon_0;$$
$$\varepsilon_2 = M_2 \mathrm{e}^{A\tau_2}\varepsilon_1 = M_2 \mathrm{e}^{A\tau_2} M_1 \mathrm{e}^{A\tau_1}\varepsilon_0;$$
$$\ldots$$
$$\varepsilon_p = M_p \mathrm{e}^{A\tau_p} M_{p-1} \mathrm{e}^{A\tau_{p-1}} \times \cdots \times M_2 \mathrm{e}^{A\tau_2} M_1 \mathrm{e}^{A\tau_1}\varepsilon_0;$$
$$\varepsilon_p = F(T)\varepsilon_0,$$

where

$$F(T) = \prod_{j=1}^{p} M_{p+1-j} e^{A\tau_{p+1-j}}.$$

The final solution to Eq. (3.33) has the form

$$\varepsilon_{pk} = \left(F(T)\right)^k \varepsilon_0, \; k = 1, 2, \ldots$$

For an autonomous system with an n-dimensional phase space one of the multipliers ρ_i, $i = \overline{1,n}$ is always equal to unity (see Andronov-Witt theorem in [70]). Thus the number of multipliers determining the local stability of a periodic motion will be equal to $n - 1$. For the periodic motion to be locally stable all the roots of Eq. (3.15) except one, must be located within the unit circle.

In the next chapter we will use the above results to perform a detailed study of the bifurcations and transitions to chaos that occur in piecewise-smooth autonomous dynamical systems. As a base model we will use a four-dimensional autonomous system of differential equations in the form (3.7) that describes a DC/DC converter with the relay control.

3.6 Model of a DC/DC Converter with Relay Feedback Control

As a concrete example of a relay system we shall consider a power converter with relay feedback control. The equivalent circuit of this system is depicted in Fig. 3.5. The power section of the circuit comprises a transistor converter of constant voltage working in switch mode and a couple of ripple LC-filters (see, e.g. [14, 16, 71–74]). Part of the output voltage σU_1, $0 < \sigma < 1$ is compared with the reference signal U_{ref} and the difference $\xi = U_{\text{ref}} - \sigma U_1$ (*deviation signal*) is applied to the relay element with hysteresis, in order to form square pulses (see, e.g. Fig. 3.1(b)) to control the switch elements of the converter. In many cases, the feedback will include a correcting circuit (compensator), for example, a PI (proportional-plus-integrating) or a PID (proportional-plus-integrating-plus-differentiating) compensator. Figure 3.5 shows an example of a converter with proportional feedback only.

In the following we will restrict our considerations to DC/DC converters. If we choose currents in the inductances L_0 and L_1 and voltages on the capacitors

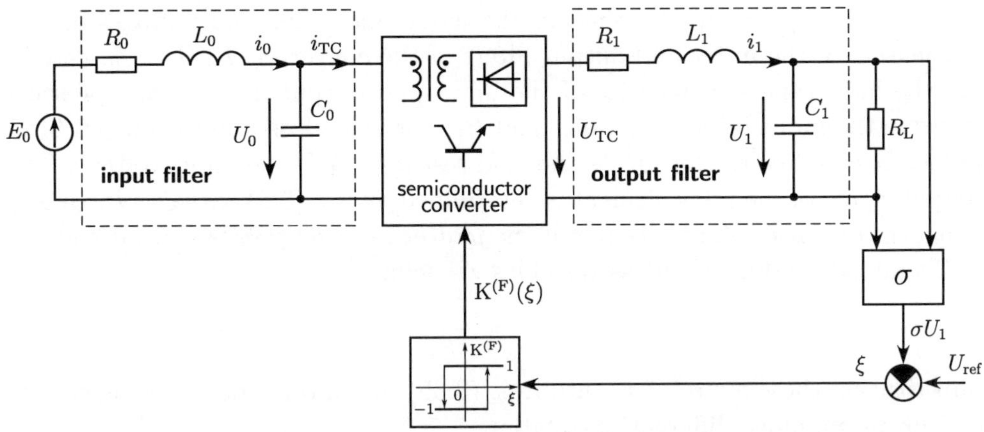

Figure 3.5. Equivalent circuit of the DC/DC converter with relay control.

C_0 and C_1 as state variables, then the behavior of the converter is described by a set of differential equations

$$C_0\frac{\mathrm{d}U_0}{\mathrm{d}t} + i_{\mathrm{TC}} = i_0; \qquad C_1\frac{\mathrm{d}U_1}{\mathrm{d}t} + \frac{U_1}{R_{\mathrm{L}}} = i_1;$$
$$L_0\frac{\mathrm{d}i_0}{\mathrm{d}t} + R_0 i_0 + U_0 = E_0; \qquad L_1\frac{\mathrm{d}i_1}{\mathrm{d}t} + R_1 i_1 + U_1 = U_{\mathrm{TC}}. \tag{3.34}$$

Here,

$$U_{\mathrm{TC}} = \begin{cases} U_0, & \mathrm{K}^{(\mathrm{F})} = 1; \\ 0, & \mathrm{K}^{(\mathrm{F})} = -1 \end{cases} \quad \text{and} \quad i_{\mathrm{TC}} = \begin{cases} i_1, & \mathrm{K}^{(\mathrm{F})} = 1; \\ 0, & \mathrm{K}^{(\mathrm{F})} = -1, \end{cases}$$

or

$$U_{\mathrm{TC}} = \frac{1}{2}\left(1 + \mathrm{K}^{(\mathrm{F})}(\xi)\right) U_0 \quad \text{and} \quad i_{\mathrm{TC}} = \frac{1}{2}\left(1 + \mathrm{K}^{(\mathrm{F})}(\xi)\right) i_1.$$

Thus, we can transform Eqs. (3.34) into the following form

$$\begin{cases} \dfrac{\mathrm{d}i_0}{\mathrm{d}t} = -\dfrac{R_0}{L_0}\, i_0 - \dfrac{1}{L_0}\, U_0 + \dfrac{E_0}{L_0}; \\ \dfrac{\mathrm{d}U_0}{\mathrm{d}t} = \dfrac{1}{C_0}\, i_0 - \dfrac{1 + \mathrm{K}^{(\mathrm{F})}(\xi)}{2C_0}\, i_1; \\ \dfrac{\mathrm{d}i_1}{\mathrm{d}t} = \dfrac{1 + \mathrm{K}^{(\mathrm{F})}(\xi)}{2L_1}\, U_0 - \dfrac{R_1}{L_1}\, i_1 - \dfrac{1}{L_1}\, U_1; \\ \dfrac{\mathrm{d}U_1}{\mathrm{d}t} = \dfrac{1}{C_1}\, i_1 - \dfrac{1}{C_1 R_{\mathrm{L}}}\, U_1, \end{cases} \tag{3.35}$$

where i_0, i_1, U_0, U_1 are currents in the inductance coils and voltages on the condensers of the input and output filters of the converter; L_0, L_1, C_0, and C_1 are the inductances, capacities of the input and output filters. R_0 represents losses in the input filter inductance and R_1 represents losses in the output filter inductance and converter. R_{L} is the load resistor. E_0 is the input voltage. The output signal of the relay element is determined by Eq. (3.8) with $\xi = U_{\mathrm{ref}} - \sigma U_1$ being the deviation signal. Here σ is the *feedback sensor gain* by output voltage.

To introduce dimensionless variables we define

$$x_1 = \frac{R_0 i_0}{E_*},\ x_2 = \frac{U_0}{E_*},\ x_3 = \frac{R_0 i_1}{E_*}, \quad \text{and} \quad x_4 = \frac{U_1}{E_*}.$$

Substituting these expressions into Eqs. (3.35), we obtain the following system of four autonomous differential equations:

$$\frac{\mathrm{d}X}{\mathrm{d}t} = G(X);\ X = (x_1, x_2, x_3, x_4)^{\mathrm{T}};\ G = (g_1, g_2, g_3, g_4)^{\mathrm{T}} \tag{3.36}$$

with

$$\begin{aligned}
g_1 &= \eta\,(-x_1 - x_2 + \Omega)\,;\\
g_2 &= \gamma\big(x_1 - \frac{1}{2}\big(1 + \mathrm{K}_k^{(\mathrm{F})}(\xi)\big)x_3\big);\\
g_3 &= \frac{\mu\big(1 + \mathrm{K}_k^{(\mathrm{F})}(\xi)\big)}{2}\,x_2 - \nu x_3 - \mu x_4;\\
g_4 &= \lambda\left(\frac{x_3}{\alpha} - \frac{x_4}{\beta}\right);\\
\mathrm{K}_k^{(\mathrm{F})} &= \mathrm{K}_{k-1}^{(\mathrm{F})}\,\mathrm{sign}\left\{\chi_0 + (-1)^{N_{k-1}}\xi(X)\right\},\ \mathrm{K}_0^{(\mathrm{F})} = -1,\\
N_k &= N_{k-1} + \frac{1 - \mathrm{sign}\big\{\mathrm{K}_k^{(\mathrm{F})}\mathrm{K}_{k-1}^{(\mathrm{F})}\big\}}{2},\ N_0 = 0,\ k = 1, 2, \ldots\\
\xi(X) &= U_{\mathrm{ref}} - UX,
\end{aligned}$$

where

$$\eta = \frac{R_0}{L_0},\ \Omega = \frac{E_0}{E_*},\ \gamma = \frac{1}{C_0 R_0},\ \mu = \frac{R_0}{L_1},\ \nu = \frac{R_1}{L_1},\ \lambda = \frac{1}{C_1 R_*},$$
$$\beta = \frac{R_{\mathrm{L}}}{R_*},\ \alpha = \frac{R_0}{R_*},\ \text{and } U = (0, 0, 0, \sigma E_*).$$

The system (3.36) can be written in matrix form

$$\frac{\mathrm{d}X}{\mathrm{d}t} = A_k X + B, \tag{3.37}$$

with

$$A_k = \begin{pmatrix} -\eta & -\eta & 0 & 0 \\ \gamma & 0 & -\frac{\gamma}{2}\left(1 + \mathrm{K}_k^{(\mathrm{F})}(\xi)\right) & 0 \\ 0 & \frac{\mu}{2}\left(1 + \mathrm{K}_k^{(\mathrm{F})}(\xi)\right) & -\nu & -\mu \\ 0 & 0 & \frac{\lambda}{\alpha} & -\frac{\lambda}{\beta} \end{pmatrix}, \quad B = \begin{pmatrix} \eta\Omega \\ 0 \\ 0 \\ 0 \end{pmatrix}$$

and

$$\begin{aligned} \mathrm{K}_k^{(\mathrm{F})} &= \mathrm{K}_{k-1}^{(\mathrm{F})} \operatorname{sign}\left\{\chi_0 + (-1)^{N_{k-1}}\xi(X)\right\}, \ \mathrm{K}_0^{(\mathrm{F})} = -1, \\ N_k &= N_{k-1} + \frac{1 - \operatorname{sign}\{\mathrm{K}_k^{(\mathrm{F})}\mathrm{K}_{k-1}^{(\mathrm{F})}\}}{2}, \ N_0 = 0, \ k = 1, 2, \ldots \\ \xi(X) &= U_{\mathrm{ref}} - UX. \end{aligned}$$

From these expressions it is easy to show that our study of the dynamical system (3.37) may be reduced to an investigation of the properties of a four-dimensional piecewise-smooth mapping of the form (3.9), (3.10).

The four-dimensional set of piecewise-smooth autonomous differential equations (3.37) displays the same dynamical phenomena as exhibited by the relay feedback system. When changing the parameters, one can observe a great variety of bifurcation phenomena including finite and infinite period-doubling cascades, Neimark–Sacker (Andronov–Hopf) bifurcations, dynamic intermittency, and hysteresis phenomena. A complete description of the complex behavior and transitions to chaos for the dynamical system (3.37) will be given in the next chapter.

3.7 DC Electric Drive with Relay Control

3.7.1 Practical realization of relay control systems

This paragraph describes in more detail the practical realization of relay systems.

Figure 3.6 shows the functional scheme of one of the basic variants of a relay system for regulating of the armature (rotor) current (rotor torque) of the motors of high-power electric drives [75]. Here, CONV is the semiconductor converter, CS is the current sensor, RE is the relay element without hysteresis,

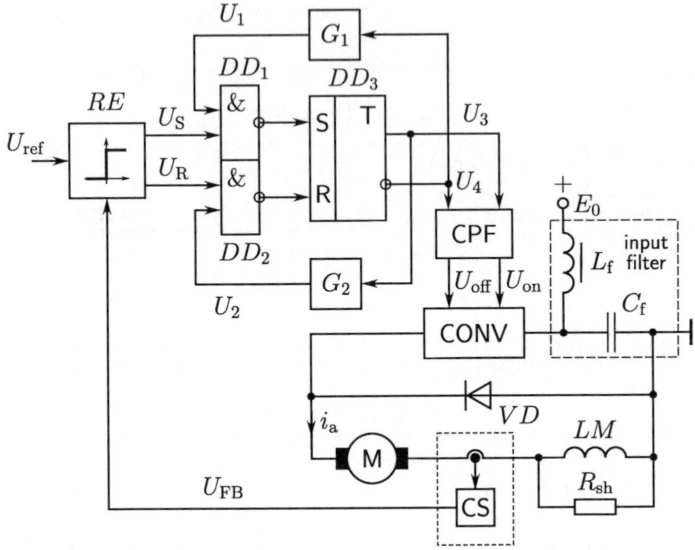

Figure 3.6. Functional scheme of a relay system without hysteresis and with time deadzone. M is the armature of the motor, LM is the excitation winding of the motor, R_{sh} is the resistor shunting the excitation winding LM in order to reduce the ripple on the magnetic flux of the motor.

G_1 and G_2 are formers of the time deadzone for the relay element, DD_3 is the control flip-flop, and CPF is the control pulse former.

Let us start by explaining some of the technical details of the system implementation. The current sensor is based on a *Hall device* and has galvanic isolation between the primary (high power) and the secondary (electronic) circuits [76]. The sensor was obtained from the LEM Group which produces a large variety of current and voltage sensors based on the Hall effect [77].

Figure 3.7 shows the functional scheme of the measuring converter (see, e.g., [76]). The measured current i_a (motor armature current), flowing through the primary winding W (power lead, Fig. 3.7), produces an excitation $i_a W$ that creates a flux Φ in the magnetic circuit. The current in the feedback winding produces an excitation $i_{FB}W_{FB}$ that creates a feedback flux Φ_{FB} directed oppositely to the flux Φ.

With the presence of currents in the windings W and W_{FB} the difference flux

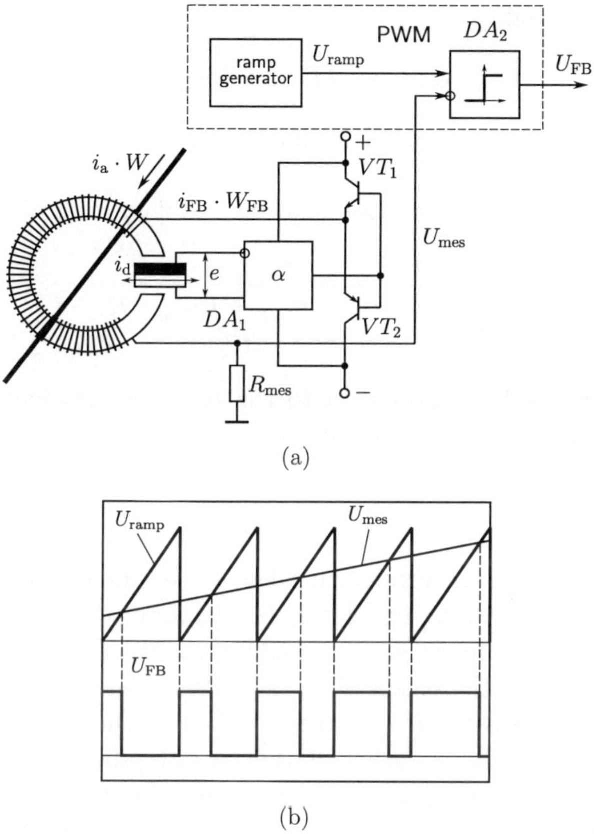

Figure 3.7. (a) Measuring converter based on the Hall device. (b) Time plots to show how the analog signal U_{mes} from the current sensor is converted into pulses by the pulse-width modulator (PWM).

$\Phi - \Phi_{FB}$ acts in the magnetic circuit. This flux produces the Hall EMF:

$$e = \frac{R_H}{d} i_d (\Phi - \Phi_{FB}) \frac{1}{S}, \tag{3.38}$$

where R_H is Hall coefficient, d is the thickness of the plate of the Hall device, i_d is the auxiliary current flowing along the plate, and S is the square of the core section.

Taking into account that $\Phi = i_a W / R_M$ and $\Phi_{FB} = i_{FB} W_{FB} / R_M$, where R_M is the magnetic resistance associated with the circuit's *magnetic path-gap*, and that the voltage gain of the amplifier equals α, we can determine the feedback current from the equality

$$\alpha e = i_{\mathrm{FB}} \left(R_{\mathrm{mes}} + R_{W_{\mathrm{FB}}} \right), \tag{3.39}$$

where $R_{W_{\mathrm{FB}}}$ is the resistance of the feedback winding, and R_{mes} is the resistance of the *measuring circuit.*

Substituting (3.38) and expressions for Φ and Φ_{FB} in (3.39) one obtains

$$i_{\mathrm{FB}} = i_{\mathrm{a}} \frac{W}{W_{\mathrm{FB}}} \cdot \frac{1}{1 + \dfrac{R_{\mathrm{mes}} + R_{W_{\mathrm{FB}}}}{\alpha W_{\mathrm{FB}}} \cdot \dfrac{R_{\mathrm{M}} S d}{R_{\mathrm{H}} i_{\mathrm{d}}}}.$$

Proceeding in the last expression to the limit of sufficiently large voltage gains α, one obtains the feedback current

$$i_{\mathrm{FB}} = i_{\mathrm{a}} \frac{W}{W_{\mathrm{FB}}}.$$

The value W/W_{FB} (transformation ratio) scales the conversion and ensures the required *current sensitivity.* Thus, for example the *transformation ratio* equals $1:5000$ for the current sensor “LEM Module LT-500-S/SP53” [77].

The output signal of the sensor U_{mes}, taken from the measuring resistor R_{mes}, is analog and transforms to the discrete form U_{FB} by the pulse-width modulator (PWM in Fig. 3.7). The duration of the output pulses of the modulator is directly proportional to the value of the measured current because the sensor characteristics is linear. The modulation frequency of the signal is determined by the requirements to the discreteness of the sensor.

When the output signal of the sensor is rectangular pulses (see Fig. 3.7), the relay element is represented by a circuit according to the scheme depicted in Fig. 3.8(a). Figure 3.8(b) shows time plots illustrating the operating principle of such a unit. This device is a threshold element with relay characteristics. It produces two outputs U_{S} and U_{R}, depending on the value of i_{a} relative to the specified value i_0

Let us also make a few comments about the operating principle of the device.

The sawtooth voltage U_{ramp} (Fig. 3.8) is produced from the ramp signal that is generated during each period of the output signal for the current sensor U_{FB}. The amplitude of the sawtooth voltage U_{ramp} is proportional to the value of the measured current i_{a} (see Fig. 3.8(b)). The output signal of the sawtooth voltage former RG is compared with the reference signal U_{ref} by the comparator

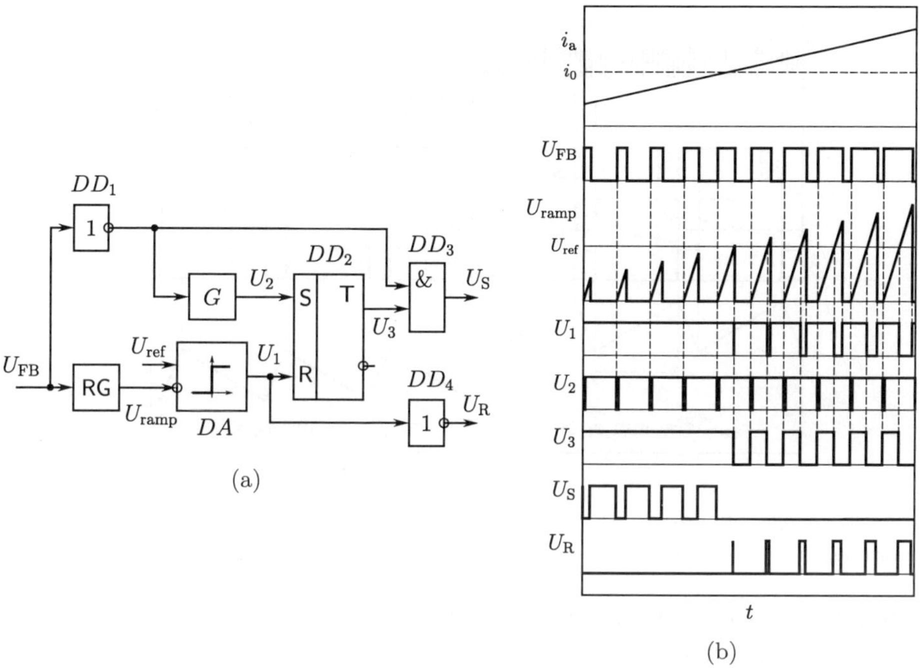

Figure 3.8. (a) Functional scheme of the relay element. RG is the ramp generator, G is the pulse former. (b) Time plots to show how the device works. i_0 is determined by the value of U_{ref}.

DA. The value of U_{ref} determines the setting of the motor armature current i_0. If the value of measured current i_a exceeds its specified value, i.e. $i_a > i_0$ (see Fig. 3.8(b)) then rectangular pulses U_1 are produced at the output of the comparator DA. The output signal of the relay element U_R is the inverted signal U_1. In the case when $i_a < i_0$, the comparator stops to produce the rectangular pulses U_1, and the selection circuit passes the rectangular pulses U_S from the output of the current sensor (inverting them). These pulses correspond to the "on" state of relay element.

Due to discreteness of the sensor signal, the relay element has a time deadzone. The magnitude of the deadzone does not exceed half of the pulse period of the measuring converter. With increasing conversion frequency this magnitude is reduced, and the relay element characteristics tends to become ideal.

An intermediate circuit for the conversion of the analog signal of the sensor to discrete form has been introduced in order to reduce the influence of the

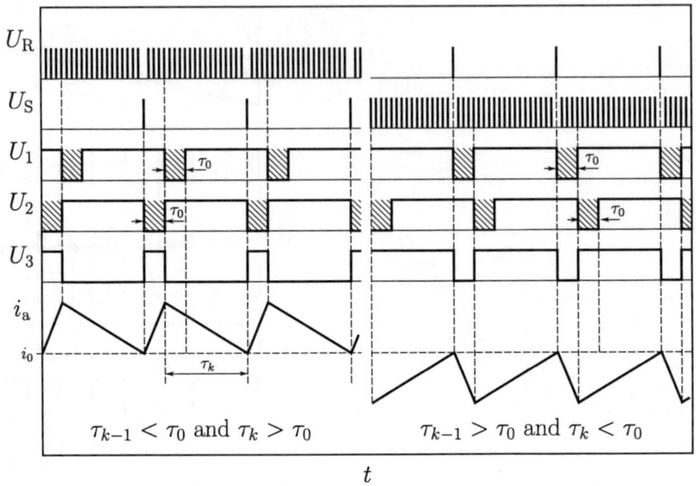

Figure 3.9. Time plots illustrating the workings of a relay system without hysteresis and with time deadzone.

feedback interference. This circuit is optional. Without this circuit the control system becomes significantly simpler. However, in this case the feedback should be amended by a passive filter to suppress high-frequency interference and by an intermediate amplifier of the sensor signal.

Now let us consider how the system presented in Fig. 3.6 works according to the time plots presented in Fig. 3.9. When the power converter CONV switches on, the current i_a in the winding of the motor armature increases as controlled by the current sensor CS. When this current exceeds the specified value i_0, the relay element switches. This causes the pulses U_R to emerge at the output of the relay element (the signal U_S equals to zero). These pulses set the control flip-flop into zero state.

We should note, that the control flip-flop may switch from one state to another with delay. This means, that the relay element for the system under consideration has time deadzone. If the time interval τ_k from the moment of the previous change in the state of the control flip-flop till the switching of the relay element is less then the duration τ_0 of the pulses of the time deadzone formers G_1 and G_2, then the flip-flop switches to another state after the delay τ_0. This happens because during the delay the former G_1 (or G_2) disables the pulses U_R (or U_S) of the relay element to pass to the corresponding flip-flop input. If $\tau_k > \tau_0$ then the control flip-flop switches without delay (see Fig. 3.9).

Introduction of a time deadzone is necessary for the system under consideration. First, the value τ_0 determines the minimum time interval between on- and off-pulses that is required for the safe operating of the system. Second, oscillations with inadmissibly high frequencies can emerge in relay systems without hysteresis. Alternatively to using time deadzones, this drawback can be avoided by introducing hysteresis.

The pulse U_{off} turning the power converter off is produced at the instant when the flip-flop resets and the armature current begins to decrease. Simultaneously, the trailing edge of the pulse U_4 starts the former G_1 that produces the pulse U_1 with the duration τ_0. Passing of the pulses U_{S} to the input S of the flip-flop is disabled during this time interval. The pulses U_{S} are produced and the signal U_{R} equals zero (see Fig. 3.8) when the current i_{a} becomes less then i_0.

The pulse U_{on} activating the power converter is produced at the instant when the flip-flop sets and the armature current increases again until the emergence of the next locking pulse. Similar to the previous case the trailing edge of the pulse U_3 starts the former G_2 disabling the passing of the pulses U_{R} to the input R of the flip-flop during the interval τ_0. The described process hereafter repeats. This results in the creation of the self-oscillatory mode in the system. The armature current oscillates around the given value i_0.

From the operating principle of the system it is seen that decreasing the value of the time deadzone of the relay element leads to a reduction of the ripple on the armature current while the oscillation frequency increases infinitely with $\tau_0 \to 0$. The minimum value of τ_0 is limited by the maximum allowable working frequency of the power converter because the switching losses rise with increasing frequency.

Figure 3.10 shows a system that differs from the above system in the construction of the relay element by having an adjustable hysteresis loop in addition to the time deadzone [78, 79]. Choosing the value of the hysteresis loop χ_0 we can limit the working frequency of the converter. On the other hand, the system ensures a higher control accuracy in comparison with the accuracy that can be obtained with a system as depicted in Fig. 3.10. This is because the oscillations of the armature current i_{a} is limited by the tolerance band determined by the thresholds $U_{\mathrm{ref}} + \chi_0$ and $U_{\mathrm{ref}} - \chi_0$ of the switching of the relay elements RE_1 and RE_2. The time deadzone causes switching delays in the transients only. Such a system can work in the mode of forced synchronization as well as in the self-oscillatory mode. Figure 3.11(a) and (b) show time plots explaining the

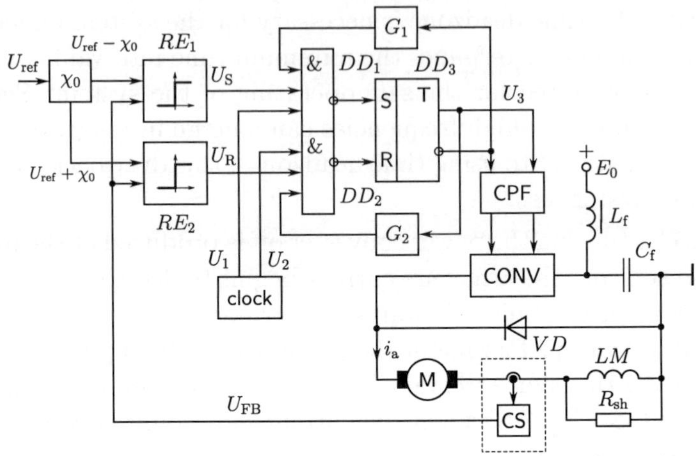

Figure 3.10. Functional scheme of a relay system with hysteresis and externally forced synchronization of the oscillations.

operating principle of the system for the modes of self-oscillations and forced synchronization, respectively. Synchronization of oscillations is performed by two pulse sequences U_1 and U_2 with equal and constant frequencies. These sequences are shifted relative to each other by half a period (see Fig. 3.11(b)). If the relative pulse duration (*pulse ratio*) is less than 0.5 then the power converter is forcedly switched on by pulses of the first sequence U_1 irrespective of the actual value of the feedback signal and it switches off when the armature current reaches the value i_{01} determined by the upper threshold of the relay element $U_{ref} + \chi_0$. As the frequency of the motor armature rotation increases, the pulse ratio grows and when it exceeds 0.5 the converter is switched off by pulses of the second sequence U_2 to switch on when the armature current reaches the value i_{01} determined by the lower threshold of the relay element $U_{ref} - \chi_0$. This allows us to combine the advantages of relay systems and systems with pulse-width modulation within a single structure.

Their simple technical implementation and small number of elements make it advantageous to use such system to control a wide variety of electrotechnological equipment involving semiconductor converters of electric energy.

Engineering solutions for the construction of relay systems shown in Figs. 3.6 and 3.10 are protected by inventor's certificates (I.C.) of USSR I.C. No. 1400920, I.C. No. 1403320, and I.C. No. 1614966 [75, 78, 79]. The system shown in

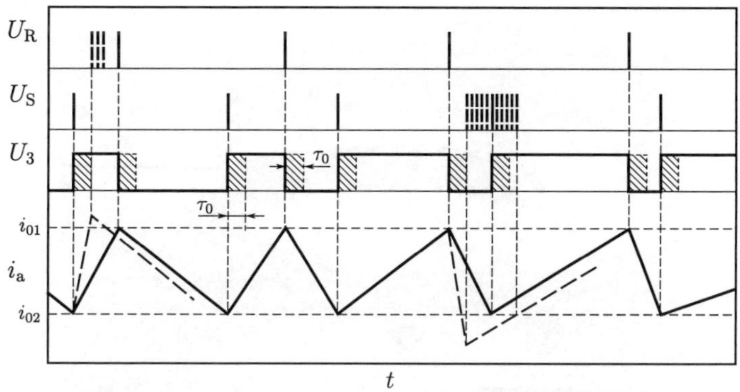

(a)

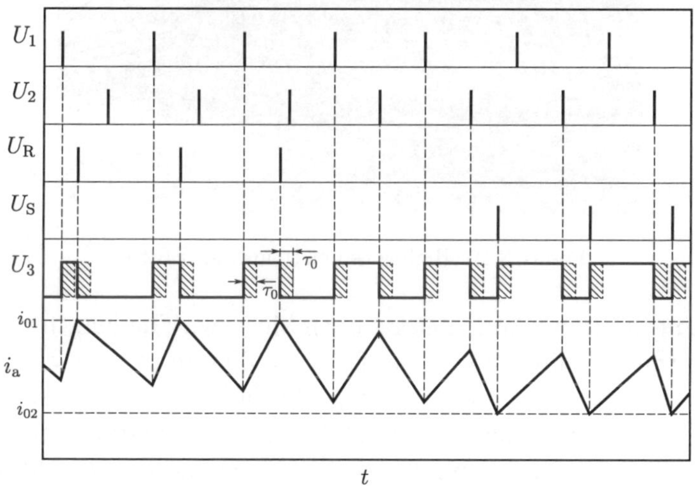

(b)

Figure 3.11. Time plots illustrating the workings of the system shown in Fig. 3.10: (a) for the mode of self-oscillations, (b) for the mode of externally forced synchronization. Tolerance band limits i_{01} and i_{02} are determined by the values $U_{ref} + \chi_0$ and $U_{ref} - \chi_0$, respectively.

Fig. 3.10 [19, 78–80] has been used to control the electric drives of the first USSR high-speed electric train ER200 (see Fig. 3.12) [81]. This electric train was developed to run on usual railway lines together with ordinary passenger and freight trains while the worldwide practice is to build special trunk-railways for high-speed trains [81]. The train was set in operation on March 1, 1984 on

Figure 3.12. High-speed electric train ER200.

the line Leningrad (now Saint-Petersburg)–Moscow. The length of the route is 650 km, and the traveling time varies from 4 h 30 min to nearly 5 h, depending on the state of the track and the presence of other traffic. In 1987, for example, the total distance over which a speed of 200 km/h was admissible was 300 km, and the traveling time according to the time-schedule was 4 h 30 min [82, 83]. However, today the train can cover the distance of 650 km in significantly shorter time. In 2000, for example, a record was established at 3 h 54 min. In 1995, the second electric train ER200-2 was set in operation and two electric trains now operate on the line Saint-Petersburg–Moscow.

The minimum moving unit of the electric train consists of two head cars H, a motor car with pantograph Mp, and a motor car without pantograph M. The head cars do not contain any traction equipment. The train is composed according to the scheme H+N(M+Mp)+H. The number N of pairs of motor cars changes depending on the required transportation work and load of the railway electric supply. Usually the train has eight cars ($N = 3$). Each motor car is equipped with a four-motor constant current electric drive, the simplified

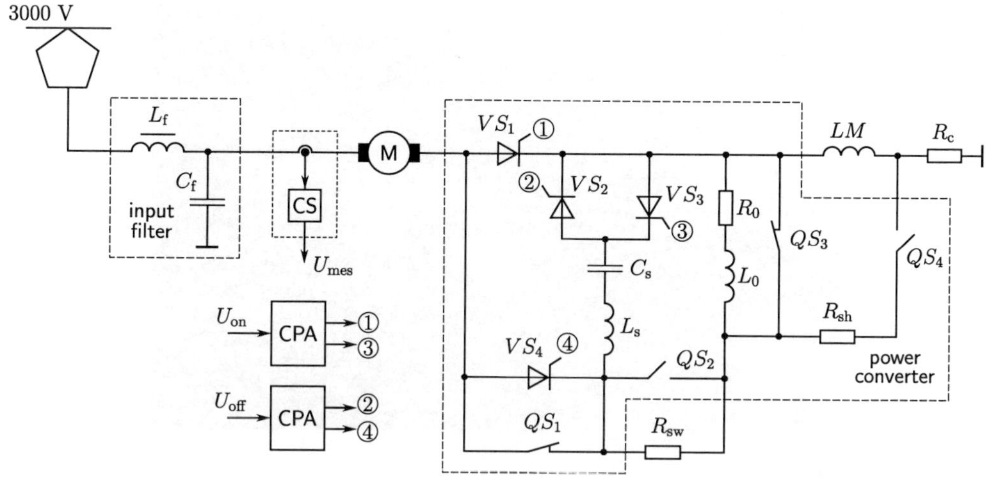

Figure 3.13. Simplified scheme of the power part of electric drive for the motor car of the electric train ER200. CPA is the control pulse amplifier. The depicted state of the switches QS_1–QS_4 corresponds to the mode of electric traction by pulse regulation of the resistance of the starting resistors.

scheme of which is shown in Fig. 3.13 [81]. The electric drives of the cars Mp and M that constitute one motor section are powered from the aerial contact wire (mains supply) via a common power filter. The total power of the traction motors of six motor cars is 10320 kW [81–83]. All high-power equipment, including the power converter of the two motor cars, is situated under the car Mp. Figure 3.14 shows the displacement of the power converters of the motor cars.

The electric drive can work in the following automatic modes:

First is the mode of the electric traction by pulse regulation of the resistance of the starting resistors R_c. Second is the mode of electric traction by pulse regulation of the magnetic flux of the traction motors. Both of these modes are implemented with serial as well as with parallel connection of groups of traction motors. Each group contains four serially connected motors. The third mode is stabilization of the train speed. The fourth mode is electrodynamic braking of the motors by regulating the magnetic flux of the traction motors, and finally the fifth mode is electrodynamic braking of the motors by regulation of the resistance of the breaking resistors. In the traction modes the electric drive is controlled by a relay system with externally forced synchronization, while the

Figure 3.14. Displacement of the power converters of the electric drives of the motor cars.

electrodynamic braking is regulated by a relay system with hysteresis and time deadzone (see Fig. 3.11). Figure 3.15(a) shows a photograph of the electronic unit of the control system for the electric drives of the motor cars Mp and M, and Fig. 3.15(b) shows the placement of the sensors (see Fig. 3.7(a)) for the armature current of the traction motors. This control system is mounted on each motor section. The electronic unit was produced by "Eksperimentalny Zavod" (Experimental Plant), Saint-Petersburg. The photographs presented in Fig. 3.14 and 3.15 were kindly granted by our collaborators from Experimental Plant in the development of the control systems Vyacheslav Boreisha, Vladimir

(a) (b)

Figure 3.15. (a) Electronic unit of the control system for the electric drives of the motor cars. (b) Placement of the sensors CS for the armature current of the traction motors.

Skobeltsyn, and Nikolay Poukhlyakov.

3.7.2 Model of electric drive with relay control

Figure 3.16 shows a block diagram for the electric drive of ER200 in the mode of electric traction by pulse regulation of resistance of the starting resistors. We will construct a mathematical model for the case when the motor torque

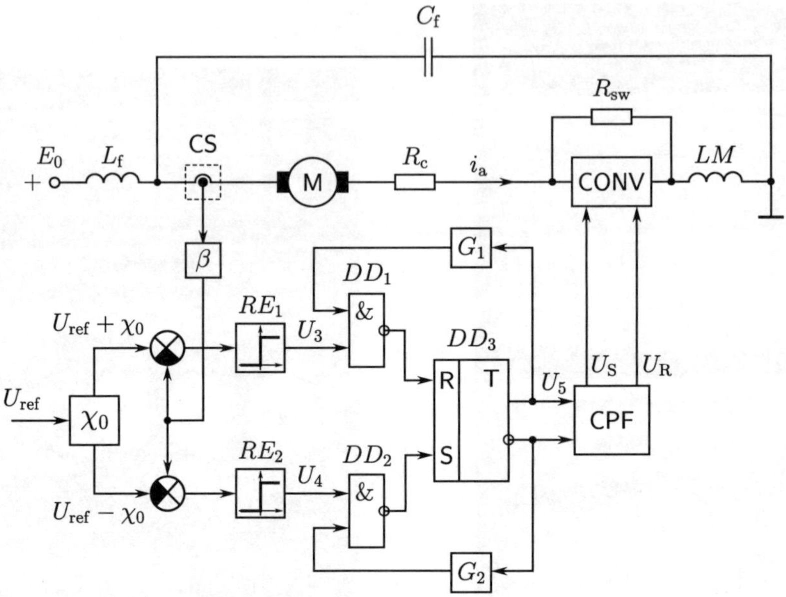

Figure 3.16. Block diagram for the DC electric drive with relay control.

(armature current) is stabilized by a relay control system with hysteresis and time deadzone. The equivalent circuit of such a system is presented in Fig. 3.17.

The mathematical model corresponding to the equivalent circuit is represented by the system of differential equations

$$\frac{\mathrm{d}X}{\mathrm{d}t} = G(X);\; X = (x_1, x_2, x_3, x_4)^{\mathrm{T}};\; G = (g_1, g_2, g_3, g_4)^{\mathrm{T}}; \tag{3.40}$$

with

$$g_1 = -\frac{R_{\mathrm{f}}}{L_{\mathrm{f}}}\,x_1 - \frac{1}{L_{\mathrm{f}}}\,x_4 + \frac{E_0}{L_{\mathrm{f}}};$$

$$g_2 = -\frac{R_{\mathrm{ef}}\big(\mathrm{K}_k^{(\mathrm{F})}(\xi)\big)}{L_{\mathrm{a}}}\,x_2 + \frac{R_{\mathrm{L}}}{L_{\mathrm{a}}}\,x_3 + \frac{1}{L_{\mathrm{a}}}\,x_4 - \frac{E_\mu(x_3)}{L_{\mathrm{a}}};$$

$$g_3 = \frac{R_{\mathrm{L}}}{L_\mu}\,(x_2 - x_3);$$

$$g_4 = \frac{1}{C_{\mathrm{f}}}\,(x_1 - x_2).$$

Here x_1, x_2, x_3, and x_4 are the filter coil, armature and magnetization currents and the filter capacitor voltage; L_{a}, L_μ, and L_{f} are the inductances of the motor

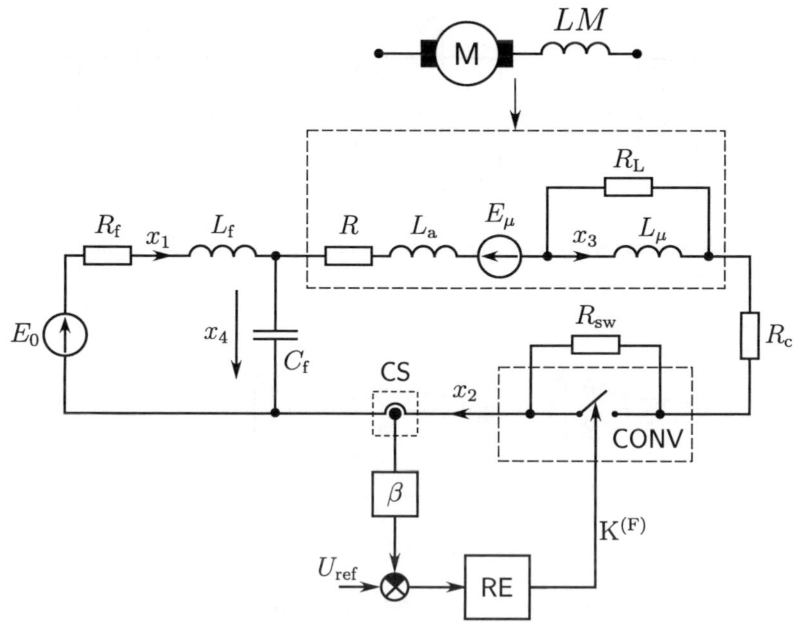

Figure 3.17. Equivalent circuit of the relay control system. In the upper part of the figure the dashed line bounds the equivalent circuit of the motor.

armature circuit, the motor magnetizing circuit and the filter, respectively; C_f is the capacity of the filter condenser; R_f and R_L are the resistances of the filter and an equivalent short-circuited eddy currents loop, respectively; $E_\mu(x_3) = \omega C_E \Phi(x_3)$ is the motor rotating EMF, where ω, $\Phi(x_3)$, C_E are the armature rotation frequency, the magnetization curve, and a design parameter of the motor, respectively. E_0 is the supply voltage.

$$R_{ef}\big(K_k^{(F)}(\xi)\big) = R + R_c + \frac{1}{2}\,(1 - K_k^{(F)})R_{sw},$$

where $R = R_e + R_a + R_L$; R_a and R_e are the resistances of the armature circuit and the motor excitation winding; R_c and R_{sw} are the values of the starting resistances. The characteristics of the relay element RE is given in Fig. 3.18.

The output signal of a relay element with time deadzone may be written as

$$K_k^{(F)} = K_{k-1}^{(F)}\,\mathrm{sign}\left\{\chi_0\,(1 + N_{k-1}^{\tau}) + (-1)^{N_{k-1}}\xi(X)\right\};$$
$$\xi(X) = U_{ref} - \beta x_2;$$

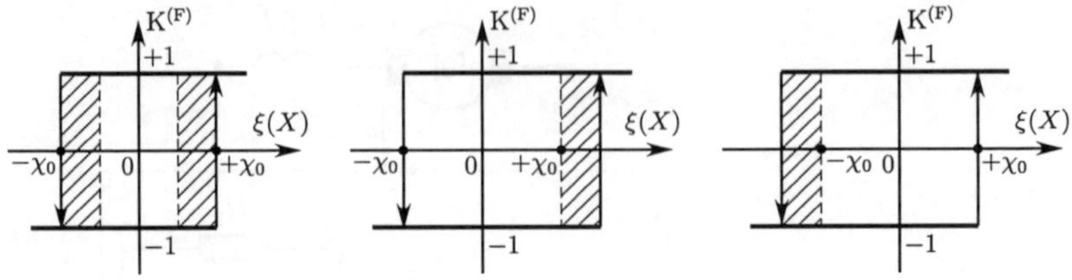

Figure 3.18. Static characteristics of the relay element. χ_0 denotes the value of the hysteresis loop.

with

$$N_k^{\tau} = \frac{\vartheta}{2}\left[1 - \text{sign}\left\{\frac{t - t_{k-1}}{\tau_0} - 1\right\}\right];$$
$$N_k = N_{k-1} + \frac{1 - \text{sign}\{\mathrm{K}_k^{(\mathrm{F})}\mathrm{K}_{k-1}^{(\mathrm{F})}\}}{2};$$
$$\mathrm{K}_0^{(\mathrm{F})} = -1;\ N_0 = 0;\ N_0^{\tau} = 0;\ \vartheta > 0 = \text{const};$$
$$t_k = \begin{cases} t_k^*, & t_k - t_{k-1} > \tau_0, \\ t_{k-1} + \tau_0, & t_k - t_{k-1} \le \tau_0; \end{cases}$$

and t_k^* being the smallest positive root of the equation

$$\chi_0 + (-1)^{N_{k-1}}\xi(X(t_k)) = 0.$$

β is the transfer constant of the current sensor. The expression for $\mathrm{K}_k^{(\mathrm{F})}$ is obtained from the time plots describing the system operation (see Fig. 3.9 and 3.11).

The experimentally determined magnetization curve $C_E\Phi(x_3)$ of the motor [13] has been interpolated by the function $\varphi(\overline{x})$:

$$C_E\Phi(x_3) = \mu\varphi(\overline{x});$$
$$\varphi(\overline{x}) = a_1 \exp(a_2\overline{x}) + a_3 \arctan(a_4\overline{x}) + a_5\overline{x}^{a_6} - a_7,$$

where the parameters $a_1 = 0.08067$, $a_2 = 0.000797$, $a_3 = 0.242$, $a_4 = 0.0108$, $a_5 = 0.01973$, $a_6 = 0.29664123$, and $a_7 = 0.0964$. $\overline{x} = x_3/x_*$ with $x_* = 1$ A and $\mu = 1$ V·min. The magnetization curve is illustrated in Fig. 3.19.

In deriving Eq. (3.40) we have made use of the following assumptions: the demagnetizing effect of the armature reaction is compensated; the converter's thyristors are ideal switching elements, except that the minimum time between

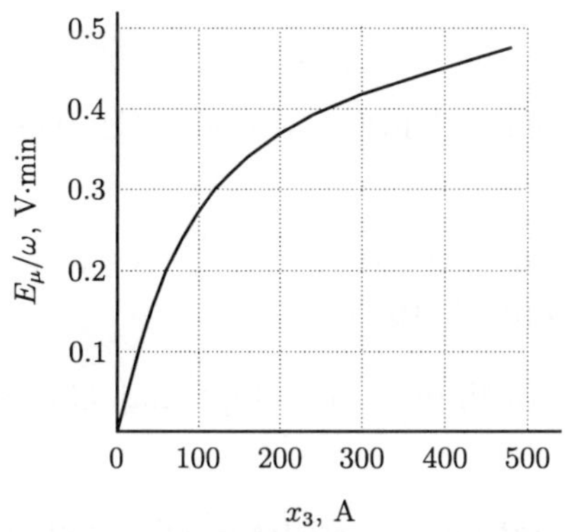

Figure 3.19. Magnetization curve for the motor. E_μ is the motor rotating EMF, ω is the armature rotation frequency, and x_3 is the magnetization current.

the subsequent changes in the converter state is determined by the value of the relay element time deadzone τ_0.

The system (3.40) should be supplemented by an equation for the armature rotation frequency. However, the class of electric drives under consideration is characterized by substantial moments of inertia which allows us to study electromagnetic and electromechanical processes separately [13]. Thus, the system (3.40) is considered in time intervals which are short with respect to the typical time periods of electromechanical processes and, at the same time, long with respect to the typical time periods of electromagnetic processes [13].

Equations (3.40) are non-integrable due to the non-linearity of the magnetization curve $C_E\Phi(x_3)$. That is why the map

$$X_k = F\left(\tau_0, \tau_k, X_{k-1}\right), \ k = 1, 2, 3, \ldots$$

with

$$\begin{aligned} \tau_k &= \begin{cases} \tau_0, & \tau_k^* - \tau_0 < 0, \\ \tau_k^*, & \tau_k^* - \tau_0 \geq 0; \end{cases} \\ \tau_k^* &\to \varphi\left(\tau_k^*, X_{k-1}\right) = 0; \\ \tau_k &= t_k - t_{k-1}; \\ X_k &= X(t_k, X_{k-1}) \end{aligned}$$

has been calculated by numerical integration of Eqs. (3.40). As previously noted, the complex behavior of this system will be analyzed in detail in the next chapter.

Bibliography

[1] Ya. Z. Tsypkin, Relay Control Systems (University Press, Cambridge, 1984).

[2] A. A. Andronov, S. E. Khaikin, and A. A. Vitt, Theory of Oscillations (Pergamon Press, Oxford, 1965).

[3] I. Flügge-Lotz, Discontinuous and Optimal Control (McGraw-Hill, New York, NY, 1968).

[4] K. J Åström, *Oscillations in Systems with Relay Feedback*, in Adaptive Control, Filtering, and Signal Processing, IMA Volumes in Mathematics and its Applications, Vol. 74, edited by Åström *et al.* (Springer–Verlag, Berlin, 1995), pp. 1–25.

[5] U. Holmberg, *Relay Feedback of Simple Systems*, Ph.D. Thesis TFRT-1034, Dept. of Automatic Control, Lund Institute of Technology, Lund, Sweden (1991).

[6] K. H. Johansson, *Relay Feedback and Multivariable Control*, Ph.D. thesis TFRT-1048, Dept. of Automatic Control, Lund Institute of Technology, Lund, Sweden (1997).

[7] I. Flügge-Lotz, Discontinuous Automatic Control (Princeton University Press, Princeton, 1953).

[8] V. I. Utkin, Sliding Modes in Control Optimization (Springer-Verlag, Berlin, 1992).

[9] R. Balasubramanian, *Stability of Limit Cycles in Feedback Systems Containing a Relay*, IEE Proc. D **128**, 24–29 (1981).

[10] D. P Atherton, *Analysis and Design of Relay Control Systems*, in CAD for Control Systems, edited by D. A. Linkens (Marcel Dekker, New York, 1993), pp. 367–394.

[11] A. Megretski, *Global Stability of Oscillations Induced by Relay Feedback*, in Preprints IFAC 13th World Congress, Vol. E (San Francisco, 1996), pp. 49–54.

[12] J. M. Goncalves, A. Megretski, and M. A. Dahleh, *Global Stability of Relay Feedback Systems*, Technical Report Preprint LIDS-P-2458, Dept. of Elec. Eng. Comp. Sci., MIT, Cambridge, MA. (1999).

[13] V. S. Baushev, Zh. T. Zhusubaliyev, and Yu. V. Kolokolov, *On Analysis of Relay Control Systems for Regulation of Electric Current in Conditions of Electrodynamic Braking of High-Speed Electric Trains*, Elektrichestvo **7**, 66–70 (1989), in Russian.

[14] Yu. K. Rozanov, Semiconductor Transformers with a Link of High Frequency (Energoatomizdat, Moscow, 1987), in Russian.

[15] V. N. Tishchenko and V. N. Kovalev, *A Study of Self-Excited Oscillations in an Asynchronous Electric Drive with a Relay-Type Thyristor Regulator for the Rotor Current*, Elektrichestvo **4**, 25–31 (1991), in Russian.

[16] B. K. Bose (Editor), Modern Power Electronics: Evolution, Technology, and Applications (IEEE Press, New York, 1992).

[17] A. V. Kobzev, G. Ya. Mikhal'chenko, and N. M. Muzhitchenko, Modulating Power Supplies of Radioelectronic Devices (Radio i Svyaz, Tomsk, 1990), in Russian.

[18] A. P. Zaytsev and V. A. Podlyagin, *Stability of Oscillation Synchronization for a Relay-Pulse System for Current Regulation*, Izv. vuzov. Elektromekhanika **7**, 94–98 (1987), in Russian.

[19] Yu. V. Kolokolov, *Development of Principle for Design of Relay-Pulse Current Regulators of DC Traction Motors*, Elektrichestvo **9**, 35–44 (1990), in Russian.

[20] G. S. Nurre, R. S. Ryan, H. N. Scofield, and J. L. Sims, *Dynamics and Control of Large Space Structures*, J. of Guidance, Control and Dynamics **7**, 514–526 (1984).

[21] N.S. Postnikov, *Stochastic Oscillations in a Nuclear Reactor with Relay Control System*, Atomnaya Energiya **76**, 3–11 (1994), in Russian.

[22] N. S. Postnikov, *Dynamic Chaos in Relay Systems with Hysteresis*, Computational Mathematics and Modeling **8**, 62–72 (1997).

[23] N. S. Postnikov, *Stochasticity of Relay Systems with Hysteresis*, Avtomatika i Telemekhanika **3**, 57–68 (1998), in Russian.

[24] K. H. Johansson and A. Rantzer, *Global Analysis of Third-Order Relay Feedback Systems*, in Preprints IFAC 13th World Congress, Vol. E (San Francisco, 1996), pp. 55–60.

[25] K. H. Johansson, A. Rantzer, and K. J. Åström, *Fast Switching in Relay Feedback Systems*, Automatica **35**, 539–552 (1999).

[26] K. H. Johansson, A. Barabanov, and K. J. Åström, *Limit Cycles with Chattering in Relay Feedback Systems*, IEEE Transactions on Automatic Control **47**, 1414–1423 (2002).

[27] A. S. Morse, *Control Using Logic-based Switching*, in Trends in Control. A European Perspective, edited by A. Isidori (Springer, 1995), pp. 69–113.

[28] R. W. Brockett, *Hybrid Models for Motion Control Systems*, in Essays in Control: Perspectives in the Theory and its Applications, edited by H. Trentelman and J. Willems (Birkhäuser, Boston, 1993), pp. 29–53.

[29] P. Kowalczyk and M. di Bernardo, *On a Novel Class of Bifurcations in Hybrid Dynamical Systems: The Case of Relay Feedback Systems*, in Proc. Hybrid Systems: Computation and Control (HSCC01), edited by M. D. di Benedetto and A. Sangiovanni-Vincentelli (Springer-Verlag, Berlin, 2001), pp. 361–374.

[30] S. R. Norsworthy, R. Schreier, and G. C. Temes, Delta-Sigma Data Converters — Theory, Design, and Simulation (IEEE Press, New York, 1997).

[31] P. M. Aziz, H. V. Sorensen, and J. van der Spiegel, *An Overview of Sigma-Delta Converters*, IEEE Signal Processing Magazine **13**, 61–84 (1996).

[32] Th. Holzhüter, *Simulation of Relay Control Systems Using MATLAB/SIMULINK*, Control Engineering Practice **6**, 1089–1096 (1998), Plenum Publishing.

[33] Z. J. Palmor, *Limit Cycles in Decentralized Relay Systems*, Int. J. Control **56**, 744 (1992).

[34] Z. J. Palmor, Y. Halevi, and T. Efrati, *A General and Exact Method for Determining Limit Cycles in Decentralized Relay Systems*, Automatica **31**, 1333–1339 (1995).

[35] Zh. T. Zhusubaliyev, *Bifurcations and Chaotic Motions in the Dynamics of Relay Automatic Control Systems*, in Proc. of the 3rd International Conf. "Recognition-97", Kursk, 25–30 (1997), in Russian.

[36] A. F. Filippov, Differential Equations with Discontinuous Right-Hand Sides (Kluwer Academic Press, Dordrecht, 1988).

[37] Yu. I. Neimark, Methods of Pointwise Mappings in the Theory of Nonlinear Oscillations (Nauka Publ., Moscow, 1972), in Russian.

[38] Yu. S. Fedosenko, *On the Structure of the Phase Portrait and Periodic Solutions of Autonomous Dynamical Systems with Impact*, Prikl. Mat. Mekh. **40**, 619–629 (1976), in Russian.

[39] M. di Bernardo, K. H. Johansson, and F. Vasca, *Self-Oscillations and Sliding in Relay Feedback Systems: Symmetry and Bifurcations*, Int. J. Bifurcation and Chaos **11**, 1121–1140 (2001).

[40] P. Kowalczyk and M. di Bernardo, *On the Existence of Stable Asymmetric Solutions in Unforced and Symmetric Relay Feedback Systems*, in Proc. European Control Conference (ECC'01) (Porto, Portugal, Sept. 2001), to appear.

[41] M. di Bernardo, K. H. Johansson, and F. Vasca, *Sliding Orbits and their Bifurcations in Relay Feedback Systems*, in Proc. IEEE Conference on Decision and Control, CDC99 (Phoenix, USA, 1999), pp. 708–713.

[42] I. N. Krutova, *Stabilization Process for a Many-Dimensional Dynamic System with Relay Control*, Avtomatika i Telemekhanika **4**, 27–43 (1999), in Russian.

[43] I. N. Krutova, *On the Stability of a Stabilization Mode for a Spacecraft Subjected to Elastic Oscillations*, Avtomatika i Telemekhanika **7**, 44–58 (1999), in Russian.

[44] Zh. T. Zhusubaliyev, E. A. Soukhoterin, V. N. Rudakov, Yu. V. Kolokolov, and E. Mosekilde, *Bifurcations and Chaotic Oscillations in an Automatic*

Control Relay System with Hysteresis, Int. J. Bifurcation and Chaos **11**, 1193–1231 (2001).

[45] Yu. I. Neimark and P. S. Landa, Stochastic and Chaotic Oscillations (Kluwer Academic Publ., Dordrecht, 1992).

[46] A. S. Alekseev, *An Electronic Analog of a Two-Position Temperature Regulator with Advance Zone*, DAN of USSR **57**, 393–396 (1952), in Russian.

[47] A. S. Alekseev, *A Two-Position Temperature Regulator with Advance Zone*, in In Memory of A. A. Andronov (USSR Acad. of Sci. Publ., 1955), pp. 45–76, in Russian.

[48] V. S. Anishchenko, Dynamical Chaos — Models and Experiments. Appearance Routes and Structure of Chaos in Simple Dynamical Systems (World Scientific, Singapore, 1995).

[49] J. Guckenheimer and P. J. Holmes, Nonlinear Oscillations, Dynamical Systems, and Bifurcations of Vector Fields (Springer, New York, 3rd printing, 1990).

[50] P. A. Cook, *Simple Feedback Systems with Chaotic Behavior*, Syst. Contr. Lett. **6**, 223–227 (1985).

[51] D. Amrani and D. P. Atherthon, *Designing Autonomous Relay Systems with Chaotic Motion*, in Proc. 28th IEEE Conf. Decision and Control, Vol. 1 (Tampa, FL, 1989), pp. 932–936.

[52] R. Genesio and A. Tesi, *Chaos Prediction in a Third Order Relay System*, Internal Report RT 29/90. Dipartimento di Sistemi ed Informatica, University of Florence, Italy (1990).

[53] Th. Holzhüter and T. Klinker, *Control of a Chaotic Relay System Using the OGY Method*, Zeitschrift für Naturforschung Section **A53**, 1029–1036 (1998).

[54] P. S. Landa, Nonlinear Oscillations and Waves in Dynamical Systems (Kluwer Academic Publ., Dordrecht, 1996).

[55] E. Mosekilde, Topics in Nonlinear Dynamics. Applications to Physics, Biology and Economic Systems (World Scientific, Singapore, 1996).

[56] A. L. Fradkov, Introduction to Control of Oscillations and Chaos (World Scientific, Singapore, 1998).

[57] G. Chen and X. Dong, From Chaos to Order: Methodologies, Perspectives and Applications (World Scientific, Singapore, 1998).

[58] Zh. T. Zhusubaliyev, E. A. Soukhoterin, and E. Mosekilde, *Border-Collision Bifurcations and Chaotic Oscillations in a Piecewise-Smooth Dynamical System*, Int. J. Bifurcation and Chaos **11**, 2977–3001 (2001).

[59] M. I. Feigin, Forced Oscillations in Systems with Discontinuous Nonlinearities (Nauka Publ., Moscow, 1994), in Russian.

[60] M. di Bernardo, M. I. Feigin, S. J. Hogan, and M. E. Homer, *Local Analysis of C-Bifurcations in n-Dimensional Piecewise-Smooth Dynamical Systems*, Chaos, Solitons and Fractals **10**, 1881–1908 (1999).

[61] S. Banerjee, P. Ranjan, and C. Grebogi, *Bifurcations in Two-dimensional Piecewise Smooth Maps — Theory and Applications in Switching Circuits*, IEEE Trans. Circ. Syst. I **47**, 633–643 (2000).

[62] Zh. T. Zhusubaliyev, *Investigation of Chaotic Regimes of a Voltage Converter with Pulse-Width Modulation*, Elektrichestvo **6**, 40–46 (1997), in Russian.

[63] R. Feldberg, M. Szymkat, C. Knudsen, and E. Mosekilde, *Iterated-Map Approach to Die Tossing*, Phys. Rev. A **42**, 4493–4502 (1990).

[64] N. V. Butenin, Yu. I. Neimark, and N. A. Fufayev, Introduction to the Theory of Nonlinear Oscillations (Nauka Publ., Moscow, 1987), in Russian.

[65] V. S. Baushev and Zh. T. Zhusubaliyev, *Indeterminable States of a Voltage Regulator with Pulse-Width Control*, Electrical Technology (3), 85–98 (1992), Elsevier Science.

[66] J. E. Dennis and R. V. Schnabel, Numerical Methods for Unconstrained Optimization and Nonlinear Equations (SIAM, 1996).

[67] M. A. Aizerman and F. R. Gantmakher, *Stability in the Linear Approximation of the Periodic Solution of a System of Differential Equations with Discontinuous Right-Hand Part*, Prikl. Mat. Mekh. **21**, 658–669 (1957), in Russian.

[68] V. S. Baushev, Zh. T. Zhusubaliyev, Yu. V. Kolokolov, and I. V. Terekhin, *Local Stability of Periodic Solutions in Sample Data Control Systems*, Automation and Remote Control **53**, part 2, 865–871 (1992), Plenum Publishing.

[69] O. A. Aleynikov, V. S. Baushev, A. V. Kobzev, and G. Ya. Mikhal'chenko, *Investigation of Stability for Periodic Modes in Nonlinear Pulse Systems*, Electrichestvo **4**, 16–21 (1991), in Russian.

[70] L. S. Pontryaguin, Ordinary Differential Equations (Nauka Publ., Moscow, 1982), in Russian.

[71] N. Mohan, T. M. Undeland, and W. P. Robbins, Power Electronics: Converters, Applications and Design (Wiley, New York, 1995).

[72] P. R. K. Chetty, Switch-Mode Power Supply Design (TAB Books Inc., 1986).

[73] R. E. Tarter, Solid-State Power Conversion Handbook (Wiley-Interscience, New York, 1993).

[74] J. G. Kasakian, M. F. Schlecht, and G. C. Verghese, Principles of Power Electronics (Addison-Wesley, Reading, MA, 1991).

[75] Zh. T. Zhusubaliyev *et al.*, I.C. No. 1400920 USSR, IPC B 60 L 15/04, *Device for Regulation of the Current of Electric Motor of Vehicle*, **21** (1988), in Russian.

[76] G. I. Katzman and I. M. Matveev, *Using of the Hall converters for Contactless Measuring of Direct Current Signals*, Pribory i Sistemy Upravleniya **10**, 33 (1986), in Russian.

[77] `http://www.lem.com`. Official Web-site of LEM Group.

[78] Zh. T. Zhusubaliyev *et al.*, I.C. No. 1403320 USSR, IPC H 02 P/175, *Device for Relay Regulation of the Current of DC Electric Motor*, **22** (1988), in Russian.

[79] Zh. T. Zhusubaliyev *et al.*, I.C. No. 1614966 USSR, IPC B 60 L 15/04, *Relay-Pulse Regulator of the Current of Traction Electric Motor*, **47** (1990), in Russian.

[80] Yu. V. Kolokolov, Zh. T. Zhusubaliyev, L. Yu. Veitsman, and R. M. Berzin, *Automatic System Controlling the Traction Electric Drives of the Second High-Speed Electric Train ER200*, Elektrotekhnika **9**, 49–52 (1990), in Russian.

[81] L. V. Gutkin, Yu. N. Dymant, and I. A. Ivanov, Electric Train ER200 (Transport Publ., Moscow, 1981), in Russian.

[82] L. V. Gutkin and V. G. Inozemtsev, *Results of Operation of the Electric Train ER200 on the Line Moscow–Leningrad*, in Investigation of High-Speed Electric Train ER200, edited by V. G. Inozemtsev (Transport Publ., Moscow, 1985), pp. 3–13, in Russian.

[83] L. V. Gutkin and D. M. Samarets, *Results of Operation of the Electric Train ER200*, Zheleznodorozhny Transport (Railway Transport) **5**, 38–41 (1994), in Russian.

Chapter 4

BIFURCATIONS AND CHAOTIC OSCILLATIONS IN RELAY SYSTEMS

4.1 Relay Control Systems with Complex Dynamics

Chapter 3 outlined some of the problems associated with the study of complex dynamics in relay control systems. In this chapter we will perform a detailed bifurcation analysis of two different types of control systems that one can meet in real-world technical applications. The aim is to investigate the structure of the two-parameter diagram of dynamical modes and to examine the regularities in the occurrence of periodic cycles. We shall also determine the main bifurcations and analyze the nature of the transitions to chaotic dynamics. Even though engineering practice at the present may not comply with such a standard, analyses of this type are required in the design of a wide variety of modern control systems.

As a first example we shall consider a constant voltage converter with relay control whose behavior is described by equations of the form (3.7). The equivalent circuit of this converter is depicted in Fig. 4.1. The power part of the circuit comprises a transistor converter of constant voltage working in switch mode, a transformer-rectifier device, and a couple of ripple LC-filters (see, e.g. [1–3]). The feedback signal ξ, carrying information about the deviation of the output voltage U_1 from its prescribed value U_{ref}, is applied to a relay element with hysteresis, in order to form the square pulses (see, e.g. Fig. 3.1(b)) used to control the switching elements of the transistor converter.

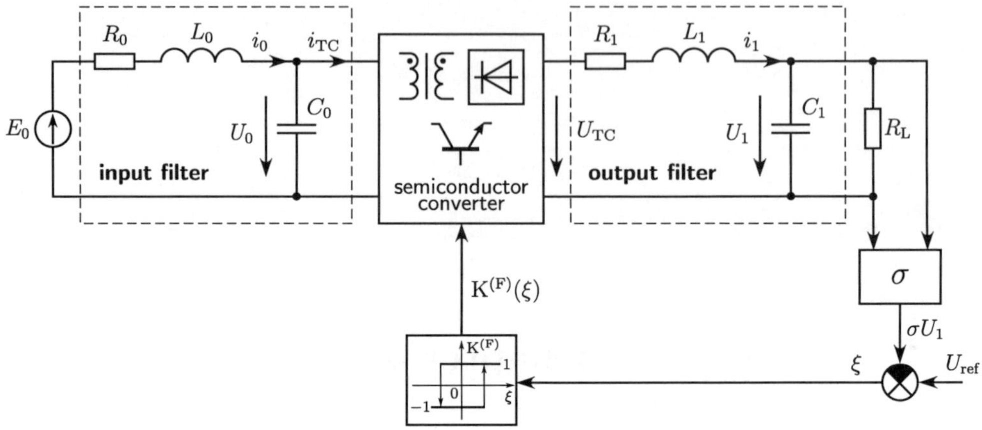

Figure 4.1. Equivalent circuit of the converter with relay control. The feedback signal ξ is applied to a relay element with hysteresis in order to form the square pulses used to control the semiconductor converter.

In power electronics this type of control is sometimes referred to as the *tolerance-band PWM technique* [4]. This technique is often used to control power electronic AC inverters for variable-frequency drives and uninterruptible power supplies [4–6]. Wasynczuk *et al.* [7] indicate that modern concepts of voltage and current control in three-phase AC inverters with decoupling of state-space variables also make use of this method. Depending on the control strategy, the feedback can be generated by the load-voltage, the load-current, or the converter (inverter) output current. The variable frequency in such converters is produced by variation of the frequency of the sinusoidal reference signal. DC/DC converters can use a similar control method. In this case the sinusoidal reference is replaced by a constant value [5, 6].

As previously noted, converters of this type are widely used as voltage regulators in power supply systems in order to obtain highly stable magnetic fields in nuclear magnetic-resonant tomographs with superconducting magnets; in energy-supply systems for space- and aircrafts [2, 3]; in test stands for experimental investigation of low- and high-temperature superconductors; and in power supply systems for radio-electronic and computer equipment [1].

As a second example of a relay system we shall consider a system applied to regulate the motor torque of a DC electric drive [8, 9]. A block diagram of this drive is shown in Fig. 4.2. Here, CONV is the power converter, (M)

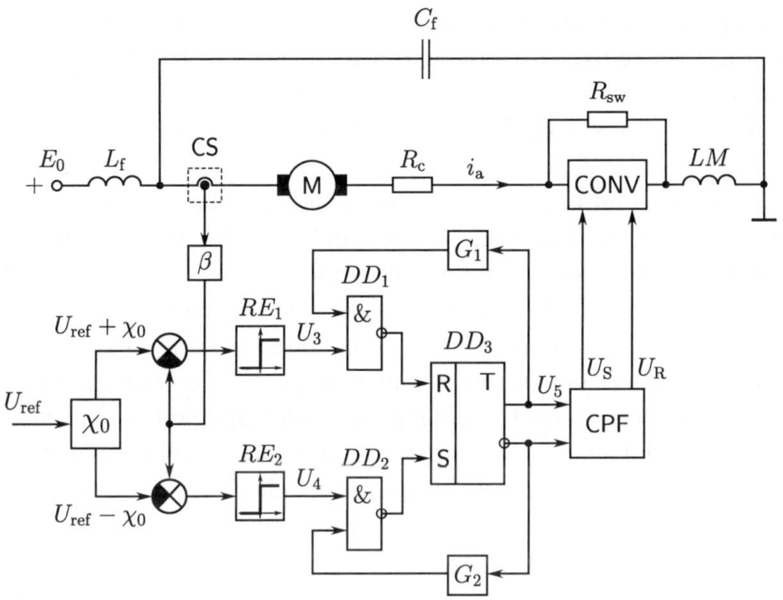

Figure 4.2. Block diagram for the DC electric drive with relay control. Unlike the system depicted in Fig. 4.1, this system has time deadzones.

is the electric DC motor, and CPF is the control pulse former. The feedback circuit contains a current sensor (CS) with the transfer constant β. A reference input-element for setting the armature current produces two reference signals $U_{ref} + \chi_0$ and $U_{ref} - \chi_0$ that determine the width of the tolerance band. Here U_{ref} determines the setting of the motor armature current. Unlike the system depicted in Fig. 4.1, the two-position relay element with hysteresis has time deadzones. The circuit of a two-position relay element with time deadzone contains two relay elements RE_1 and RE_2 without hysteresis, the logic elements DD_1 and DD_2, the control flip-flop DD_3, and the deadzone formers G_1 and G_2.

In Chapter 3 we considered several possible schemes for relay systems of this kind. We also described their practical realization and constructed the mathematical model for the system shown in Fig. 4.2. In Sec. 4.5 we shall present a comprehensive study of the regularities in the development of complex behavior when the operating conditions change.

The output voltage (or current) requirements of semiconductor converters vary significantly, but the following requirements are typical: The converter must provide the prescribed control accuracy combined with a low level of output-voltage (or -current) ripple and a high speed of regulation (see, e.g. [1, 2]). As a rule, these requirements can be fulfilled for a given class of systems only when the systems work in a mode of periodic oscillations, corresponding to a period-1 cycle [1, 2].

Under realistic conditions of operation, smaller or larger parameter changes will always take place. The basic factors affecting the working characteristics of the converters are usually input voltage variations or changes of the load working modes [1, 2]. Such variations often lead to qualitative changes of the dynamics. Sometimes this shows up as a sudden boosting of the amplitude of the alternating component in the output voltage or as a transition to irregular motion [12]. In other cases the changes are accompanied with dynamical alternation or intermittency, where aperiodic oscillations arise from time to time on the background of an apparently stable periodic motion, or one can observe a random switching between motions with different dynamical characteristics. Such transitions may be induced by external noise of very small amplitude [10].

4.2 Two-Parameter Analysis of Dynamical Modes

In accordance with the equivalent circuit given in Fig. 4.1, the mathematical model of a converter with relay control is represented as a system of autonomous differential equations [11]

$$\frac{\mathrm{d}X}{\mathrm{d}t} = G(X);\ X = (x_1, x_2, x_3, x_4)^{\mathrm{T}};\ G = (g_1, g_2, g_3, g_4)^{\mathrm{T}}. \tag{4.1}$$

Here

$$\begin{aligned}
g_1 &= \eta\,(-x_1 - x_2 + \Omega)\,; \\
g_2 &= \gamma\big(x_1 - \frac{1}{2}\big(1 + \mathrm{K}_k^{(\mathrm{F})}(\xi)\big)x_3\big); \\
g_3 &= \frac{\mu\big(1 + \mathrm{K}_k^{(\mathrm{F})}(\xi)\big)}{2}\,x_2 - \nu x_3 - \mu x_4; \\
g_4 &= \lambda\left(\frac{x_3}{\alpha} - \frac{x_4}{\beta}\right);
\end{aligned}$$

$$\begin{aligned}
\mathrm{K}_k^{(\mathrm{F})} &= \mathrm{K}_{k-1}^{(\mathrm{F})}\,\mathrm{sign}\left\{\chi_0 + (-1)^{N_{k-1}}\xi(X)\right\},\ \mathrm{K}_0^{(\mathrm{F})} = -1, \\
N_k &= N_{k-1} + \frac{1 - \mathrm{sign}\{\mathrm{K}_k^{(\mathrm{F})}\mathrm{K}_{k-1}^{(\mathrm{F})}\}}{2},\ N_0 = 0,\ k = 1, 2, \ldots
\end{aligned}$$

$$\xi(X) = U_{\text{ref}} - UX,$$

where

$$x_1 = \frac{R_0 i_0}{E_*},\ x_2 = \frac{U_0}{E_*},\ x_3 = \frac{R_0 i_1}{E_*},\ x_4 = \frac{U_1}{E_*};$$
$$\eta = \frac{R_0}{L_0},\ \Omega = \frac{E_0}{E_*},\ \gamma = \frac{1}{C_0 R_0},\ \mu = \frac{R_0}{L_1},\ \nu = \frac{R_1}{L_1},\ \lambda = \frac{1}{C_1 R_*},$$
$$\beta = \frac{R_{\text{L}}}{R_*},\ \alpha = \frac{R_0}{R_*},\ U = (0, 0, 0, \sigma E_*).$$

i_0, i_1, U_0 and U_1 are currents in the inductance coils and voltages on the condensers of the input and output filters of the converter; L_0, L_1, C_0, C_1, R_0, R_1 are the inductances, capacities, and resistances of the input and output filters; R_{L} is the load resistor; σ is the feedback sensor gain by output voltage; E_0 is the input voltage; and ξ is the deviation signal. Note that if the reference signal varies in time, then Eq. (4.1) becomes non-autonomous.

In order to study the dynamical system (4.1) we will vary the input voltage E_0 and the load resistor R_{L}. In the mathematical model they are represented by the dimensionless parameters Ω and β, respectively. The remaining parameters are supposed to be constant and have the following values: $R_1 = 2.1\ \Omega$, $L_0 = 10^{-3}$ H, $L_1 = 6 \cdot 10^{-3}$ H, $C_0 = 3 \cdot 10^{-6}$ F, $C_1 = 2.5 \cdot 10^{-5}$ F, $U_{\text{ref}} = 2.4$ V, $\sigma = 0.1$, and $\chi_0 = 0.25$ V. For the normalizing voltage and resistance we use $E_* = 1$ V and $R_* = 1\ \Omega$, respectively.

In this way the investigation of the dynamical system (4.1) has been reduced to the study of the properties of the mapping (3.9), (3.10). Hence, we can obtain limit cycles by solving Eq. (3.12), and analyses of the local stability of the periodic solutions can be performed by means of the methods introduced in Sec. 3.4.

Let Π be the set of parameters of the dynamical system. Π spans a q-dimensional space. Every point $P \in \Pi$ will correspond to a definite set of fixed parameter values. Let $X_{\text{c}}(t, P)$ be a locally stable periodic motion of the dynamical system corresponding to the point P, and let $\Pi_h^m \subset \Pi$, $h = 0, 1, 2, \ldots, m \geq 1$ [13] be simply connected sets such that there exists a locally stable periodic motion $X_{\text{c}}(t, P)$ (an m-cycle) that varies continuously with the parameters in the sets Π_h^m for any $P \in \Pi_h^m$.

The sets Π_h^m correspond to different coexisting stationary motions for a particular P. The index h is introduced to distinguish sets which are responsible for motions with the same period m.

One may ask if the following condition is true:

$$\Pi \setminus \bigcup_{h,m} \Pi_h^m = \emptyset? \tag{4.2}$$

If this is the case, the system only displays stable periodic motions (m-cycles) (or aperiodic motions associated with a narrowing of the *inner basin of attraction* and the presence of noise [14] (see also [15])). The concept of an inner basin of attraction will be discussed in detail in Sec. 4.4 below. If (4.2) is not fulfilled, then there exist parameters for which other types of motion occur, e.g., chaotic or quasiperiodic oscillations. Let us denote the domains of such modes of oscillations as $\tilde{\Pi}$ ($\tilde{\Pi} = \Pi \setminus \bigcup_{h,m} \Pi_h^m$).

Let us appoint one of the sets Π_h^m to be the basic set and denote it as Π_0. We shall refer to the periodic motion, which is determined on Π_0, as the basic mode of operation (see also [1]). The choice of the basic set Π_0 reflects the requirements to the functional quality of the system. It is possible in certain cases that a domain that corresponds to chaotic oscillations, may be used as the basic set. For example, Postnikov [16, 17] describes such a case, where the use of chaotic oscillations as the operating condition substantially improves the quality performance of a control system.

If one of the following conditions is fulfilled

$$\Pi_0 \cap \left[\bigcup_{h,m} \Pi_h^m \setminus \Pi_0\right] = \emptyset,\ \Pi \setminus \bigcup_{h,m} \Pi_h^m = \emptyset$$

or (4.3)

$$\Pi_0 \cap \left[\Pi \setminus \bigcup_{h,m} \Pi_h^m\right] = \emptyset,\ \Pi \setminus \bigcup_{h,m} \Pi_h^m = \emptyset,$$

then there exists only one stable motion for any $P \in \Pi_0$, and this motion is globally stable.

To distinguish the structure of the objects by their dynamical properties, as a generalization of previous investigations [14, 18] Baushev *et al.* [13] introduced the notions of *normal* and *abnormal structures*. The structures of dynamical objects which have the property (4.3) are called normal ([13], see also [19]). Baushev *et al.* [13] emphasized that the notions of normal and abnormal structures are relative and compared these notions to the notions of *sensitivity* and *robustness*. Catastrophic phenomena typically arise from abnormal structures [13, 14, 19]. This is connected with the non-uniqueness of the motions and with the possibility of hard intersections of the sets $\tilde{\Pi}$ and Π_h^m with the basic set Π_0.

A procedure for normalization of the structure of a control system with pulse-width modulation was considered by Baushev *et al.* [13]. The dynamics of a similar system was studied in detail in [13, 14, 19]. A more precise definition of the notion of a "catastrophe" [20] was also given by Baushev *et al.* [13]. We cite this definition to show what the term catastrophe, as used in the present book, means.

Definition 4.1. Let a dynamical system (and the mathematical model that describes it) have the stationary motions $X_c^{(1)}$, $X_c^{(2)}$,... of static or periodic type. Some quantitative characteristics Θ_1, Θ_2,... may be associated with each motion. The inequality $\Theta_i < \Theta_j$ is then considered to imply that the motion $X_c^{(i)}$ is better than the motion $X_c^{(j)}$ in terms of some qualitative feature, which is determined by the characteristics of the two motions. By a *catastrophe* we mean the fulfilling of the following set of conditions:

(a) the trajectory in phase space crosses from the basin of attraction of one motion $X_c^{(i)}$ to the basin of attraction of another motion $X_c^{(j)}$. This transfer may occur due to the loss of stability for $X_c^{(i)}$ under a small parameter variation or as a result of a minute disturbance whose value is comparable with the radius of the basin of attraction;

(b) $\Theta_i \ll \Theta_j$, this includes the possibility of self-destruction of the object;

(c) the transition from the motion $X_c^{(i)}$ to the motion $X_c^{(j)}$ occurs suddenly.

Abnormal structures may be said to be *unsound* when these conditions are satisfied. Thus, the problem of qualitative design is closely associated with the construction of normal structures [13, 14].

The above notions will now be used to analyze the dynamic peculiarities of a relay system with hysteresis. The choice of control parameters out of the collection of all relevant parameters will be based on common experience from engineering practice. The set Π can be represented as $\Pi = \Pi_{\text{int}} \oplus \Pi_{\text{ext}}$ [13]. This means that if

$$\begin{aligned} P_{\text{int}} &= (p_1^{\text{int}}, p_2^{\text{int}}, \ldots, p_d^{\text{int}}) \in \Pi_{\text{int}}, \\ P_{\text{ext}} &= (p_1^{\text{ext}}, p_2^{\text{ext}}, \ldots, p_s^{\text{ext}}) \in \Pi_{\text{ext}}, \end{aligned} \qquad d + s = q,$$

then

$$P = (P_{\rm int}, P_{\rm ext}) \in \Pi.$$

Here, $\Pi_{\rm int}$ and $\Pi_{\rm ext}$ are the sets of *internal* and *external parameters*, respectively. A distinction between these types of parameters was applied, e.g., by Krasovski [21] and by Baushev *et al.* [13]. It suffices here to note that the external parameters $P_{\rm ext} = (p_1^{\rm ext}, p_2^{\rm ext}, \ldots, p_s^{\rm ext}) \in \Pi_{\rm ext}$ represent the working conditions and the properties of the surroundings, and that the range of variation of these parameters often is much larger than the same range for the internal parameters.

As suggested by Baushev *et al.* [13] we chose the external conditions as control parameters. The remaining parameters are regarded as fixed, and it is supposed that their values are chosen so as to provide the required control quality at the nominal values of the external parameters. So from now on by Π we mean the set $\Pi_{\rm ext}$.

Let the normalized input voltage Ω $(= E_0/E_*)$ and the normalized load resistance β $(= R_{\rm L}/R_*)$ be the external parameters of the dynamical system (4.1) with a range of variation $\Pi = \big\{(\Omega, \beta) : \Omega_{\min} < \Omega < \Omega_{\max}, \beta_{\min} < \beta < \beta_{\max}\big\}$. The division of parameter space for our relay control system (4.1) into domains of different modes of oscillations is shown in Fig. 4.3. The figure presents only those domains which are relatively large. The bifurcation curves divide Π into parts in a complex manner, forming an original mosaic of the domains $\tilde{\Pi}$ and Π_h^m, $h = 0, 1, 2, \ldots$, $m \geq 1$ and their intersections. The sets $\tilde{\Pi}$ and Π_h^m and their intersections correspond to domains of different shadings in Fig. 4.3.

For the sake of simplicity we shall consider only the domains $\tilde{\Pi}$ and Π_h^m. To make the analysis easier we first consider Π_h^m individually and then their collections. The sets Π_h^1, $h = 0, 1, 2, 3$ correspond to the regions of parameter space, where a locally stable 1-cycle exists (Fig. 4.4) and the sets Π_j^2, $j = 0, 1, 2$ correspond to the presence of a 2-cycle (Fig. 4.5). Let Π_0^1 be the basic set. Figure 4.6 indicates the collection Π_h^1, $h = 0, 1, 2, 3$; Π_j^2, $j = 0, 1, 2$, where A, B, C, D denote bifurcation nodes, i.e., points of intersection of the bifurcation curves. The basic set Π_0 is bounded by the bifurcation curve N_+ on the left side, where stable 1-cycle merges with an unstable cycle and disappears in a saddle-node bifurcation. The period-doubling bifurcation curves N_- form the parts AB and CD of the boundary of the set Π_0, and the bifurcation curves N_φ form the parts BC and DE. When crossing N_φ, the 1-cycle bifurcates into a two-dimensional torus. The domain Π_1^1 is bounded by the curve N_φ on the left

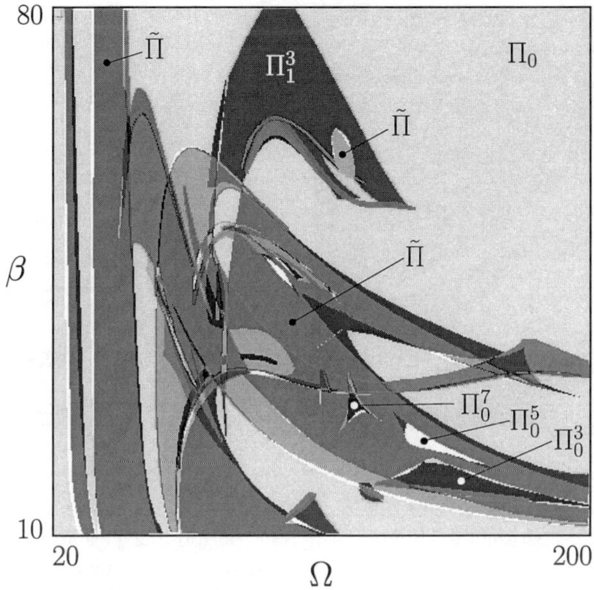

Figure 4.3. Regions with different dynamical behaviors for a voltage converter. The parameters Ω and β represent the input voltage and the load resistance, respectively. Π_0 is the basic set where the system normally operates. Complex behavior typically arises when the load resistance or input voltage become relatively low.

and by the curve N_- on the right.

The basic set Π_0 and the set Π_0^2 are separated by a period-doubling bifurcation curve N_-. The same is the case for Π_1^1 and Π_1^2. It is not difficult to see that all sets Π_j^2, $j = 0, 1, 2$ intersect with Π_0, where $\left(\Pi_0 \cup \Pi_0^2 \cup \Pi_2^2\right) \cap \left(\Pi_1^1 \cup \Pi_1^2\right) \neq \emptyset$ and $\Pi_1^1 \cap \Pi_1^2 = \emptyset$. The domain, that corresponds to the intersections of Π_0^2 and Π_1^2, includes other sets, which correspond to motions different from the motions in Π_h^1 and Π_j^2 ($j = 0, 1, 2$; $h = 0, 1, 2, 3$) such as 3-, 6-cycles, chaotic and quasiperiodic oscillations.

The set $\tilde{\Pi}$ consists of the domains with quasiperiodic and chaotic modes and their intersections (Fig. 4.7). In Sec. 4.3 we shall consider the set that is denoted by $\hat{\Pi}$ in Fig. 4.7 in more detail. The internal structure of $\hat{\Pi}$ proves to be very complicated, it has many windows with regular dynamics, some of which begin with m-cycles, arising in saddle-node bifurcations.

Figure 4.7 also presents the set $\hat{\Pi}$ consisting of the domains which correspond to stable 5-, 7-, 9-, 10-, 11-, 14-, 21-, 25-, 27-, 50-...-cycles and the domains

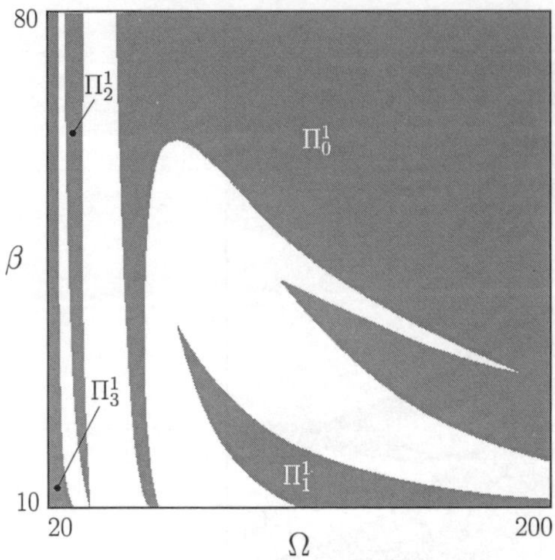

Figure 4.4. Domains of existence for different period-1 cycles in the parameter plane.

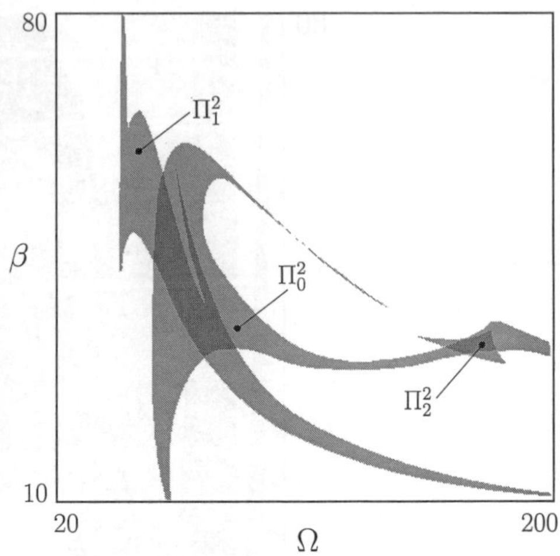

Figure 4.5. Domains of existence for different period-2 cycles in the parameter plane. The darker regions are regions of coexistence of two different 2-cycles.

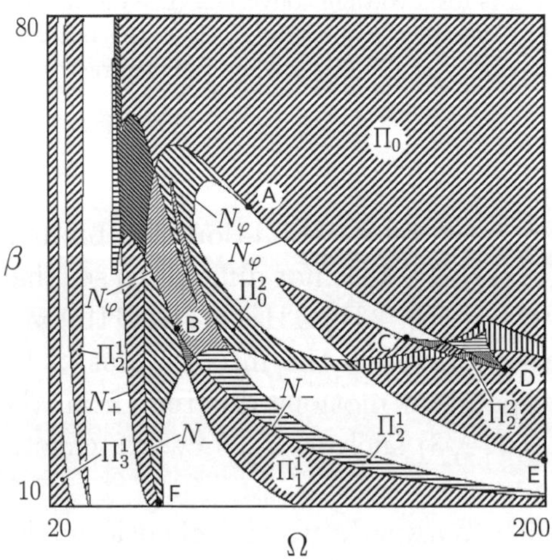

Figure 4.6. Collection of the sets that represent the domains of existence of period-1 and period-2 cycles.

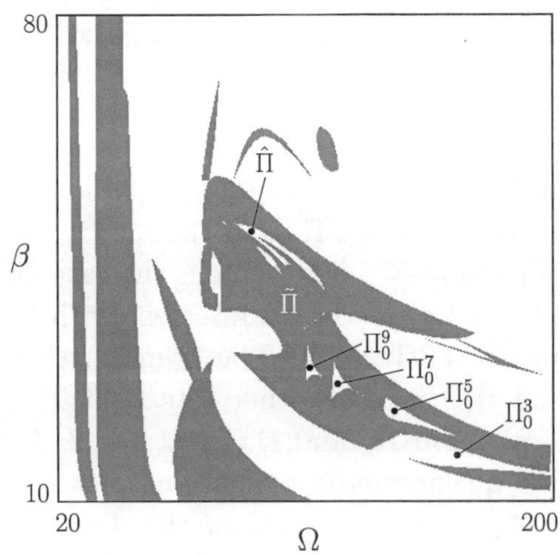

Figure 4.7. Collection of the sets that correspond to the domains of existence of period-3, -5, -7, -9 cycles, and of quasiperiodic and chaotic oscillations.

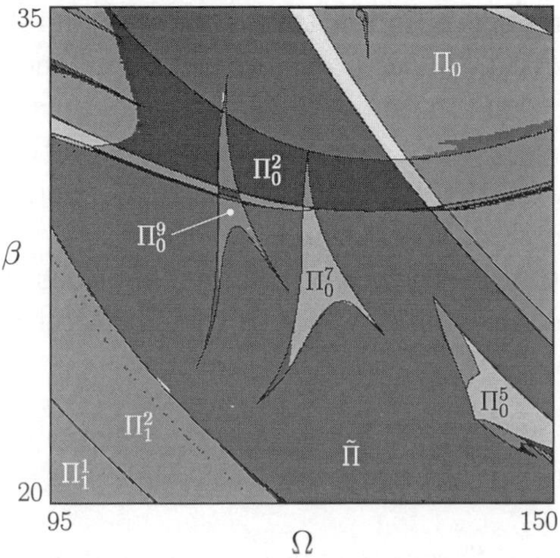

Figure 4.8. Magnified view of that part of the diagram in Fig. 4.3 where we find the sets Π_0^5, Π_0^7, and Π_0^9.

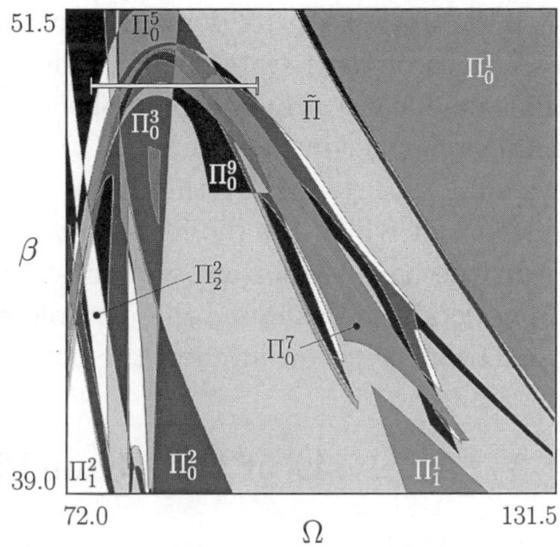

Figure 4.9. Magnified view of the set $\hat{\Pi}$ as denoted in Fig. 4.7. This set is seen to have a very complicated structure.

of their intersections. The sets Π_0^3, Π_0^5, Π_0^7, Π_0^9, correspond to such values of the parameters where a stable 3-, 5-, 7-, 9-cycle, respectively, exists. Figure 4.8 gives a magnified view of the part of Fig. 4.3 that contains the domains Π_0^5, Π_0^7, Π_0^9. Another part of Fig. 4.3 is shown magnified in Fig. 4.9. This region is considered in detail below.

Now let us make the distinction

$$\Pi_0 = \Pi_{\text{conv}} \cup (\Pi_0 \setminus \Pi_{\text{conv}}),$$

where Π_{conv} is that part of Π_0 in which a unique motion exists. The set Π_{conv} is referred to as the *domain of convergence* [13]. It may be simply connected, not simply connected (as in Fig. 4.3), or an empty domain. Non-uniqueness of behavior occurs when the parameters correspond to the set $\Pi_0 \setminus \Pi_{\text{conv}}$.

Let Π_{conv} be a non-empty set. Π_{conv} coincides with Π_0 for normal structures. In practice it is often more important to know only the radius of the domain of convergence rather than the actual form of the domain. Let $P_0 \in \Pi_0$ be the basic (nominal) values of the parameters [13]. It should be noted that cases may arise where $P_0 \notin \Pi_{\text{conv}}$.

Definition 4.2. The *radius of the domain of convergence* r_{conv} is the minimal distance from the basic point P_0 to the boundary of Π_{conv} [13].

If the radius of the domain of convergence is small, then nearly negligible variations of the parameters or minute disturbances of the operational conditions may lead to catastrophic phenomena and manifest themselves in terms of sudden changes in the dynamics. This can be very dangerous for an operating system. As revealed by the above analysis, the structure of the relay control system, for which the model (4.1) was derived, is abnormal.

In the next section we will consider the occurrence of different motions in more detail and examine the possible scenarios for changes in the character of oscillations with variation of the parameters.

4.3 Analysis of the Branching Patterns

Our study of the dynamical system (4.1) in different regions of parameter plane has shown that the complication of oscillations with variation of the parameters takes place according to the following scenarios:

(a) a cascade of period-doubling bifurcations leading to aperiodic motions;

(b) hard transitions from some periodic motions to others;

(c) the sudden emergence of multiple stable motions with subsequent period-doubling bifurcations;

(d) incomplete cascades of "direct" and "reverse" period-doubling bifurcations;

(e) hard transitions from periodic motions to chaotic or quasiperiodic oscillations and vice versa;

(f) the appearance and destruction of quasiperiodic oscillations.

The concrete sequence of changes in the nature of oscillations with variation of the parameters depends on the choice of section in Π or, in other words, on the trajectory of deformation. To illustrate this, let us make an analysis for the two different sections:

1. $\{(\Omega, \beta) : 56.0 \leq \Omega \leq 132.0; \beta = 31.0\}$;
2. $\{(\Omega, \beta) : 75.0 \leq \Omega \leq 95.0; \beta = 49.5\}$.

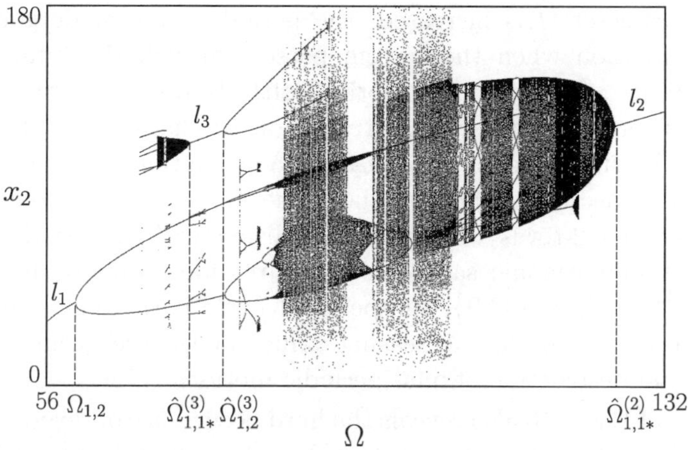

Figure 4.10. Bifurcation diagram for $\beta = 31$ and $56.0 \le \Omega \le 132.0$. l_1, l_2, and l_3 denote different domains of existence for a period-1 cycle.

Consider first section 1. One of the interesting peculiarities of the transition of our relay system from motions of one type to motions of another type under continuous variation of Ω is illustrated in the bifurcation diagram of Fig. 4.10. Here, the value of the control parameter Ω is plotted along the abscissa and the normalized voltage x_2 $(= U_0/E_*)$ along the vertical axis. The figure shows not only the transition from regular motion to chaos through a cascade of period-doubling bifurcations, which is typical of many dissipative systems, but also some nontrivial properties of the dynamical system (4.1).

Inspection of the diagram reveals three intervals where a stable 1-cycle exists (see Figs. 4.3 and 4.4):

$$\begin{aligned} l_1 &= \{\Omega : 56.0 < \Omega < \Omega_{1,2}\}; \\ l_2 &= \{\Omega : \hat{\Omega}^{(2)}_{1,1*} < \Omega < 132.0\}; \qquad l_1, l_2 \in \Pi_0,\ l_3 \in \Pi_1^1. \\ l_3 &= \{\Omega : \hat{\Omega}^{(3)}_{1,1*} < \Omega < \hat{\Omega}^{(3)}_{1,2}\}, \end{aligned}$$

Complication of the basic motion (existing in the interval l_1) with increasing Ω takes place via an accumulating sequence of period-doubling bifurcations that leads to aperiodic motion at some threshold value of the parameter Ω $(\Omega \approx 84.2)$. This is followed by a domain of chaoticity. A quite different pattern is observed in the evolution of the 1-cycle which exists in the range $l_3 = \{\Omega : \hat{\Omega}^{(3)}_{1,1*} < \Omega < \hat{\Omega}^{(3)}_{1,2}\}$.

As the parameter Ω is increased, the period-1 cycle undergoes a period-doubling bifurcation when the system passes through the bifurcation value $\Omega = \hat{\Omega}^{(3)}_{1,2}$. Next, a quasiperiodic motion softly branches off from the 2-cycle when $\Omega \approx 90$, and the subsequent transition to chaos occurs through torus destruction. In the same way, a quasiperiodic motion arises from from the 1-cycle when Ω decreases through the value $\Omega^{(3)}_{1,1*}$. This is followed by the destruction of the 2-torus through the collision with a period-5 saddle cycle born in a subharmonic saddle-node bifurcation. Finally, in the interval $l_2 = \{\Omega : \hat{\Omega}^{(2)}_{1,1*} < \Omega < 132.0\}$, the period-1 cycle is the only stable solution. The transition to chaos for decreasing values of Ω here occurs through the appearance and destruction of quasiperiodic motions.

Inspection of Fig. 4.10 also reveals the hard occurrence of several other stable motions, such as the period-3, and -6 cycles, etc. The subsequent bifurcations of these motions complicate the character of oscillations. So that coexistence of motions takes place over wide ranges of Ω.

Let us dwell a little on these observations. The results of our numerical calculations for the existence of stable m-cycles are reported in Table 4.1. In accordance with the presentation we introduced in Chapter 2, the first column contains a designation of the branch to which the domain of stability of given m-cycle belongs; the second column gives the periodicity of the m-cycle; and the last column contains the ranges of Ω where the locally stable m-cycle exists.

It is easy to check that the condition (3.24) is fulfilled for the dynamical system (4.1). Hence the updating matrix M_k in the expression for the basic matrix does not depend on the values of ϑ_1 and ϑ_2, but is uniquely determined by Eq. (3.22) whatever choice is made for the values of ϑ_1 and ϑ_2.

It is convenient to represent the data of Table 4.1 as a branching diagram. In this diagram the value of the active parameter Ω is plotted on the horizontal axis, and a value that characterizes the complexity of the motion is plotted on the vertical axis. As a measure of complexity it is natural to regard either the period T of a periodic motion (of m-cycle) or the value m. In our original work [14, 18, 19], the value m was used as such a measure for the non-autonomous dynamical system .

When presenting the results of the investigations in such a way, many peculiarities in the change of character of the oscillations, which may seem mysterious in the two-parameter diagrams (Figs. 4.3–4.7), attain a more transparent character.

Branch	m	Ω
$V_{1,1}$	1	$54.198 < \Omega < 59.740$
	2	$59.740 < \Omega < 78.261$
	4	$78.261 < \Omega < 82.840$
	8	$82.840 < \Omega < 83.911$
	16	$83.911 < \Omega < 84.155$
	32	$84.155 < \Omega < 84.209$
$V_{2,1}$	2	$101.159 < \Omega < 120.921$
	4	$97.805 < \Omega < 101.159$
	8	$96.606 < \Omega < 97.805$
$V_{6,1}$	6	$73.852 < \Omega < 75.451$
	12	$75.451 < \Omega < 75.774$
$V_{3,1}$	3	$80.857 < \Omega < 82.215$
	6	$80.263 < \Omega < 80.857$
	12	$80.103 < \Omega < 80.263$
$V_{8,1}$	8	$70.849 < \Omega < 71.366$

Table 4.1. List of intervals with periodic solutions for $\beta = 31.0$ and $54.0 \leq \Omega \leq 132.0$.

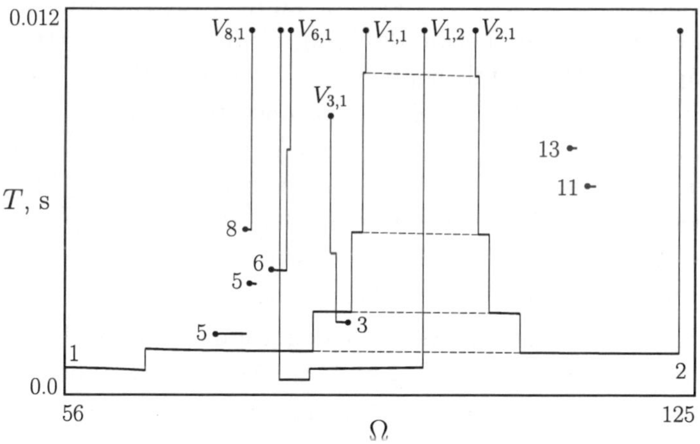

Figure 4.11. Branching pattern for $\beta = 31$ and $56.0 \leq \Omega \leq 132.0$.

The designation $V_{i,j}$ is introduced for the branches such that the index i indicates that the i-cycle is minimum for $V_{i,j}$, and that is there are no k-cycles within $V_{i,j}$ with $k < i$. The maximum value of j denotes the number of observed branches, having the same i.

The branch $V_{1,1}$ begins with the 1-cycle, which is born in a saddle-node bifurcation when $\Omega = \Omega_{1,1}$ ($\Omega_{1,1} \approx 54.198$). At $\Omega = \Omega_{1,2}$ ($\Omega_{1,2} \approx 59.74$) the stable 1-cycle becomes unstable and a stable 2-cycle softly arises. The stable 2-cycle becomes unstable in a new period-doubling bifurcation at $\Omega = \Omega_{1,4}$ ($\Omega_{1,4} \approx 78.261172$). This leads to the soft occurrence of a stable 4-cycle, and so on. The branch $V_{1,1}$ consists of the intervals $\{\Omega : \Omega_{1,2^{i-1}} < \Omega < \Omega_{1,2^i}\}$, $i = 1, 2, \ldots$, $\Omega_{1,1} < \Omega_{1,2} < \cdots < \Omega_{1,2^i} < \cdots < \Omega_{1,*}$.

The parameters $\Omega_{1,2^i}$, $i = 1, 2, \ldots$ correspond to the loss of stability of the 2^{i-1}-cycle and to a soft occurrence of the 2^i-cycle (as one of the multipliers of the 2^{i-1}-cycle emerges from the circumference of the unit circle through -1). When i increases a geometric narrowing of the intervals of existence of the 2^i-cycle occurs, so that $V_{1,1}$ has the limiting value $\Omega_{1,*} \approx 84.2$, which corresponds to aperiodic motion. This is the so-called *accumulating point*. When $\Omega > \Omega_{1,2^i}$, all the 2^{i-1}-cycles still exist, but only as unstable cycles of saddle type.

A more detailed analysis of the evolution of the saddle-shaped 2^i-cycles within the limits of local stability theory shows that along with $\Omega_{1,2^i}$, $i = 1, 2, \ldots$ there exist parameter values $\Omega_{2,2^i} > \Omega_{1,2^i}$, ($\Omega_{2,2^{i+1}} < \Omega_{2,2^i}$, $\Omega_{2,2^i} \notin V_{1,1}$), $i = 2, 3 \ldots$ such that with $\Omega > \Omega_{2,2^i}$ all saddle 2^i-cycles become locally stable as a result of "reverse" period-doubling bifurcations [22] (the numerically largest multiplier of the saddle-shaped 2^i-cycle enters the unit circle through -1). Here, $\Omega_{2,2^i}$, $i = 2, 3 \ldots$ are the parameters of the period-doubling bifurcations, and $\{\Omega : \Omega_{2,2^{i+1}} < \Omega < \Omega_{2,2^i}\}$, $i = 1, 2, \ldots$ are the intervals of parameter values where the 2^i-cycles are locally stable. The branch $V_{2,1}$ is an aggregate of the intervals $\{\Omega : \Omega_{2,2^{i+1}} < \Omega < \Omega_{2,2^i}\}$, $i = 1, 2, \ldots$, $\Omega_{2,*} < \cdots < \Omega_{2,2^i} < \cdots < \Omega_{2,4} < \Omega_{2,2}$, with $\Omega_{2,*} \approx 96.65$ being the accumulating point of the branch. This phenomenon is illustrated in Fig. 4.12.

Figure 4.13 presents the results of a calculation of the local stability for the 4-cycle. At the bifurcation point $\Omega_{1,8} \approx 82.840544$ the multiplier $\rho_4(\Omega)$ of the 4-cycle becomes equal to -1. Later, as the parameter Ω increases, $\rho_4(\Omega)$ emerges from the unit circle through -1, a stable 8-cycle arises softly, and the 4-cycle becomes a saddle. When the parameter Ω passes through the value $\Omega_{2,8} \approx 97.80592$, $\rho_4(\Omega)$ enters the unit circle through -1, and the 4-cycle becomes stable again. The intervals $\{\Omega : \Omega_{1,2^{i+1}} < \Omega < \Omega_{2,2^{i+1}}\}$, $i = 1, 2, \ldots$, where the 2^i-cycles are unstable, are shown on the branching pattern by dashed lines (see also Fig. 4.12).

Let us examine another phenomenon displayed by the solutions to the dynamical system (4.1). This phenomenon is associated with the period-1 and -2 motions. Let us start with an examination of the evolution of the 1-cycle that

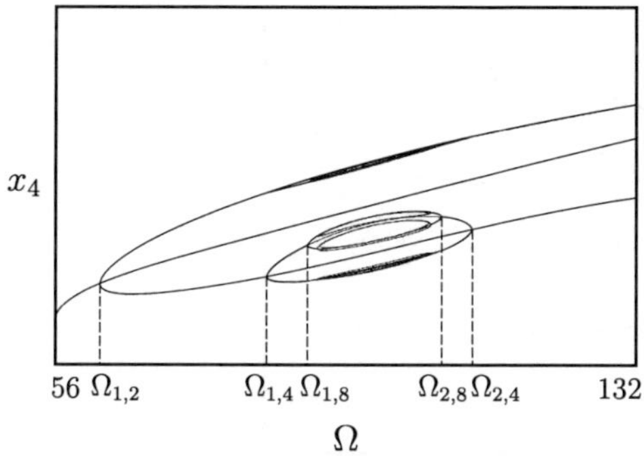

Figure 4.12. Cascade of direct and reverse period-doubling bifurcations for the diagram shown in Fig. 4.10.

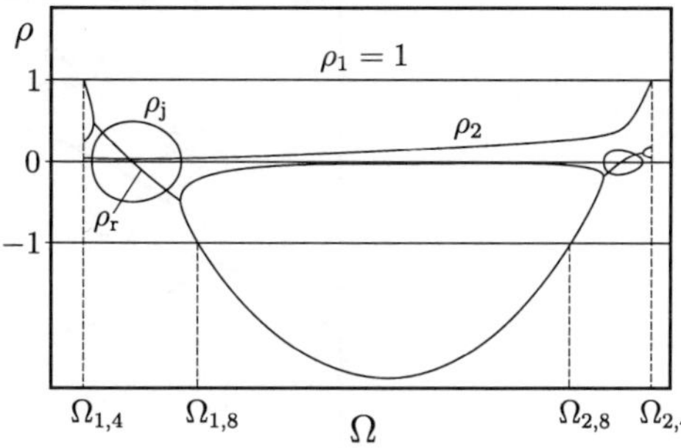

Figure 4.13. Variation of the multipliers of the 4-cycle which undergoes direct and reverse period-doubling bifurcation with the parameter Ω.

arises in a hard bifurcation at $\Omega = \Omega_{1,1}$ ($\Omega_{1,1} \approx 54.198$). With $\Omega > \Omega_{1,2}$, one of the multipliers of the 1-cycle emerges from the unit circle through -1, and the 1-cycle becomes a saddle. With increasing Ω, a pair of complex-conjugate roots of the Eq. (3.15) appears $\rho_{3,4} = \rho_{\mathrm{r}} \pm j\rho_{\mathrm{i}}$ ($|\rho_{3,4}| > 1$). At the point $\Omega = \hat{\Omega}^{(2)}_{1,1*}$ the multipliers $\rho_{3,4} = \rho_{\mathrm{r}} \pm j\rho_{\mathrm{i}}$ enter the unit circle, and the 1-cycle is stabilized (see Fig. 4.10). As previously noted, $\hat{\Omega}^{(2)}_{1,1*}$ corresponds to a value of the parameter Ω, where the 1-cycle bifurcates into a two-dimensional torus.

With further increase in Ω, namely when $\Omega = \hat{\Omega}^{(2)}_{1,1}$, the 1-cycle (see $X^{(1)}_{\mathrm{c}}$ in Fig. 4.14) undergoes a new period-doubling bifurcation. This corresponds to the point of intersection of the trajectory of deformation with the bifurcation curve connecting the nodes C and D (Fig. 4.6). The period-1 cycle remains a saddle until $\Omega = \hat{\Omega}^{(2)}_{1,2}$ (see $X^{(1)}_{\mathrm{c}*}$ in Fig. 4.14), at which point the largest (in absolute value) multiplier of the 1-cycle enters the unit circle through -1, and the 1-cycle again becomes stable. The transition that takes place at $\Omega = \hat{\Omega}^{(2)}_{1,2}$ is referred to as a *subcritical period-doubling bifurcation.*

When $\Omega = \hat{\Omega}^{(2)}_{1,2\oplus}$, a 2-cycle $X^{(2)}_{\mathrm{c}}$ (Fig. 4.14), which has appeared from the 1-cycle as a result of a period-doubling bifurcation, meets with the unstable 2-cycle $X^{(2)}_{\mathrm{c}*}$ and disappears in a saddle-node bifurcation as the largest (in absolute value) multiplier of the 2-cycle becomes equal $+1$.

In this way an interval $\{\Omega : \Omega^{(2)}_{1,2} < \Omega < \Omega^{(2)}_{1,2\oplus}\}$ exists, where two stable motions $X^{(1)}_{\mathrm{c}}$ and $X^{(2)}_{\mathrm{c}}$ coexist (Fig. 4.14). On the boundaries of this interval one may observe a hard change in the character of the oscillations with hysteresis

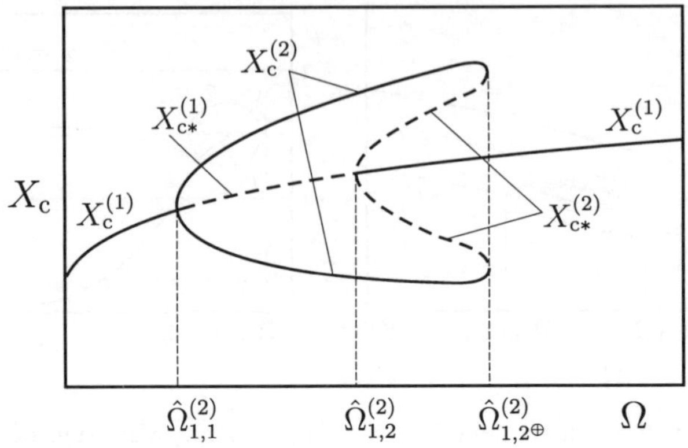

Figure 4.14. Subcritical period-doubling leading to the coexistence of the stable period-1 and period-2 cycles near the boundaries of the domain $\Pi_0 \cap \Pi_2^2$. This coexistence can give rise to hysteretic transitions.

as it is typical for subcritical transitions. For example, if one moves along the parameter from $\hat{\Omega}_{1,2}^{(2)}$ in the direction of its increase, then with $\Omega > \hat{\Omega}_{1,2\oplus}^{(2)}$ the system jumps to the mode, corresponding to the 1-cycle. Analogously, if the system follows the stable 1-cycle with decreasing Ω and passes through the bifurcation value $\hat{\Omega}_{1,2}^{(2)}$, where the stable 1-cycle joins with the unstable 2-cycle, one may observe the reverse transition from the 1-cycle to the 2-cycle.

It is evident, that before a periodic motion loses its stability (or disappears by combining with an unstable motion), the basin of attraction of this mode or, more exactly, the radius of this basin becomes quite small, and random disturbances, which are always present, can switch the system into another mode even before the radius of the basin of attraction vanishes [20]. This is similar to the phenomenon discussed for transonic flutter in aircraft wings in Chapter 1, except that the phenomenon in that case was associated with a subcritical Hopf bifurcation of the equilibrium point.

Now let us examine the evolution of the 2-cycle that arises softly from the 1-cycle as a result of a period-doubling bifurcation when $\Omega = \Omega_{1,2}$. When the parameter Ω increases the loss of stability of the 2-cycle is accompanied by the soft emergence of a 4-cycle. In this case the 2-cycle continues to exist but only as a saddle. When passing through the value $\Omega = \Omega_{2,4}$, the 2-cycle becomes stable again. This takes place via a reverse bifurcation (Fig. 4.15). When Ω continues to increase, the stable 2-cycle again loses its stability. A

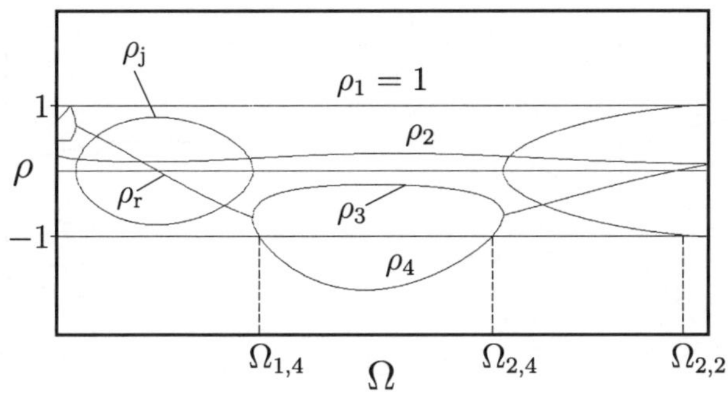

Figure 4.15. Multiplier diagram illustrating the appearance of quasiperiodic oscillation from a period-2 cycle.

two-dimensional torus arises softly when, at $\Omega = \Omega_{2,2} \approx 120.9214$, the pair of complex-conjugate multipliers of the 2-cycle leaves the unit circle (see Fig. 4.15). The 2-cycle continues to exist as an unstable focus for all $\Omega > \Omega_{2,2}$.

Figures 4.16 and 4.17 show three-dimensional projections of the phase trajectories and power spectra of the various oscillations that arise through the period-doubling bifurcations $1 \to 2 \to 4 \to 8 \to 16 \to 32$ within the branch $V_{1,1}$.

Like the branch $V_{1,1}$, other branches start with periodic cycles that are born in hard bifurcations or they correspond to the domains of existence of resonance cycles that arise by synchronization of two oscillatory modes. For example, the branch $V_{6,1}$ starts with the 6-cycle. A pair of 6-cycles (a stable cycle and an unstable cycle) are born in a saddle-node bifurcation (see Table 4.1). When the parameter Ω increases, the stable 6-cycle undergoes a cascade of period-doubling bifurcations accumulating at the value $\Omega_{6,*} \approx 75.7738$.

We have also observed domains where an *incomplete cascade* of period-doubling bifurcations is realized. At $\Omega = \Omega_{1,3} \approx 80.8577$, the 3-cycle is born in a hard bifurcation. When the parameter Ω decreases the 3-cycle undergoes two period-doubling bifurcations: $3 \to 6 \to 12$. At $\Omega = \Omega_{1,12^{\oplus}}$, one of the multipliers of the 12-cycle becomes equal $+1$, the stable 12-cycle and the unstable 12-cycle join and disappear. The intervals $\{\Omega : \Omega_{1,12^{\oplus}} < \Omega < \Omega_{1,12}\}$, $\{\Omega : \Omega_{1,12} < \Omega < \Omega_{1,6}\}$, $\{\Omega : \Omega_{1,6} < \Omega < \Omega_{1,3}\}$, $\Omega_{1,12^{\oplus}} < \Omega_{1,12} < \Omega_{1,6} < \Omega_{1,3}$ form the branch $V_{3,1}$.

As previously noted, an incomplete cascade of period-doubling bifurcations is a characteristic phenomenon for piecewise-smooth systems. Where piecewise-

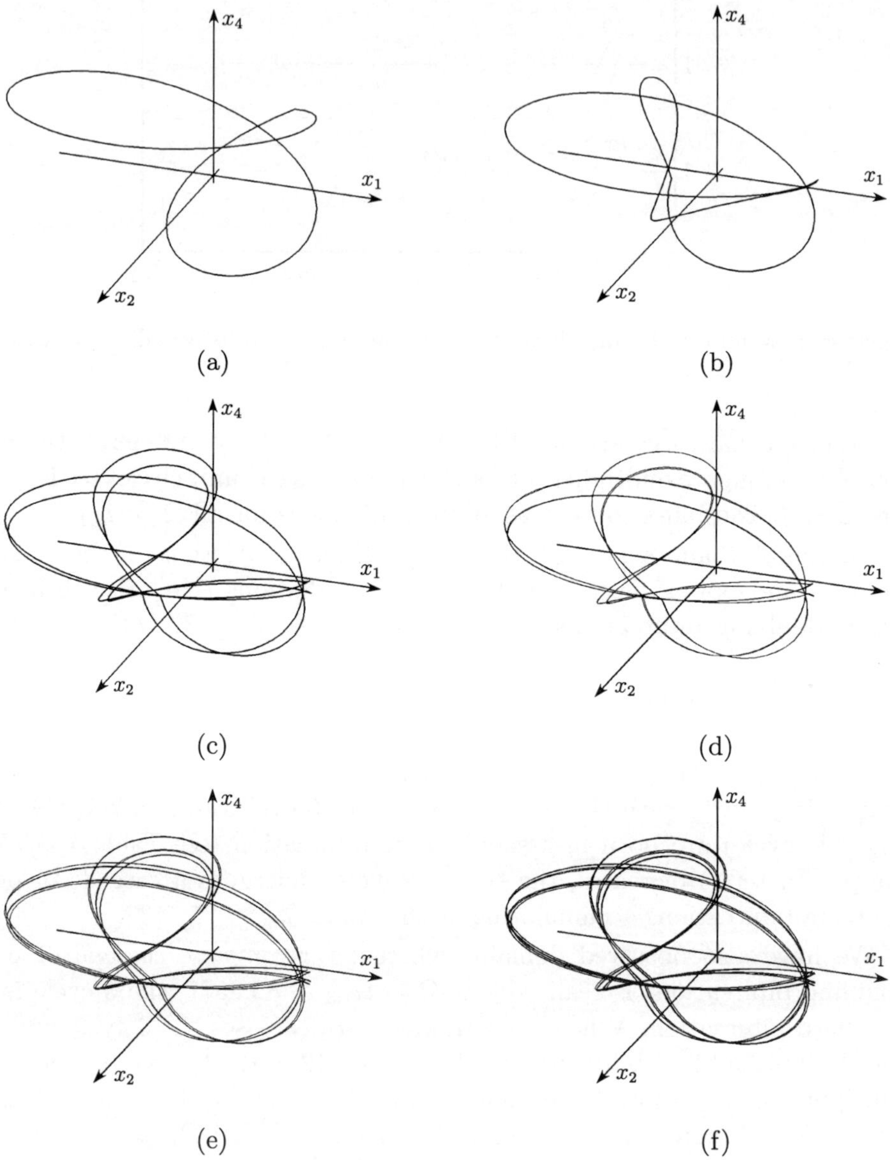

Figure 4.16. Three-dimensional projections of the phase trajectories of the various oscillations. The figure illustrates the period-doubling cascade $1 \to 2 \to \cdots \to 32$ ((a)–(f), respectively) for the branch $V_{1,1}$.

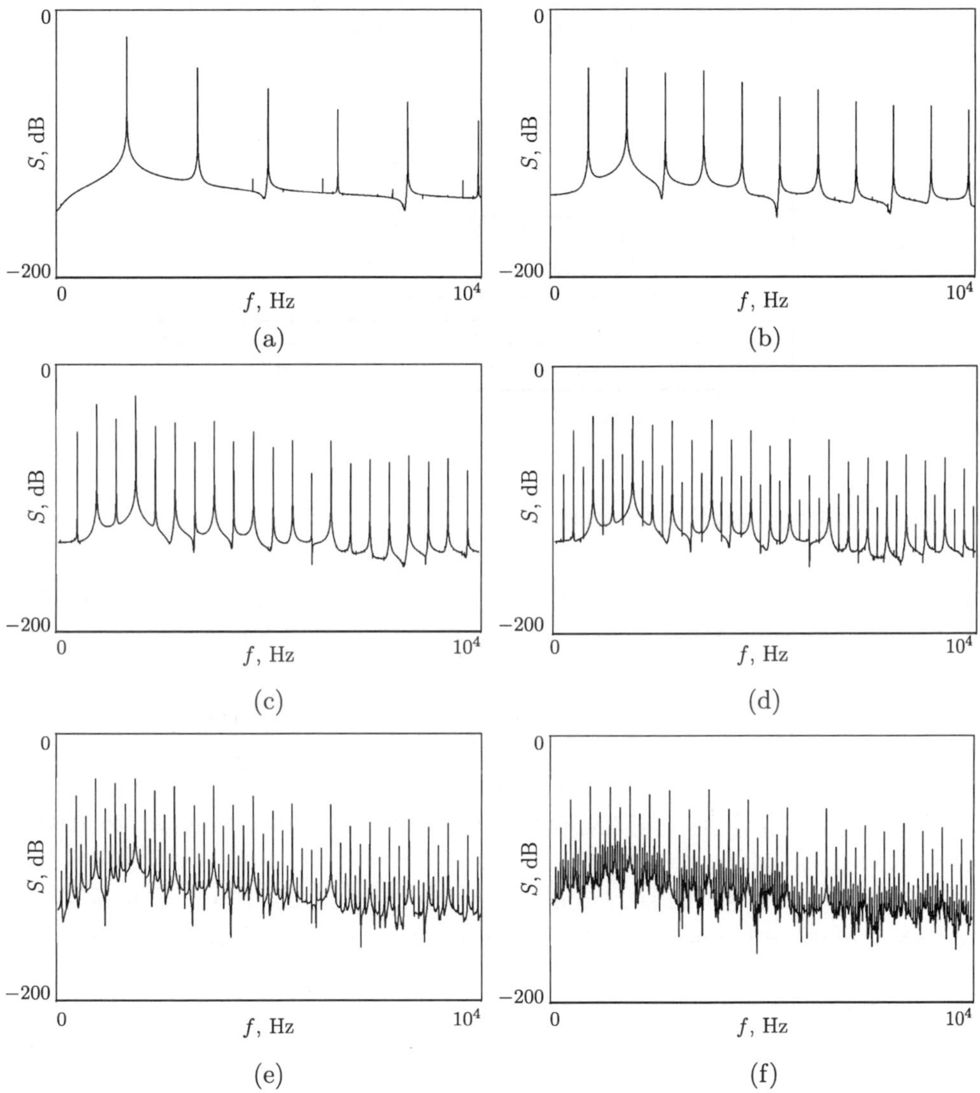

Figure 4.17. Power spectra of the various oscillations. Each period-doubling bifurcation leads to the appearance of a new subharmonic component in the spectrum $1 \to 1/2 \to 1/4 \to 1/8 \to 1/16 \to \ldots$, etc.

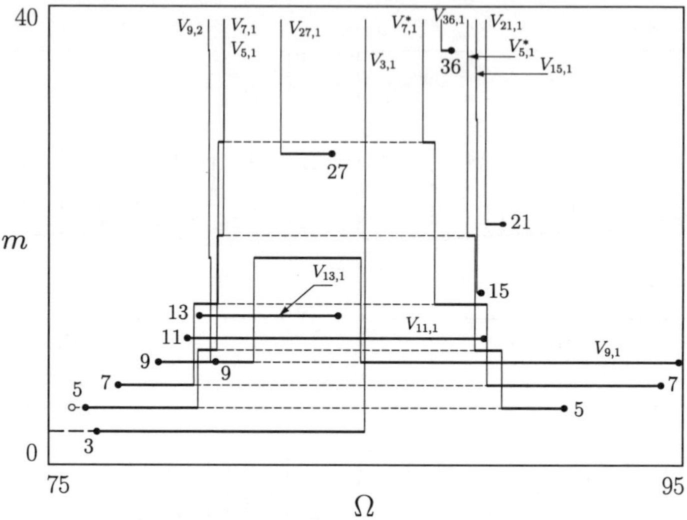

Figure 4.18. Branching pattern for $\beta = 49.5$ and $75.0 \leq \Omega \leq 95.0$.

linear systems only display a single period-doubling and smooth systems display an infinite cascade, a finite sequence is often observed for piecewise-smooth systems.

The range of existence of some branches is so narrow that these branches can be illustrated in Fig. 4.17 only by vertical lines. There are many windows with narrow ranges of Ω in the domain of quasiperiodicity where resonance cycles are realized. The branching pattern shows only those windows that are relatively wide.

Let us now turn to the second section, defined on page 178 and depicted in Fig. 4.9 as a line segment. The results of our numerical calculations are reported in Table 4.2 and are presented in Fig. 4.18 as a branching pattern. Unlike the previously considered section (Fig. 4.11) where we used the periodicity of the motion as a measure of complexity, we here use the value of m as such a measure. Table 4.2, in addition to the columns composing Table 4.1, contains a column indicating the nature of the loss of stability for the m-cycle at the boundary points of the branch. This column consists of two subcolumns corresponding to the left (with the minimum value of the parameter Ω) and the right (with the maximum value of the parameter Ω) branch boundaries. The designations used in this column are:

Branch	m	Ω	Boundary point nature	
			Left	Right
$V_{3,1}$	3 6	$67.36 < \Omega < 84.97732$ $84.97732 < \Omega < 84.97745$	$*$	$+$
$V_{5,1}$	5 10 20 40	$76.01 < \Omega < 79.76$ $79.76 < \Omega < 80.39$ $80.39 < \Omega < 80.57$ $80.57 < \Omega < 80.61$	$*$	$-$
$V^*_{5,1}$	5 10 20	$89.21 < \Omega < 91.17$ $88.36 < \Omega < 89.21$ $88.14 < \Omega < 88.36$	$-$	$+$
$V_{7,1}$	7 14 28	$77.31 < \Omega < 79.65$ $79.65 < \Omega < 80.41$ $80.41 < \Omega < 80.59$	$+$	$-$
$V^*_{7,1}$	7 14 28	$88.73 < \Omega < 94.18$ $87.11 < \Omega < 88.73$ $86.74 < \Omega < 87.11$	$-$	$+$
$V_{9,1}$	9 18 9	$84.84 < \Omega < 94.92$ $81.53 < \Omega < 84.84$ $78.54 < \Omega < 81.53$	$+$	$+$
$V_{9,2}$	9 18 36	$80.17 < \Omega < 80.32$ $80.11 < \Omega < 80.17$ $80.09 < \Omega < 80.11$	$-$	$+$
$V_{11,1}$	11	$79.44 < \Omega < 88.68$	$+$	$+$
$V_{13,1}$	13	$79.87 < \Omega < 84.13$	$+$	$+$
$V_{15,1}$	15 30	$88.43 < \Omega < 88.49$ $88.40 < \Omega < 88.43$	$-$	$+$
$V_{21,1}$	21	$88.68 < \Omega < 89.24$	$-$	$+$
$V_{27,1}$	27	$82.34 < \Omega < 83.96$	$-$	$+$
$V_{36,1}$	36	$87.32 < \Omega < 87.62$	$-$	$+$

Table 4.2. List of the main intervals with periodic solutions along the line segment shown in Fig. 4.9.

"+" for disappearance (or birth) via a saddle-node bifurcation, and

"−" for a period-doubling bifurcation and part of a cascade accumulating at a parameter value corresponding to aperiodic motion.

The symbol "$*$" indicates two specific cases considered in detail below.

This section consists of thirteen branches all completely included in the range $75.0 \leq \Omega \leq 95.0$. There are some branches that intersect this region without being entirely included in it. Such branches are not considered. All branches composing this pattern have some common properties, namely:

- Each branch $V_{m,k}$ is an aggregate of ranges where stable $2^n m$-cycle exists, for $n = 0, 1, \ldots, r$. The value of r may be finite or infinite.
- If the branch consists of more than one such range, i.e. $r \geq 1$, then in all internal boundary points between the ranges where the stable $2^n m$-cycle and $2^{n+1}m$-cycle exist, the $2^n m$-cycle loses its stability as the $2^{n+1}m$-cycle softly arises via a period-doubling bifurcation.

Besides by the quantitative parameters m and k, the branches $V_{m,k}$ differ qualitatively by the nature of the stability loss at the external boundary points, i.e., the point of appearance for the limit cycle with the smallest m and the point of disappearance for the cycle with the largest m within the limits of the branch. On this basis all the branches can be divided into the following classes [11]:

(i) The branches $V_{9,2}$, $V_{15,1}$, $V_{21,1}$, $V_{27,1}$ and $V_{36,1}$ begin from cycles arising via saddle-node bifurcation. When decreasing the parameter Ω the m-cycle undergoes a cascade of period-doubling bifurcations leading to aperiodic motion.

(ii) The branches $V_{11,1}$ and $V_{13,1}$ both begin and end with parameter values at which stable and unstable cycles join and disappear in a saddle-node bifurcation.

(iii) The branch $V_{9,1}$ (similar to the branches $V_{11,1}$ and $V_{13,1}$) begins and ends with parameter values $\Omega_{9,1} \approx 78.5385$ and $\Omega^*_{9,1} \approx 94.9232$ where the 9-cycle arises (disappears) via a saddle-node bifurcations. But, unlike the previous case, this mode undergoes only two period-doubling bifurcations with changing Ω, first a direct bifurcation (at $\Omega = \Omega_{18,1} \approx 81.5327$), and then a reverse one (at $\Omega = \Omega^*_{18,1} \approx 84.8455$). The bifurcation and multiplier diagrams shown in Fig. 4.19(a,b) clearly illustrate this structure. The branch $V_{9,1}$ consists of two symmetrical half-branches (symmetrical not by the values of parameter Ω but by their structure). These half-branches consist of a single segment where a 9-cycle exists and are connected by a segment corresponding to the domain of existence of an 18-cycle.

Connected with this branch is another peculiarity: The branches $V_{9,1}$ and $V_{9,2}$ have non-empty intersections, so that two 9-cycles with different dy-

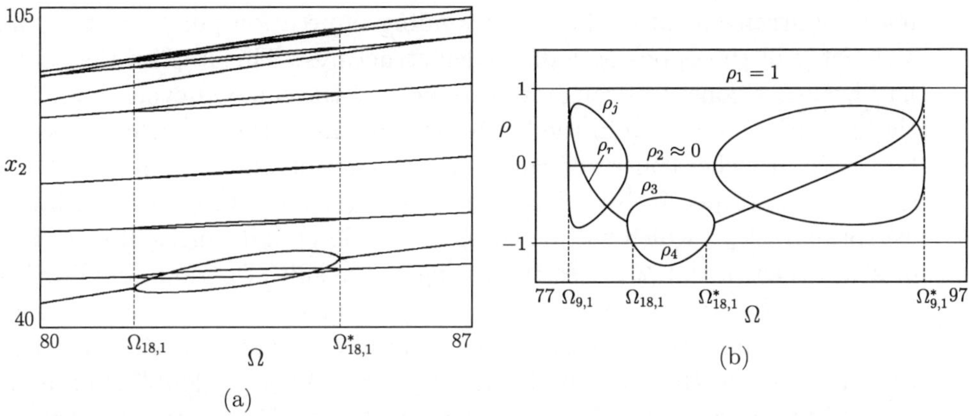

Figure 4.19. Bifurcation (a) and the multiplier (b) diagrams illustrating the bifurcations within the branch $V_{9,1}$.

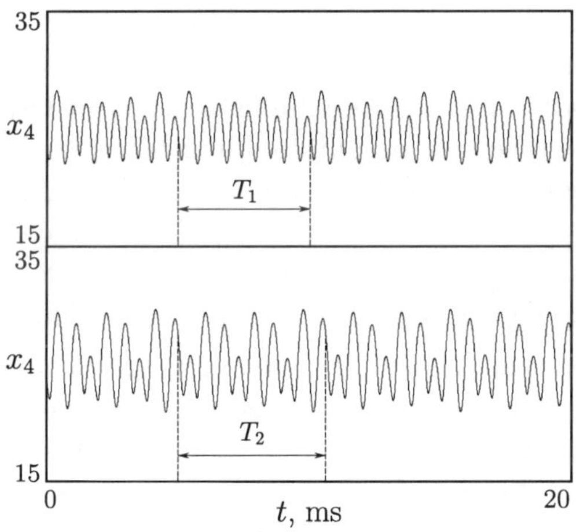

Figure 4.20. Temporal variations of two different period-9 cycles existing for $\beta = 49.5$ and $80.1733 \leq \Omega \leq 80.3162$.

namic characteristics exist within the range $80.1733 < \Omega < 80.3162$. Figure 4.20 shows the temporal variation of the first (top curve) and the second (bottom curve) 9-cycle.

(iv) The branches $V_{5,1}$, $V^*_{5,1}$ and $V_{7,1}$, $V^*_{7,1}$ may pairwise be considered to constitute a single branch consisting of two symmetrical half-branches (see Fig. 4.18). Here again the symmetry is to be understood, not as a symmetrical arrangement of the corresponding bifurcation points, but as the symmetry of the opposing half-branch structures. This implies that both half-branches consist of ranges of existence of the same m-cycles and that the transitions involved in the births and deaths of the m-cycles are similar. Both branches consist of ranges of existence of a basic m-cycle ($m = 5$ and $m = 7$, respectively) arising via a saddle-node bifurcation (with the exception of $V_{5,1}$, which will be considered below) and ranges of existence of $2^n m$-cycles, $n = 1, 2, \ldots$ softly arising as a result of period-doubling bifurcations. It is worth mentioning, that the corresponding m-cycles from the left and the right half-branch not only have the same periodicity, but they are one and the same mode, undergoing two period-doubling bifurcations (direct and reverse) in its range of existence. The ranges of existence of unstable cycles are shown by dashed lines in the branching pattern of Fig. 4.18. The evolution of each m-cycle is a replica of the evolution observed for the branch $V_{9,1}$ and described above. It is seen that the branch $V_{9,1}$ can be considered as a degenerate case of a junction of the half-branches with a structure similar to that just described, only with an incomplete cascade of period-doubling bifurcations. Since an infinite cascade of period-doubling bifurcations occurs within the branches $V_{5,1}$, $V^*_{5,1}$ and $V_{7,1}$, $V^*_{7,1}$, it can be said that the half-branches $V_{5,1}$, $V^*_{5,1}$ and $V_{7,1}$, $V^*_{7,1}$ are connected by the range of existence of aperiodic oscillations in the same way as the half-branches of $V_{9,1}$ are connected by a period-18 range.

The branch $V_{5,1}$ complies with the above description with a single exception, which is related to the mechanism of the onset of the basic motion. The multiplier diagram shown in Fig. 4.21 illustrates the nature of birth for the basic motion. The top diagram is meant to show the point of the stability loss for the 5-cycle $\Omega_{5,1} \approx 76.2485$. This cycle arises at $\Omega^q_{5,1} \approx 76.0104$. At this point, the maximum multiplier (in absolute value) is greater than one ($\rho_{\max} \approx 1.0685$). With an increase of the parameter Ω, this multiplier becomes complex and, with a pair of complex-conjugate multipliers located outside the boundaries of the unit circle, the motion remains locally unstable right up to $\Omega_{5,1}$.

(v) The branch $V_{3,1}$ begins with a 3-cycle arising in a hard bifurcation. The nature of appearance of this motion has certain peculiarities in comparison with the usual saddle-node bifurcation. The multiplier diagrams are presented in Fig. 4.22(a) (near the point of birth) and in Fig. 4.22(b) (in

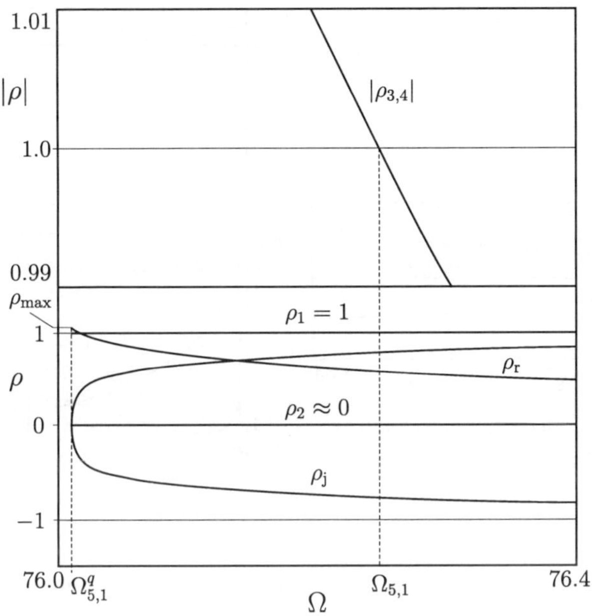

Figure 4.21. Mechanism of birth for the 5-cycle that starts the branch $V_{5,1}$.

the whole region of existence for the 3-cycle). The pair of unstable 3-cycles arise in the point $\Omega^u_{3,1} \approx 67.3586$. With increasing Ω one of the 3-cycles becomes stable (its multipliers are plotted as solid lines in Fig. 4.22(a)), with the multipliers of the other 3-cycle (dashed line in Fig. 4.22(a)) moving further and further away from the unit circle, such that this motion remains unstable in its whole range of existence. When the motion arises all the multipliers are real but after a small increase of the parameter Ω two multipliers become complex conjugate, remaining outside of the boundary of the unit circle up to the value $\Omega_{3,1} \approx 76.8338$, where the other motion becomes locally stable. Figure 4.22(b) shows the Ω-dependence of the multipliers (bottom diagram) and the absolute values of the multipliers (top diagram) for the 3-cycle. The points of formation and stabilization are clearly seen. The first 3-cycle at $\Omega_{6,1} \approx 84.9773$ undergoes a period-doubling bifurcation. This leads to the soft appearance of a 6-cycle which then disappears via a saddle-node bifurcation at $\Omega \approx 84.9774$.

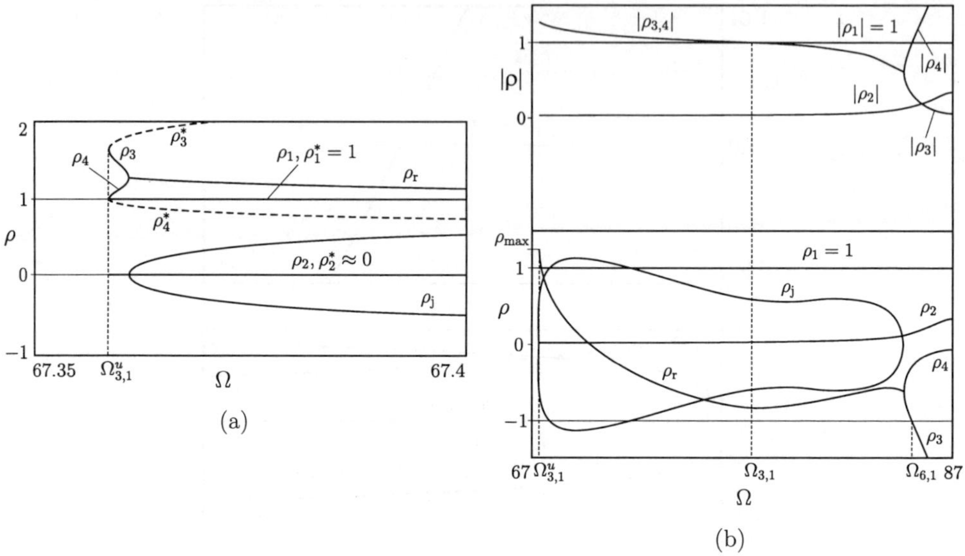

Figure 4.22. Multiplier diagram showing the transitions of the 3-cycle that belongs to the branch $V_{3,1}$.

Let us consider the inner structure of domains Π_h^m to understand this complexity of branches. Most of the domains Π_h^m in Fig. 4.9 look like stripes (Π_0^2, Π_1^2, Π_2^2) or tongues (Π_0^5, Π_0^7, Π_0^9). A sketch of the structure of the latter is presented in Fig. 4.23. As shown, this domain is the infinite aggregate of the domains of existence of the periodic modes $\hat{\Pi}_h^{2^{n-1}m}$, $n = 1, 2, \ldots$ The "external" boundary of the domain is the saddle-node bifurcation curve N_+. The "internal" boundary of the domain of existence of the basic m-cycle is the period-doubling curve N_-. The domains of existence of $2^n m$-cycles, $n = 1, 2, \ldots$ are structurally similar to that of the basic m-cycle. With increasing n the domain rapidly becomes narrower until it reaches the accumulation curve N_* which is the "internal" boundary of the domain Π_h^m. This is an example of the simple organization of the period-doubling transition in two-parameter systems discussed in Chapter 2.

Figure 4.23(a,b,c) shows the possible arrangement of the section and the tongue-like domain. If the section intersects the accumulation curve N_* (Fig. 4.23(a)) then the branch corresponding to this domain consists of two half-branches each including a complete period-doubling cascade. Hence, it belongs to the class **(iv)**. The mutual arrangement of the considered section and the domains Π_0^5 and Π_0^7 (Fig. 4.24(b,e)) causes all the above mentioned features

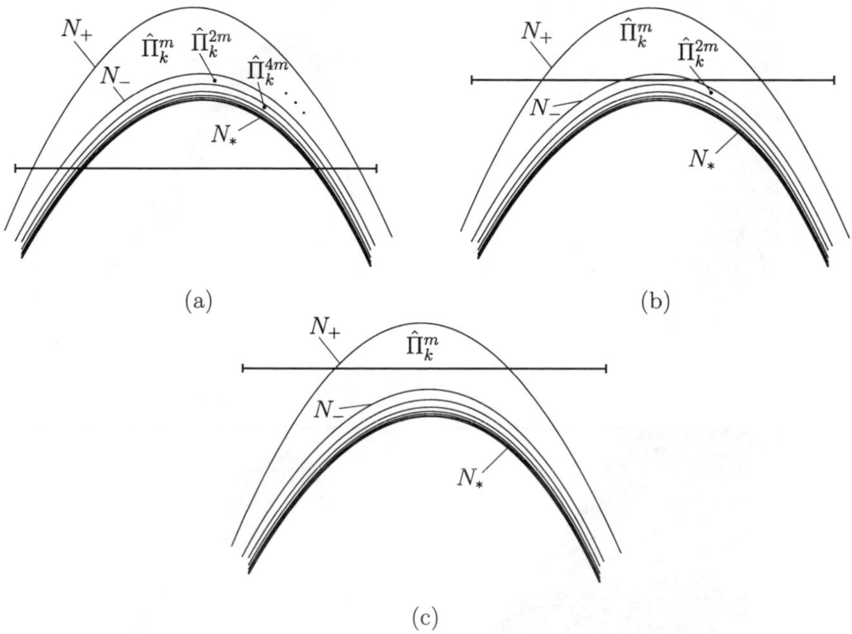

Figure 4.23. Sketch of the structure of the two-parameter period-doubling cascades with possible sections.

for the branches of class **(iv)**.

Figure 4.23(b) represents the case when the section does not intersect the accumulation curve and passes not only through the domain of existence of the basic motion $\hat{\Pi}_h^m$ but also through a finite number of the domains of existence of the derived motions $\hat{\Pi}_h^{2^n m}$, $n = 1, 2, r$; $r < \infty$ (in shown case $r = 1$). With such mutual arrangement the corresponding branch belongs to class **(iii)**. Figure 4.24(d) shows the mutual arrangement of the considered section and the domain Π_0^9 corresponding to this case.

Finally, if the section intersects only the domain of existence of the basic motion $\hat{\Pi}_h^m$ as it is shown in Fig. 4.23(c), then the corresponding branch begins and ends with the hard appearance of this motion (via a saddle-node bifurcation). Hence, the branch does not include other modes and it belongs to class **(ii)**.

The remaining domains that look like stripes have the following features:

- the domain Π_0^3 (by contrast to the above description) consists of a finite number of domains of existence for the periodic motions, namely of two, $\hat{\Pi}_0^3$ and $\hat{\Pi}_0^6$. The left boundary of $\hat{\Pi}_0^3$ (the left boundary of Π_0^3) and the

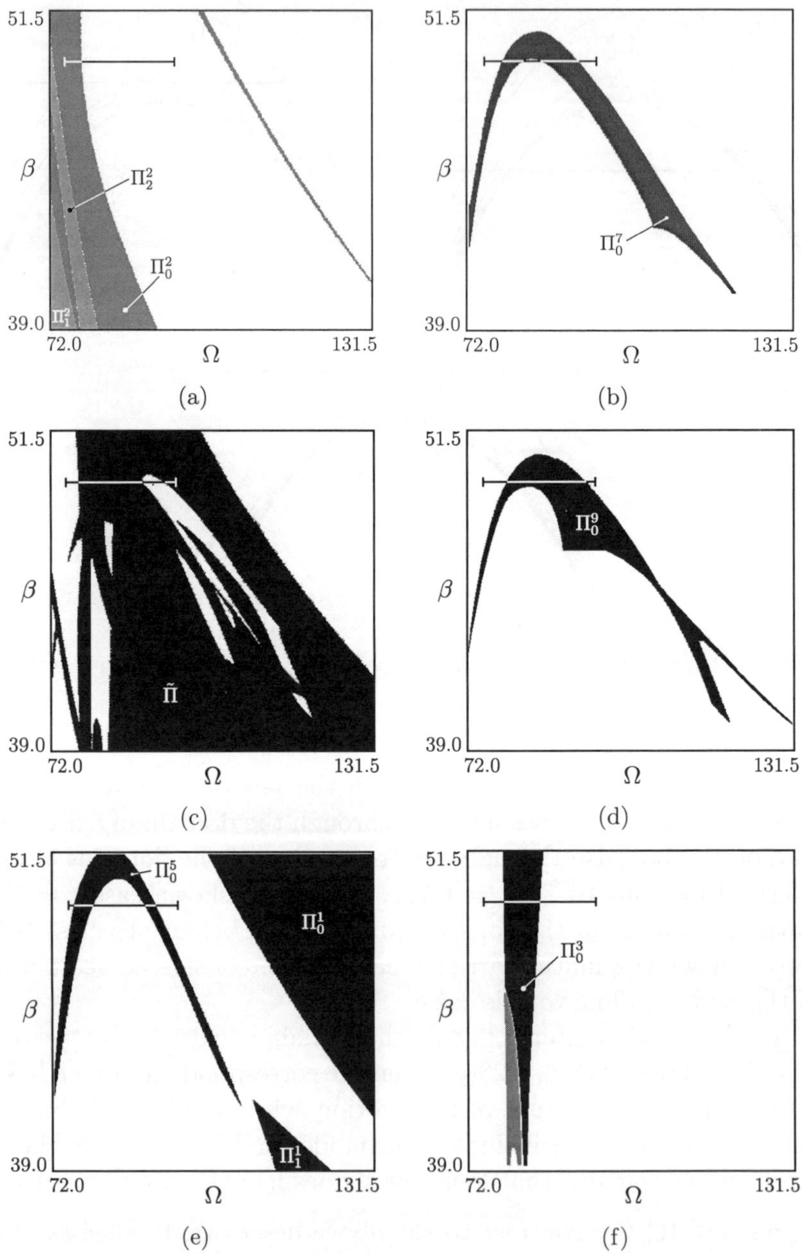

Figure 4.24. Mutual arrangement of the section $75.0 \leq \Omega \leq 95.0$, $\beta = 49.5$ and the domains shown in Fig. 4.9.

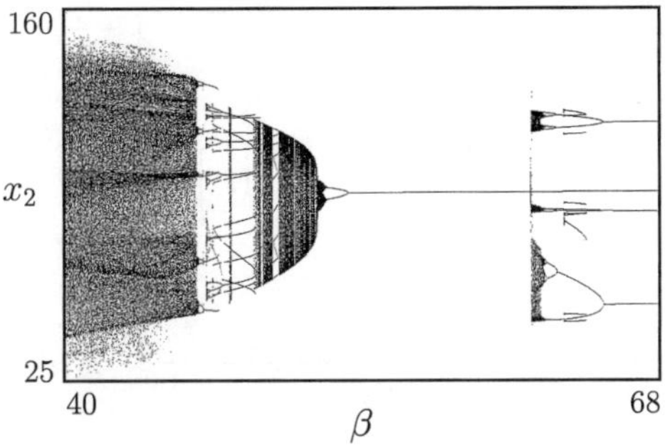

Figure 4.25. Bifurcation diagram for Ω = 95.77228 and $40.0 \leq \beta \leq 68.0$ (emergence and destruction of a quasiperiodic motion).

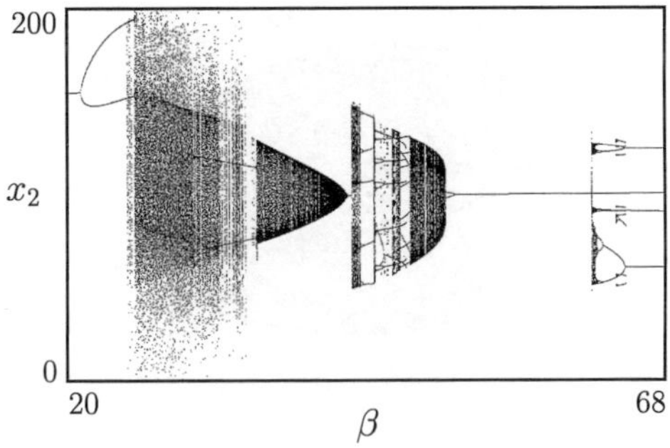

Figure 4.26. Bifurcation diagram for Ω = 101.37129 and $20.0 \leq \beta \leq 68.0$ (emergence and destruction of a quasiperiodic motion).

right boundary of $\hat{\Pi}_0^6$ (the right boundary of Π_0^3) are the curves N_+. The domain $\hat{\Pi}_0^6$ is very narrow. In the section corresponding to Fig. 4.18, for instance, it is about 0.0013 wide.

- the right boundary of Π_0^2 corresponds to the formation of a quasiperiodic motion (curve N_φ).

Figures 4.25 and 4.26 illustrate a possible scenario for the transition to chaos via the emergence of a quasiperiodic motion corresponding to a two-dimensional torus in phase space and, with change of a parameter, its subsequent destruction.

The results of our analyses of the branching patterns and Figs. 4.11 and 4.18 show that there are intervals of parameter Ω, where quasiperiodic and chaotic oscillations coexist with periodic motions. The forms of phase trajectories projected into a three-dimensional space and the spectra of oscillations corresponding to modes of chaotic and quasiperiodic oscillations are shown in Figs. 4.27 and 4.28. In this domain the motion of the dynamical system may be periodic, quasiperiodic or chaotic in character.

4.4 Chaotization of Oscillations in Relay Systems

Neimark and Landa [15] introduced such notions as "*transformer of stochasticity*" and "*amplifier of stochasticity*", by which they referred to the response of dynamical systems with random impacts. The distinction between a trans-

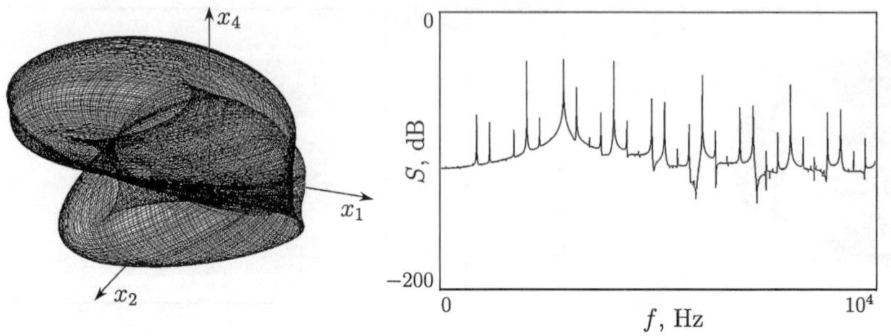

Figure 4.27. Three-dimensional projection and power spectrum of quasiperiodic oscillations.

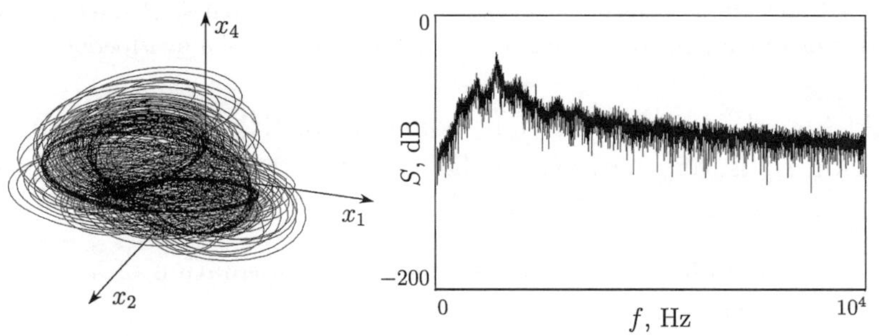

Figure 4.28. Three-dimensional projection and power spectrum of chaotic oscillations.

former of stochasticity and an amplifier of stochasticity lies in the degree of amplification, which may be significantly higher for amplifiers of stochasticity than for transformers of stochasticity. As noted by Neimark and Landa [15, p. 57], nonregularity and chaoticity of motions may be presented "as the combined effect of regular and random components, i.e., as a disturbance of a more or less expressed regular component by some random value." One cause of amplification of stochasticity could be the existence of several asymptotically stable limit cycles in phase space. A motion along one of the cycles might then switch to another cycle because of fluctuations [10, 15].

Numerous results of experiments show that this type of "quasistochastic" [22] oscillations are common in real systems [15, 22–24]. Here the transition to irregular (or non-determinate) dynamics is determined by small random disturbances, which nevertheless exceed some threshold value [14, 15, 18, 22, 25, 26].

Let $X_c(t)$ be the stable m-cycle and let $\varepsilon(t) = X(t) - X_c(t)$ be the perturbation, which is determined by the action of a disturbing factor and by the properties of the dynamical system (4.1) itself. As in [18] we assume that the disturbing factor acts at discrete moments, so that $\varepsilon(t)$ represents the dynamic characteristics of the system with the exception of these moments. As in the papers by Baushev *et al.* [14, 18], by the disturbance magnitude we mean

$$\delta = \max_t \|\varepsilon(t)\|.$$

There is non-uniqueness of m-cycles in the range of existence of $V_{1,1}$ on Ω. The analysis of the dynamic characteristics of the control system with pulse-width modulation [14, 18] has shown that at some definite relations between the radii of the basins of attraction of the stable m-cycles and δ, *non-determinate dynamics* occurs. In the paper by Baushev and Zhusubaliyev [18] this was demonstrated for parameters of a system with pulse-width modulation corresponding to the coexistence of a stable 1-cycle and a stable 3-cycle.

Let us assume that there exist p stable periodic motions having the radii of the basins of attraction R_1, R_2, …, R_p for the chosen parameters.

Figure 4.29 gives a phase portrait of the dynamical system (4.1) projected onto a three-dimensional space for a case where several locally stable periodic motions exist for the same parameters. Integral curves corresponding to 3-, 11-, 33-, 36-cycles are denoted by numbers 3, 11, 33 and 36, respectively.

Different relations between the radii of the basins of attraction and δ will lead to motions of different types.

For example, if

$$\max_t \|\varepsilon(t)\| > R_i,\ i = 1, 2, \ldots, p,$$

then randomness of the disturbing factor may lead to non-determinate dynamics.

If

$$R_i < \max_t \|\varepsilon(t)\| < R_1,\ i = 2, 3, \ldots, p,$$

then it is possible to obtain either determinate or non-determinate dynamics depending on the initial conditions.

If with

$$\max_t \|\varepsilon(t)\| > R_i,\ i = 1, 2, \ldots, p,$$

$$R_1 \gg R_i,\ i = 2, 3, \ldots, p,$$

then a catastrophic transition is possible from determinate to non-determinate dynamics and vise versa.

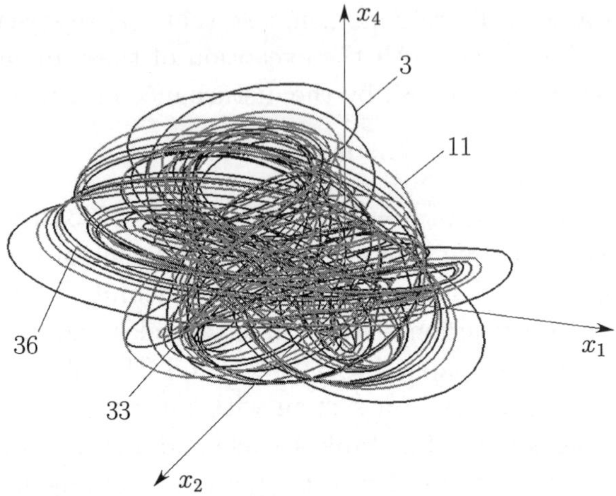

Figure 4.29. Phase portrait of the dynamical system (4.1) projected onto a three-dimensional space when several locally stable periodic motions exist with the same parameters.

A second cause of "amplification of stochasticity" is less evident as it is connected with the possibility of noise-induced jumps from one part of the integral curve to another for the same periodic motion [10, 15]. A detailed analysis of the properties of high-period motions of one-dimensional maps preceded consideration of this mechanism [15] (see also [22]), and it proved quite difficult to evaluate the mechanism quantitatively. This evaluation was made by Baushev *et al.* [14] for the one-dimensional logistic map and subsequently generalized to the dynamics of control systems with pulse-width modulation.

To formulate the results we need some background from the original publication [14]. As mentioned above, Baushev *et al.* [14] started with an analysis of the one-dimensional logistic map

$$\begin{aligned} w_k &= f(\lambda, w_{k-1}),\ k = 1, 2, \ldots,\ 0 < w_k < 1; \\ f(\lambda, w) &= \lambda w(1 - w),\ 1 < \lambda < 4. \end{aligned} \tag{4.4}$$

A branching pattern for the logistic map was built similarly to the construction we described in the previous section of this chapter for the dynamical system (4.1).

Along with well-known concepts such as "basins of attraction", new concepts such as *inner basin of attraction* D_{int} and *radius* R_{int} *of the inner basin of attraction* were introduced [14]. It was shown that non-determinate dynamics arises at a definite relation between R_{int} and the magnitude of the random disturbance. These concepts are analogous to the concepts "global basin of attraction" and "local basin of attraction" applied by Neimark and Landa [15]. Let us consider the results in more detail.

Assume that w_{ck}, $k = \overline{1,m}$ is a stable m-cycle of the logistic map, i.e., the roots of equation

$$w_{\text{c}} - f^{(m)}(w_{\text{c}}) = 0,\ f^{(m)} = \underbrace{f \circ f \circ \cdots \circ f}_{m\ times},$$

where w_{c} is any of w_{ck}, $k = \overline{1,m}$. Let D_w be the basin of attraction of the m-cycle w_{ck} with the radius R_w.

It is evident that

$$R_w \leq \min\bigl(\min_k(w_{ck}), \min_k(1 - w_{ck})\bigr).$$

Now assume that some disturbance ε_0 occurs at an initial node corresponding to $k = 0$ as a result of the action of some disturbing factor, that is

$$w_0 = w_{\text{c}} + \varepsilon_0. \tag{4.5}$$

The full equation with respect to the disturbance ε_0 is written as

$$\varepsilon_k = \lambda\,(1 - 2w_{\text{c}(k-1)} - \varepsilon_{k-1})\,\varepsilon_{k-1},\ k = 1, 2\ldots$$

Hence

$$\varepsilon_m = \varphi(\varepsilon_0)\,\varepsilon_0,\ \varphi(\varepsilon_0) = \lambda^m \prod_{k=1}^{m}(1 - 2w_{\text{c}(k-1)} - \varepsilon_{k-1}),$$

and

$$\varepsilon_{nm} = \left[\prod_{k=1}^{n} \varphi(\varepsilon_{(n-k)m})\right] \varepsilon_0.$$

Let ε_0 be such that $w_0 \in D_w$. In this case $\varepsilon_k \to 0$ when $k \to \infty$. But the character of convergence depends on ε_0. When $|\varepsilon_0|$ is small enough, the disturbance will decay monotonically. It follows from the fact that $\varphi(0) = \rho_m$, where ρ_m is the multiplier of a m-cycle, which is $|\rho_m| < 1$ for a locally stable m-cycle.

If $|\varphi(\varepsilon_0)| > 1$, then at first the disturbance will increase (in absolute value) from cycle to cycle.

k	w_{ck}	k	w_{ck}
0	0.50	10	0.85059
1	0.92054	11	0.46972
2	0.26933	12	0.91675
3	0.72461	13	0.2810
4	0.73476	14	0.74395
5	0.71760	15	0.70140
6	0.74618	16	0.77117
7	0.69737	17	0.64976
8	0.77709	18	0.83795
9	0.63782		

Table 4.3. Superstable period-19 cycle that exists at $\lambda \approx 3.68216782318268$.

Definition 4.3. The collection of $w_0 \in D_w$, which ensures a monotonical diminution of $|\varepsilon_{nm}|$, $n = 1, 2, \ldots$ is called the *inner basin of attraction* D_{int}.

It is evident that $D_{\text{int}} \subset D_w$. Let R_{ik} be the smallest root of the following equation

$$|\varphi(\varepsilon_0)| - 1 = 0. \tag{4.6}$$

The index k is determined by the point w_{ck} chosen in (4.5).

Definition 4.4. By the *radius of an inner basin of attraction* we mean the value

$$R_{\text{int}} = \min_k(R_{ik}),\ k = \overline{1, m}.$$

The value of R_{int} for any known m-cycle may be obtained numerically from the solution of the Eq. (4.6). In particular, the radius for one of the stable 19-cycles (given in Table 4.3) was computed to be $R_{\text{int}} \approx 2.6045 \cdot 10^{-9}$ for $\lambda \approx 3.68216782318268$ (at this parameter value cycle is superstable) [14].

Some numerical experiments were carried out by Baushev *et al.* [14] with the above mentioned λ for a mapping with the disturbance

$$w_k = \lambda w_{k-1}(1 - w_{k-1}) + \delta\gamma_k. \tag{4.7}$$

Here, $-1.0 < \gamma_k < 1.0$ are uniformly distributed random numbers, and the term $\delta\gamma_k$ plays the role of the disturbing factor. Let us note that the disturbing factor and the disturbance ε_k are different. But near the m-cycle we may assume that $\varepsilon_k \approx \delta\gamma_k$. So the magnitude of the disturbance is determined by the choice of δ.

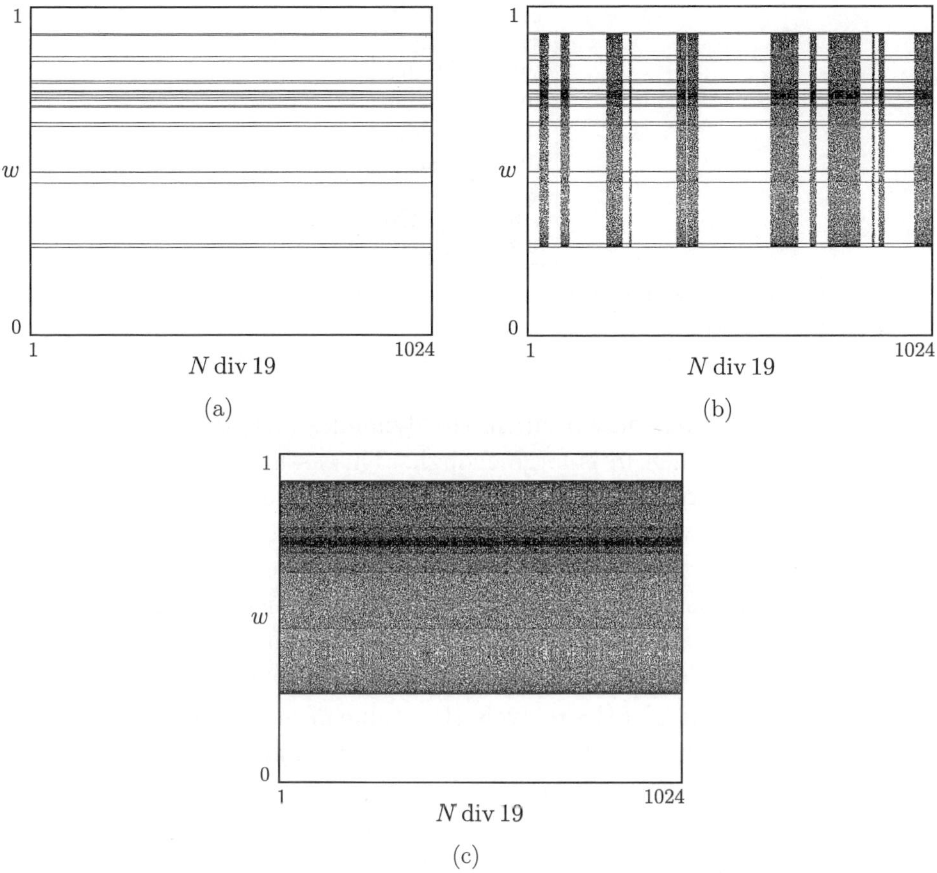

Figure 4.30. Dynamics of the logistic map with noise (4.7) for different values of δ: (a) $\delta \approx 2.5 \cdot 10^{-9}$, (b) $\delta \approx 4 \cdot 10^{-9}$, (c) $\delta \approx 10^{-7}$.

Inspection of Fig. 4.30 clearly reveals the indeterminate (chaotic) dynamics for $\delta \approx 10^{-7}$ (c) and the convergence to a 19-cycle when $\delta \approx 2.5 \cdot 10^{-9}$ (a). The alternation between determinate and indeterminate dynamics is observed when $\delta \approx 4 \cdot 10^{-9}$ (Fig. 4.30(b)).

The radius of the inner basin of attraction is generally a function both of the period of the cycle and of the bifurcation parameter, $R_{\text{int}} = R_{\text{int}}(m, \lambda)$. For the same m-cycle, existing in the range $\lambda_{m\text{L}} < \lambda < \lambda_{m\text{R}}$, the value of R_{int} depends slightly on λ. However, R_{int} generally drops significantly as m is increased [14].

If we assume as in [14] that the mapping (3.9), (3.10) has the same property,

m	Ω
52	90.8425
52	90.84409
104	90.8449
104	90.8452
208	90.8453
208	90.8454
416	90.845426
832	90.84544

Table 4.4. Period-doubling of a 52-cycle as the parameter Ω is increased. List of stable cycles existing for different values of Ω.

then chaotic oscillations may occur in the dynamics (4.1) due to a small inner basin of attraction when m is large enough. This assumption is supported by results of some numerical experiments [14]. These computations were performed with $\Omega = 90.8425$ which corresponds to the existence of a stable 52-cycle. Subsequently, the evolution of this motion was considered as a function of Ω. The parameter Ω was increased in small steps. The Cauchy problem was solved for each value of Ω by the mapping (3.9), (3.10) with different initial conditions which was chosen randomly. Convergence (or nonconvergence) to a periodic motion and the features of the m-cycle (the value of m and X_{ck}, $k = \overline{1,p}$) were determined by a special algorithm, which is not examined here.

Our attempts to discover non-uniqueness of the stable m-cycle over the range $90.8425 < \Omega < 90.87$ were not successful. Then the calculations were made by using (3.9), (3.10) until $k = Nm$ under the initial conditions corresponding to the observed m-cycle, and the values X_{ck} were plotted in a diagram. The value N was varied from $N = 1000$ to $N > 5000$.

The calculation of m-cycles for several values of Ω are reported in Table 4.4. As seen in the table, the 52-cycle undergoes several period-doubling bifurcations when Ω increases. We could observe four period-doubling bifurcations in the numerical calculations. However, using the numerical approach it was impossible to find motions with $m > 832$ due to limitations in the precision of the calculations. Noise of increasing intensity occurs against the background of the periodic dynamics when Ω increases, and the fuzziness of the trajectories of the motion becomes more noted. Beginning with some $\Omega > \Omega_*$, an alternation in the dynamics of the system is observed, with some short-term "turbulent" splashes [10, 22] appearing against a background of relatively long-term regular

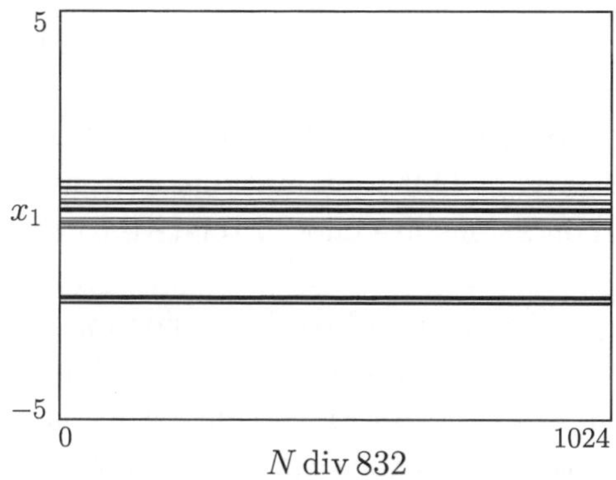

Figure 4.31. Iterations of the mapping (3.9), (3.10) with parameter values corresponding to the existence of a period-832 cycle.

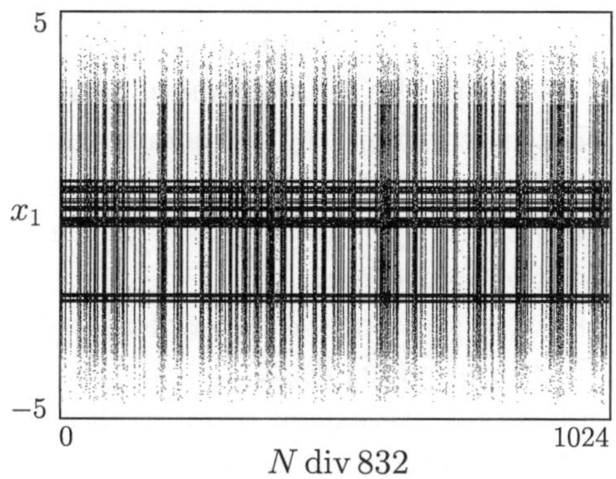

Figure 4.32. The alternation phenomenon in the dynamics of the mapping (3.9), (3.10).

oscillations ("laminar phase" [10, 22]). When moving from Ω_* the duration of the turbulent phase becomes longer and the duration of the laminar phase becomes shorter. At some $\Omega > \tilde{\Omega}_*$ the motion assumes a chaotic character.

The results of our numerical calculations are presented in Figs. 4.31 and 4.32. Figure 4.31 shows the process of appearance of a stable periodic motion, which corresponds to an 832-cycle with $\Omega = 90.84544$. The dynamics of the mapping (3.9), (3.10) in Fig. 4.32, which corresponds to $\Omega = 90.84544$, illustrates the alternation (or intermittency) phenomenon.

The occurrence of non-determinate modes may be explained by a small inner basin of attraction when m is large. In this case the existence of even small random noise sources associated with truncations in the numerical calculations, lead to a non-monotonical behavior of the disturbance.

The results of these investigations allow us to make some summarizing remarks concerning the causes of non-determinate modes.

Let the dynamical system operate in the presence of random noise and let the parameters tend to fluctuate. Let an empty intersection of the branches $V_{i,j}$ exist, that is, let the structure on which the dynamical system (4.1) was determined be normal. Assume that the state of the dynamical system corresponds to the 1-cycle existing within the range of $V_{1,1}$ (Fig. 4.11). The degree of nondeterminacy will be defined by the value of the parameter Ω. The nearer the value of Ω is to the boundary of the branch, which corresponds to the accumulating point $\Omega_{1,*}$, the higher the degree of nondeterminacy. The degree of non-determinate motion is explained by the relation of the noise magnitude to the radius of the inner basin of attraction.

Let us consider the branching pattern shown in Fig. 4.11 again. The basic branch $V_{1,1}$ is intersected by some branches $V_{i,j}$. Different motions may here be realized, depending on how the initial conditions are chosen relative to the basins of attraction.

Let the state of the dynamical system (4.1) correspond to the basic motion, as in the previous case. If the noise magnitude does not exceed the radius of the basin of attraction of basic motion (and of other motions that occur as a result of plural bifurcations of the basic motion) then the degree of increase of nondeterminacy when changing the parameter Ω towards the boundary of the branch will take place as in the previous case.

If the noise magnitude becomes comparable to the radius of the basin of attraction for the motion that the dynamical system performs, then an abrupt change in the degree of nondeterminacy is possible. In this case, either chaotic dynamics or a catastrophic transition from determinate to chaotic dynamics, and vise versa, occur at definite relations of the radii of the basins of attraction and the magnitude of the noise [14, 18].

4.5 Complex Oscillations in a DC Electric Drive

The present section is devoted to investigation of the complex behavior of a DC electric drive in which the motor torque is stabilized by a relay control system [8, 9]. Both the realization and the principles of operation for this system were described in Sec. 3.7. The two-position relay element in the system is considered to have time deadzone and hysteresis. In practice the value of the hysteresis for the relay element is chosen in such a way that the time deadzone does not cause delay in the switching of the power converter for the nominal operating mode. For the system under consideration this mode corresponds to periodic oscillations with two pulses of different polarity during the period T (i.e., to a period-1 cycle). In this mode, the time deadzone influences the dynamics only during the transient. However, when more complex dynamics arises, the time deadzone becomes essential.

In this section we shall:

- examine how the dynamical characteristics of the system depend on the

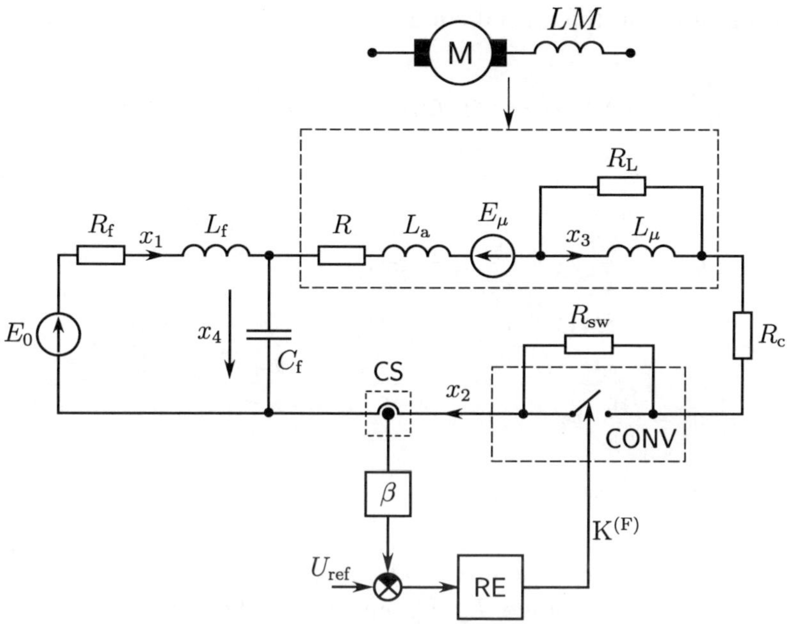

Figure 4.33. Equivalent circuit of the relay control system.

parameters of the relay element, the rotation frequency of the motor armature, and the parameters of the input filter.

- study the type and organization of the bifurcations through which the oscillatory modes change when varying the parameters.

- determine the range of parameters for the relay element and the input filter which minimize the oscillation amplitude, widen the domain of convergence, and up-grade the control accuracy within the limits set by the operating frequency of the converter.

4.5.1 Mathematical model of a relay system with hysteresis and time deadzone

The equivalent circuit of the system is presented in Fig. 4.33.

The mathematical model corresponding to the equivalent circuit is represented as a system of differential equations

$$\frac{\mathrm{d}X}{\mathrm{d}t} = G(X); \; X = (x_1, x_2, x_3, x_4)^{\mathrm{T}}; \; G = (g_1, g_2, g_3, g_4)^{\mathrm{T}}; \tag{4.8}$$

with

$$\begin{aligned}
g_1 &= -\frac{R_{\mathrm{f}}}{L_{\mathrm{f}}} x_1 - \frac{1}{L_{\mathrm{f}}} x_4 + \frac{E_0}{L_{\mathrm{f}}};\\
g_2 &= -\frac{R_{\mathrm{ef}}\big(\mathrm{K}_k^{(\mathrm{F})}(\xi)\big)}{L_{\mathrm{a}}} x_2 + \frac{R_{\mathrm{L}}}{L_{\mathrm{a}}} x_3 + \frac{1}{L_{\mathrm{a}}} x_4 - \frac{E_\mu(x_3)}{L_{\mathrm{a}}};\\
g_3 &= \frac{R_{\mathrm{L}}}{L_\mu}(x_2 - x_3);\\
g_4 &= \frac{1}{C_{\mathrm{f}}}(x_1 - x_2).
\end{aligned}$$

Here x_1, x_2, x_3, and x_4 are the filter coil, armature and magnetization currents, and the filter capacitor voltage; L_{a}, L_μ, and L_{f} are the inductances of the motor armature circuit, the motor magnetizing circuit and the filter, respectively; C_{f} is the filter condenser capacity; R_{f} and R_{L} are the resistances of the filter and an equivalent short-circuited eddy currents loop, respectively; $E_\mu(x_3) = \omega C_E \Phi(x_3)$ is the motor rotating EMF, where ω, $\Phi(x_3)$, C_E are the armature rotation frequency, the magnetization curve and a design parameter of the motor; E_0 is the supply voltage.

$$R_{\mathrm{ef}}\big(\mathrm{K}_k^{(\mathrm{F})}(\xi)\big) = R + R_{\mathrm{c}} + \frac{1}{2}\big(1 - \mathrm{K}_k^{(\mathrm{F})}\big) R_{\mathrm{sw}},$$

where $R = R_{\mathrm{e}} + R_{\mathrm{a}} + R_{\mathrm{L}}$; R_{a}, R_{e} are the resistances of the armature circuit and the motor excitation winding, respectively; R_{c} and R_{sw} are the values of the starting resistances. The characteristics of the relay element RE are given in Fig. 4.34.

The output signal of a relay element with time deadzone may be written as [27]

$$\begin{aligned}
\mathrm{K}_k^{(\mathrm{F})} &= \mathrm{K}_{k-1}^{(\mathrm{F})} \operatorname{sign}\big\{\chi_0 \big(1 + N_{k-1}^{\tau}\big) + (-1)^{N_{k-1}} \xi(X)\big\};\\
\xi(X) &= U_{\mathrm{ref}} - \beta x_2;
\end{aligned}$$

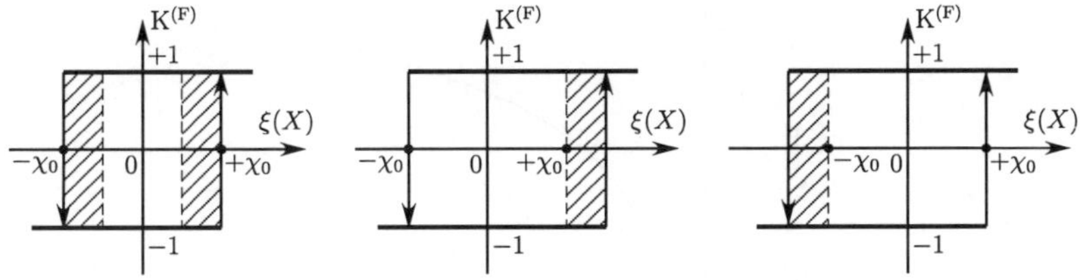

Figure 4.34. Characteristics of the relay element. χ_0 denotes the value of the hysteresis loop.

with

$$N_k^\tau = \frac{\vartheta}{2}\left[1 - \operatorname{sign}\left\{\frac{t - t_{k-1}}{\tau_0} - 1\right\}\right];$$

$$N_k = N_{k-1} + \frac{1 - \operatorname{sign}\{\mathrm{K}_k^{(\mathrm{F})}\mathrm{K}_{k-1}^{(\mathrm{F})}\}}{2};$$

$$\mathrm{K}_0^{(\mathrm{F})} = -1;\ N_0 = 0;\ N_0^\tau = 0;\ \vartheta > 0 = \text{const};$$

$$t_k = \begin{cases} t_k^*, & t_k - t_{k-1} > \tau_0, \\ t_{k-1} + \tau_0, & t_k - t_{k-1} \le \tau_0; \end{cases}$$

and with t_k^* being the smallest positive root of the equation

$$\chi_0 + (-1)^{N_{k-1}}\xi(X(t_k)) = 0.$$

β is the transfer constant of the current sensor. The expression for $\mathrm{K}_k^{(\mathrm{F})}$ is determined from the time plots describing the system's operation (see Fig. 3.9 and 3.11).

The experimentally obtained magnetization curve $C_E\Phi(x_3)$ for the motor [8] has been interpolated by the function $\varphi(\overline{x})$:

$$C_E\Phi(x_3) - \mu\varphi(\overline{x});$$
$$\varphi(\overline{x}) = a_1 \exp(a_2\overline{x}) + a_3 \arctan(a_4\overline{x}) + a_5\overline{x}^{a_6} - a_7,$$

where $a_1 = 0.08067$, $a_2 = 0.000797$, $a_3 = 0.242$, $a_4 = 0.0108$, $a_5 = 0.01973$, $a_6 = 0.29664123$, $a_7 = 0.0964$, $\overline{x} = x_3/x_*$, $x_* = 1$ A, $\mu = 1$ Vmin. This dependence is given in Fig. 4.35.

The Eqs. (4.8) are non-integrable due to the nonlinearity of the magnetization curve $C_E\Phi(x_3)$. That is why the map

$$X_k = F(\tau_0, \tau_k, X_{k-1}),\ k = 1, 2, 3, \ldots$$

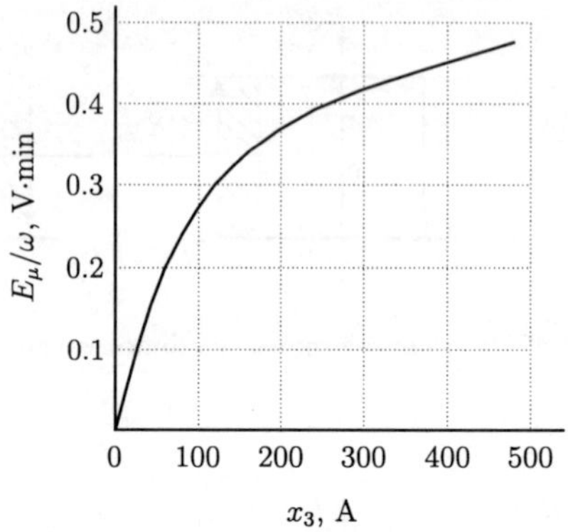

Figure 4.35. Magnetization curve for the motor.

with

$$\begin{aligned} \tau_k &= \begin{cases} \tau_0, & \tau_k^* - \tau_0 < 0, \\ \tau_k^*, & \tau_k^* - \tau_0 \geq 0; \end{cases} \\ \tau_k^* &\rightarrow \varphi\left(\tau_k^*, X_{k-1}\right) = 0; \\ \tau_k &= t_k - t_{k-1}; \\ X_k &= X(t_k, X_{k-1}) \end{aligned}$$

in our investigations has been solved by numerical integration of the Eqs. (4.8).

4.5.2 Analysis of the complex behavior

The basic parameters that determine the dynamic behavior of our system (4.8) include the value of the hysteresis loop χ_0 and the time deadzone of the relay element τ_0, the value of R_c, and the parameters of the input LC-filter. The duration τ_0 is determined by the inertial properties of the power thyristors and by the parameters of the converter's forced switching unit. For the converter in question the value τ_0 is equal to $4.5 \cdot 10^{-4}$ s. The dynamical system (4.8) has been studied with the following values of the parameters: $R_a = 0.0618\ \Omega$, $L_a = 2.8 \cdot 10^{-3}$ H, $R_e = 0.1\ \Omega$, $R_L = 4\ \Omega$, $L_\mu = 3 \cdot 10^{-3}$ H, $R_f = 0.5\ \Omega$, $L_f = 7.2 \cdot 10^{-3}$ H, $C_f = 0.28 \cdot 10^{-3}$ F, $E_0 = 3000$ V, $R_{sw} = 2.2\ \Omega$, $R_c = 1.8\ \Omega$.

Let us begin the analysis of the dynamical system (4.8) by defining the conditions of existence for oscillations.

The system (4.8) has a definite form for each state of the relay element:

$$\frac{\mathrm{d}X}{\mathrm{d}t} = G^{+}(X) \qquad \text{for } \mathrm{K}^{(\mathrm{F})} = 1 \tag{4.9}$$

and

$$\frac{\mathrm{d}X}{\mathrm{d}t} = G^{-}(X) \qquad \text{for } \mathrm{K}^{(\mathrm{F})} = -1. \tag{4.10}$$

Between two subsequent switchings of the relay element the phase trajectories of the dynamical system (4.8) can be determined by solving either the system (4.9) or the system (4.10).

A stable fixed point (equilibrium point) exists in phase space for each of the systems (4.9) and (4.10). These fixed points can be determined as a solution to the equations

$$G^{+}(X) = 0, \tag{4.11}$$

$$\begin{cases} R_{\mathrm{f}}x_1 + x_4 = E_0; \\ -(R + R_{\mathrm{c}})x_2 + R_{\mathrm{L}}x_3 + x_4 = E_{\mu}(x_3); \\ x_1 = x_2;\ x_2 = x_3; \end{cases}$$

for $\mathrm{K}^{(\mathrm{F})} = 1$ and to the equations

$$G^{-}(X) = 0, \tag{4.12}$$

$$\begin{cases} R_{\mathrm{f}}x_1 + x_4 = E_0; \\ -(R + R_{\mathrm{c}} + R_{\mathrm{sw}})x_2 + R_{\mathrm{L}}x_3 + x_4 = E_{\mu}(x_3); \\ x_1 = x_2;\ x_2 = x_3; \end{cases}$$

for $\mathrm{K}^{(\mathrm{F})} = -1$.

Let us first show that these fixed points are unique.

Equations (4.11) and (4.12) can be rewritten as

$$\begin{cases} R_{\mathrm{f}}x_1 + x_4 = E_0; \\ -(R + R_{\mathrm{c}} + \frac{1}{2}(1 - \mathrm{K}^{(\mathrm{F})})R_{\mathrm{sw}})x_2 + R_{\mathrm{L}}x_3 + x_4 = E_{\mu}(x_3); \\ x_1 = x_2;\ x_2 = x_3; \end{cases}$$

Solving of (4.13) is reduced to obtaining a root of the nonlinear equation

$$\begin{gathered} (R_{\mathrm{f}} + R_{\mathrm{a}} + R_{\mathrm{e}} + \tfrac{1}{2}(1 - \mathrm{K}^{(\mathrm{F})})R_{\mathrm{sw}})x_3 = E_0 - E_{\mu}(x_3); \\ x_1 = x_3;\ x_2 = x_3;\ x_4 = E_0 - R_{\mathrm{f}}x_3,\ E_0 > E_{\mu}(x_3). \end{gathered}$$

For the specific value of $\mathrm{K}^{(\mathrm{F})}$ this equation has the unique solution (see Fig. 4.35) that can be numerically obtained by means of iteration methods.

We conclude that two fixed points $X^+ = (x_1^+, x_2^+, x_3^+, x_4^+)$ and $X^- = (x_1^-, x_2^-, x_3^-, x_4^-)$ exist in the phase space of the dynamical system (4.8). One of these points corresponds to the "on" state of the relay element ($\mathrm{K}^{(\mathrm{F})} = +1$) and another corresponds to the "off" state ($\mathrm{K}^{(\mathrm{F})} = -1$).

The local nature of these fixed points, i.e., the behavior of the integral curves of the systems (4.11) and (4.12) in their small neighborhoods is determined by the eigenvalues λ_k, $k = \overline{1,4}$ of the Jacobian matrix:

$$\det\left(\frac{\mathrm{d}C_{\pm}(X)}{\mathrm{d}X}\bigg|_{X=X^{\pm}} - \lambda E\right) = 0.$$

In practice, one of the main requirements when choosing the value of the resistances R_c and R_sw is to ensure the stability of fixed points, i.e., all eigenvalues of the Jacobian matrix must be situated in the left part of the complex λ-plane with respect to the imaginary axis.

In this case the solution of the Cauchy problem for (4.8) for the given state of relay element will tend towards the appropriate equilibrium point due to attracting nature of the fixed points. Change of the state of the relay element reverses the direction of motion for the phase point because the position of the attracting point is changed.

Oscillations arises only when the following conditions are satisfied:

$$x_3^+ > \frac{1}{\beta}(U_\mathrm{ref} + \chi_0) \qquad \text{and} \qquad x_3^- < \frac{1}{\beta}(U_\mathrm{ref} - \chi_0).$$

Indeed, if

$$x_3^- > \frac{1}{\beta}(U_\mathrm{ref} - \chi_0)$$

then from (4.8) it follows that

$$\lim_{t\to\infty} X(t) = X^-$$

for any initial conditions.

In the same way, if

$$x_3^+ < \frac{1}{\beta}(U_\mathrm{ref} + \chi_0)$$

then

$$\lim_{t\to\infty} X(t) = X^+.$$

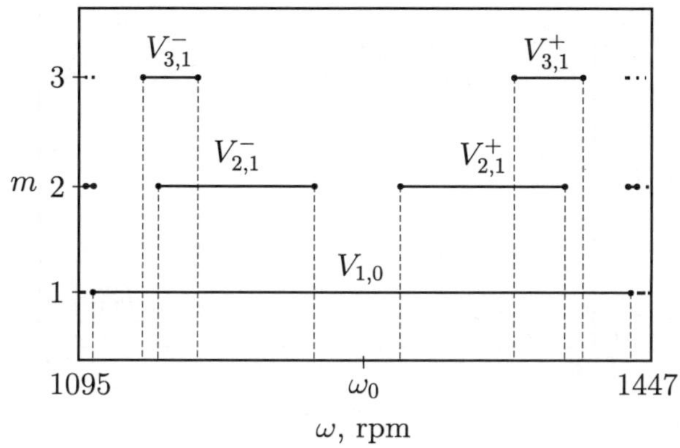

Figure 4.36. Branching pattern calculated for $U_{\text{ref}} = 3.4745$ V and $\chi_0 = 0.1865$ V. 1095 rpm < ω < 1447 rpm.

Figure 4.36 is a branching pattern calculated for $U_{\text{ref}} = 3.4745$ V and $\chi_0 = 0.1865$ V. The assumed values of U_{ref} and χ_0 correspond to a motor armature current setting of 347.45 A and an armature current oscillation amplitude of 37.3 A [27].

The corresponding range for controlling the rotation frequency of the traction motors is $\Omega_1 = \{\omega : 1095 \text{ rpm} < \omega < 1447 \text{ rpm}\}$. This means that oscillations do not occur when $\omega < 1095$ rpm or $\omega > 1447$ rpm. At these values of ω there exists a single fixed point determined by Eq. (4.12) or by Eq. (4.11), respectively. The relative duration of the pulse of the converter (pulse ratio) for the 1-cycle changes from 0 to 1 through the range Ω_1.

Figure 4.36 makes it clear that the dynamics is symmetrical with respect to $\omega = \omega_0$. At this value of ω the pulse ratio of the converter equals to 0.5. It is also seen that within the range $\Omega_1 = \{\omega : 1095 \text{ rpm} < \omega < 1447 \text{ rpm}\}$ together with a 1-cycle there are other stable motions such as 2- and 3-cycles, all of which appear in hard transitions. The domains of existence of these limit cycles are symmetrically situated around ω_0. For the sake of convenience let us denote the branch which corresponds to the parameter-ω domain where the 1-cycle exists as $V_{1,0}$; while the other branches are denoted: $V^-_{m,1} \subset \{\omega : \omega_{\min} < \omega < \omega_0\}$ and $V^+_{m,1} \subset \{\omega : \omega_0 < \omega < \omega_{\max}\}$, $m = 2, 3$. Figure 4.36 shows that all the branches $V^{\pm}_{m,1}$, $m = 2, 3$ have intersections (or regions of coexistence) with $V_{1,0}$.

Figure 4.37 displays the ω-dependences of the oscillation frequencies $f_{m,k}(\omega)$, $m = 2, 3$; $k = 0, 1, 2, \ldots$ that correspond to the 1-, 2-, and 3-cycles. Under vari-

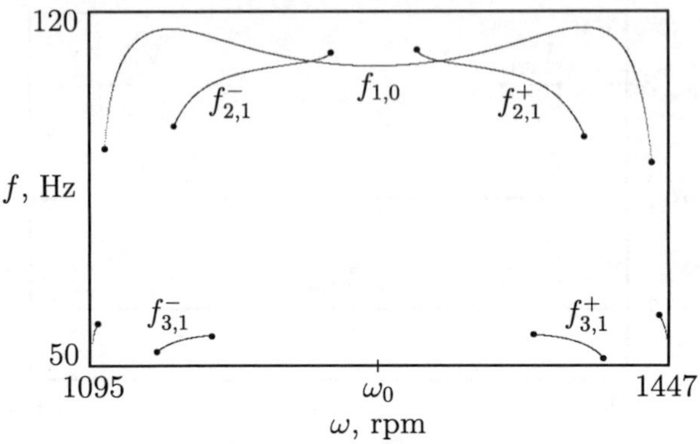

Figure 4.37. Variations of the oscillation frequency with ω for $U_{\text{ref}} = 3.4745$ V and $\chi_0 = 0.1865$ V. 1095 rpm $< \omega <$ 1447 rpm.

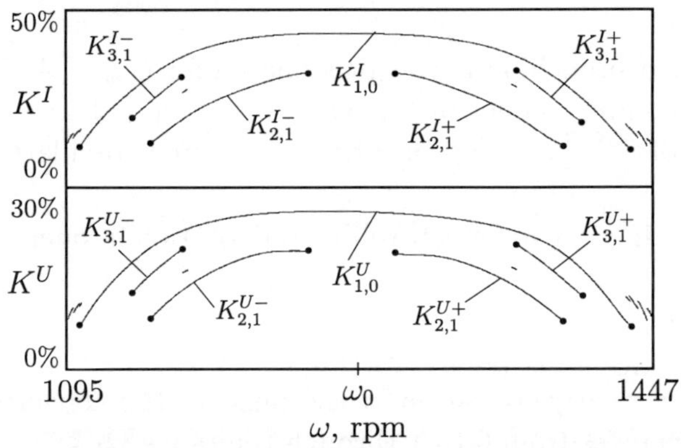

Figure 4.38. Current and voltage ripple factors $\left(K_{m,k}^{I\pm}(\omega)\right.$ and $\left.K_{m,k}^{U\pm}(\omega)\right)$ as functions of ω for different m-cycles, $m = 1, 2, 3$; $k = 0, 1, 2, \ldots$

ation of ω the oscillation frequency of the 1-cycle changes non-monotonously. The curve $f_{1,0}(\omega)$ has two maxima symmetrical around ω_0. The first maximum corresponds to a frequency of about 116.5 Hz, the second one to 117 Hz. Between these maxima, when ω increases, the oscillation frequency of the 1-cycle at first goes down and reaches the value of $f_{1,0}(\omega) = 109.5$ Hz at ω_0 to subsequently increase up to the second maximum and finally decrease again.

Figure 4.38 shows a family of curves for the input filter's current and voltage ripple factors $K_{m,k}^{I\pm}(\omega)$, $K_{m,k}^{U\pm}(\omega)$, $m = 1, 2, 3$; $k = 0, 1, 2, \ldots$. The ripple factor is a measure of the amplitude of the periodic variations of a signal in terms of its mean value. The curves in Fig. 4.38 correspond to different m-cycles ($m = 1, 2, 3$) within the range $\Omega_1 = \{\omega : 1095 \text{ rpm} < \omega < 1447 \text{ rpm}\}$. Here the subscript is introduced to mark to which motion the corresponding values, $K^{I\pm}_{m,k}(\omega)$ or $K^{U\pm}_{m,k}(\omega)$, are referred. The ripple factors for a cycle are determined as

$$K^x = \frac{1}{\overline{x}} \max_{0<t<T} |x(t) - \overline{x}| \cdot 100\%,$$
$$\overline{x} = \frac{1}{T} \int_0^T x(t)\, dt,$$

where T is the period of the limit cycle. With increasing ω the values of $K^I_{1,0}(\omega)$ and $K^U_{1,0}(\omega)$ for the 1-cycle grow up monotonically and reach their maximum at those values of ω where $f_{1,0}(\omega)$ has its maxima. Between these points, the values of $K^I_{1,0}(\omega)$, $K^U_{1,0}(\omega)$ remain practically constant. Our calculations show that the ripple factors of the 1-, 2-, and 3-cycles are almost the same:

$$10\% < K^I_{1,0}(\omega) < 40\%; \qquad 5\% < K^U_{1,0}(\omega) < 25\%;$$
$$10\% < K^{I\pm}_{2,1}(\omega) < 30\%; \qquad 6\% < K^{U\pm}_{2,1}(\omega) < 19\%;$$
$$16\% < K^{I\pm}_{3,1}(\omega) < 31\%; \qquad 9\% < K^{U\pm}_{3,1}(\omega) < 22\%.$$

The armature current ripple factor is constant over the whole control range for all the oscillation modes because the hysteresis value is chosen so that the relay element time deadzone should have no delaying effect upon the switching of the converter elements under the established modes. Otherwise the switching frequency would increase over the maximum allowed value, which for the converter under investigation is equal to 400 Hz.

Figures 4.39(a)–4.39(c) show the temporal variations of the filter coil current (x_1), the armature current (x_2), the magnetization current (x_3), and the filter capacity voltage (x_4) for different values of ω. The figures demonstrate that the profiles of the motors current curves are greatly distorted due to superposition of the current and voltage harmonics of the input filter.

Let us denote the range of ω where only one stable period-1 limit cycle exists as the domain of convergence Ω_{conv}.

The domain of convergence Ω_{conv} does not coincide with the range $\Omega_1 = \{\omega : 1095 \text{ rpm} < \omega < 1447 \text{ rpm}\}$. As it follows from Fig. 4.36, Ω_{conv} is not simply connected and it occupies rather narrow ranges of ω. As a result of this, when ω changes continuously, the mode characteristics will not change

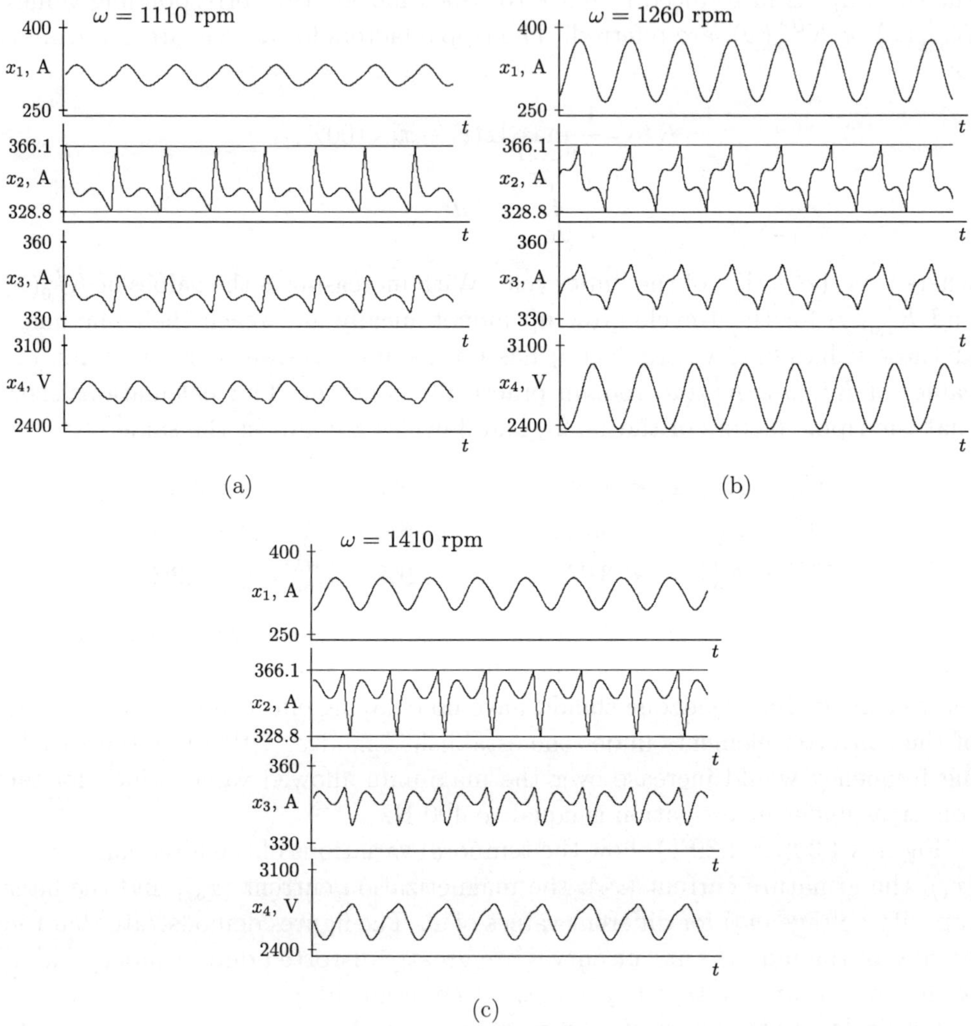

Figure 4.39. Shape of the oscillations in x_1, x_2, x_3, x_4 for different types of period-1 limit cycles with different values of ω. Notice the considerable second and third harmonic content in the temporal variation of x_2 and x_3.

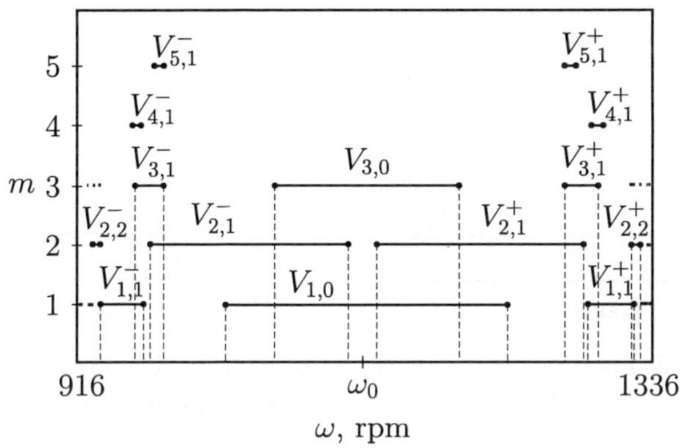

Figure 4.40. Branching pattern calculated for $U_{\text{ref}} = 4.164$ V and $\chi_0 = 0.2$ V. 916 rpm $< \omega <$ 1336 rpm.

smoothly, but abruptly (catastrophically). This is related to the fact, that an abrupt transition from one oscillation mode to another can take place while increasing the motor's armature rotation frequency because of non-uniqueness of the motions within the wide range of ω values and because of hard intersections of the branches $V^{\pm}_{m,1}$, $m = 2, 3$ with $V_{1,0}$. However, such transitions will not entail any considerable increase in the ripple factors and will not degrade the static control accuracy, because, as it has been demonstrated above, the characteristics of the modes occurring under the same values of the parameters show little difference. Yet, the dynamic control error can be considerable, and when a system is operating in a high-period oscillation mode one observes an essential deterioration of the current and voltage spectra of the input filter.

When $U_{\text{ref}} = 4.164$ V and $\chi_0 = 0.2$ V the dynamics of the system becomes substantially more complicated (see Fig. 4.40). The continuity of the 1-cycle upon the parameter ω is broken. The domain of convergence $\Omega_{\text{conv}} = \emptyset$. This means that coexistence of motions occurs over the whole control range 916 rpm $< \omega <$ 1336 rpm. The symmetry of the branching pattern with respect to ω is preserved (see Fig. 4.40). The branching pattern clearly reveals three relatively wide intervals of ω wherein the stable 1-cycle exists. The branch corresponding to the biggest of these intervals is denoted as $V_{1,0}$. The boundaries of $V_{1,0}$ are symmetrical with respect to ω_0. A branch $V_{3,0}$ appears. It corresponds to the domain of existence of a 3-cycle. As before, all other branches are denoted $V^{\pm}_{m,k}$, ($m = 1, 2, 3$; $k = 1, 2, 3, \ldots$). Here $V^-_{m,k} \subset \{\omega : \omega_{min} < \omega < \omega_0\}$,

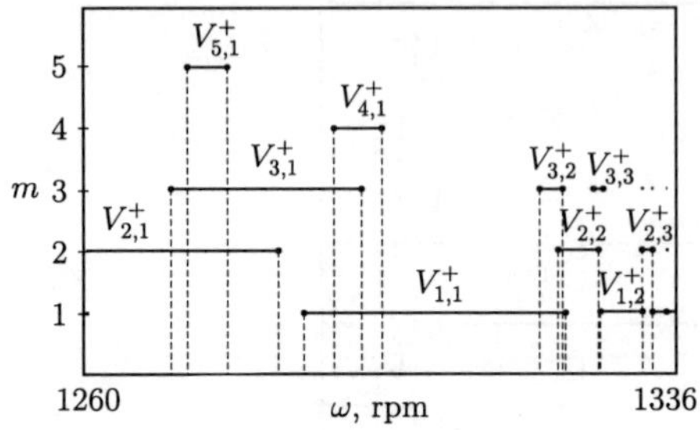

Figure 4.41. Magnified portion of the branching pattern shown in Fig. 4.40.

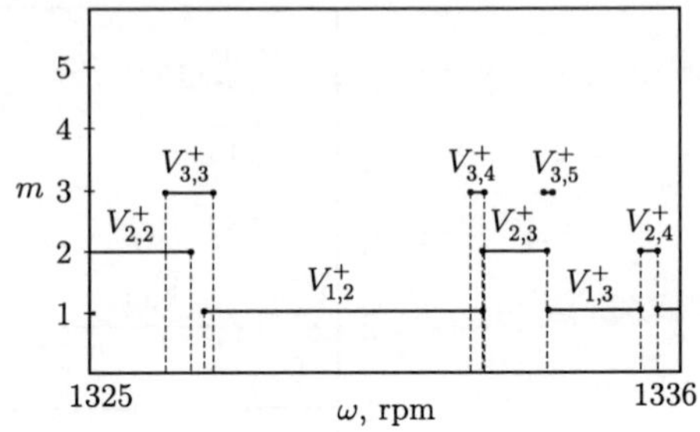

Figure 4.42. Magnified portion of the branching pattern shown in Fig. 4.41.

$V^+_{m,1} \subset \{\omega : \omega_0 < \omega < \omega_{max}\}$, $k = 1, 2, 3, \ldots$. The first subscript indicates the cycle period (the value of m), and the second one is introduced to distinguish between branches with the same m.

The number of branches $V^{\pm}_{m,k}$, $m = 1, 2, 3$, $k = 1, 2, 3, \ldots$ increases dramatically when moving along the parameter ω towards the boundaries of the control range $\Omega_2 = \{\omega : 916 \text{ rpm} < \omega < 1336 \text{ rpm}\}$ (see Figs. 4.40–4.42).

Let $\delta^{\pm}_{m,k} = \left[\omega^{\pm}_{P(m,k)}; \omega^{\pm}_{L(m,k)}\right]$ be the range of ω where the branch $V^{\pm}_{m,k}$ exists. We have found that $\lim\limits_{k\to\infty} \left|\delta^{\pm}_{m,k}\right| = 0$, $m = 1, 2, 3$ (see Figs. 4.40–4.42). Exceptions are the branches corresponding to the domains of existence of a 4- and a 5-cycle. There are four such domains in Ω_2. The branches which correspond to the domains of existence of 4- and 5-cycles in Fig. 4.40 are denoted as $V^{\pm}_{m,1}$, $m = 4, 5$.

Figure 4.43 shows the curves $f(\omega)$ for $m = 1, 2, 3$. The character of these variations is essentially the same as when $U_{\text{ref}} = 3.4745$ V and $\chi_0 = 0.1865$ V with the exception that there are discontinuities in the characteristics $f_{m,k}(\omega)$, $m = 1, 2, 3$; $k = 0, 1, 2, \ldots$. In the intervals of ω between these discontinuities other oscillatory modes exist, different from the 1-cycle. By virtue of the above stated properties of the branching pattern, the number of such intervals will be infinite.

That is why in the further analysis we shall restrict ourselves to studying the motion characteristics that correspond to 1-cycles determined only upon those branches whose domains of existence are relatively large, for example, the branches $V_{1,0}$, $V^-_{1,1}$, $V^+_{1,1}$. When increasing ω within the domain of existence of $V_{1,1}$, the frequency $f^-_{1,1}(\omega)$ increases monotonically and reaches its maximum at the bifurcation value of the parameter which corresponds to the right boundary

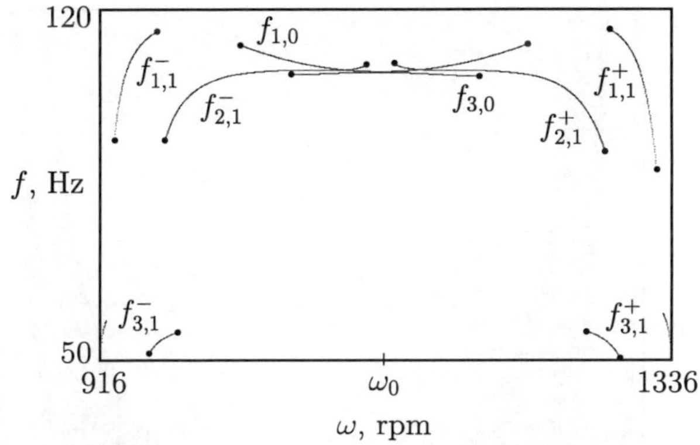

Figure 4.43. Variations of the oscillation frequency with ω for $U_{\text{ref}} = 4.164$ V and $\chi_0 = 0.2$ V. 916 rpm $< \omega <$ 1336 rpm.

of $V_{1,1}^-$. Within $V_{1,0}$ the frequency $f_{1,0}(\omega)$ first decreases and reaches its minimum at ω_0, and then increases up to the value of about 112 Hz on the boundary of $V_{1,0}$. The dependences $f_{1,1}^+(\omega)$ and $f_{1,1}^-(\omega)$ are symmetrical with respect to ω_0.

Ripple factors will change close to the same values as at $U_{\text{ref}} = 3.4745$ V and $\chi_0 = 0.1865$ V:

$$
\begin{aligned}
&9\% < K_1^I(\omega) < 45\%; && 5\% < K_1^U(\omega) < 35\%;\\
&9\% < K_{2,1}^{I\pm}(\omega) < 35\%; && 5\% < K_{2,1}^{U\pm}(\omega) < 25\%;\\
&15\% < K_{3,1}^{I\pm}(\omega) < 25\%; && 9\% < K_{3,1}^{U\pm}(\omega) < 20\%.
\end{aligned}
$$

The maximum static control error does not exceed 2%.

Then an attempt was made to improve the dynamics of the system at the expense of diminishing the value of the relay element hysteresis loop, because, as it is seen from Fig. 4.37 and Fig. 4.43 there is a rather big reserve in the value of the maximum allowed operation frequency of the power converter. However, as our investigations have shown, the decrease of the hysteresis value of the relay element does not result in an up-grading of the control quality, because in this case the domain of existence of the 1-cycle becomes considerably narrower, and modes of larger period will prevail in the dynamics of the system.

We have found a strong dependence of the dynamic properties of a control system on the parameters of the power filter. The results of a two-parameter analysis of the dynamical system (4.8) in the parameter plane $\Pi = \{(\omega, C_{\text{f}}) : 916 \text{ rpm} < \omega < 1336 \text{ rpm}; 70\ \mu\text{F} < C_{\text{f}} < 910\ \mu\text{F}\}$ are given in Fig. 4.44 [27].

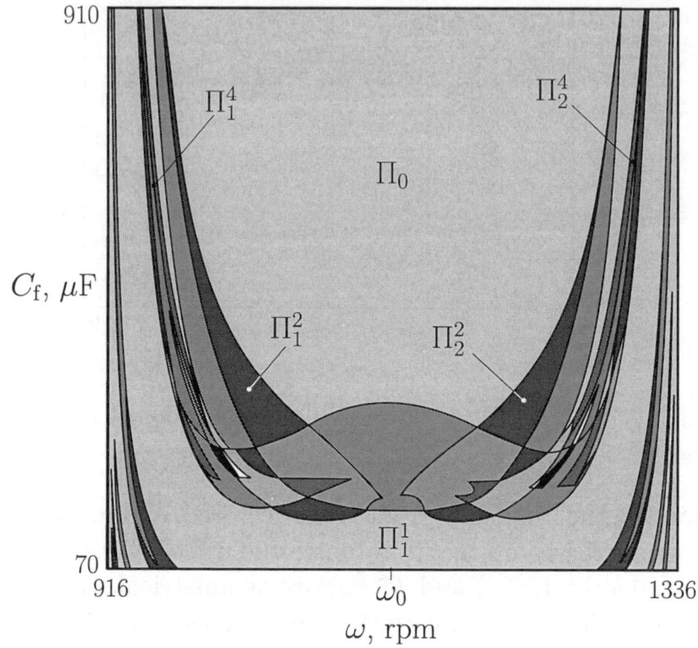

Figure 4.44. Division of the system parameter plane $\Pi = \{(\omega, C_f) : 916 \text{ rpm} < \omega < 1336 \text{ rpm}; 70\ \mu\text{F} < C_f < 910\ \mu\text{F}\}$ into domains with different periodic modes. This plot has been calculated for $L_f = 3.6 \cdot 10^{-3}$ H, $U_{ref} = 4.164$ V and $\chi_0 = 0.2$ V.

The calculations have been performed with $L_f = 3.6 \cdot 10^{-3}$ H, $U_{ref} = 4.164$ V, $\chi_0 = 0.2$ V, while all other parameters were assumed to be fixed.

To simplify the analysis, Figs. 4.45–4.48 show separately some of the sets Π_k^m ($m = 1, 2, 3, 4, 5$; $k = 0, 1, 2, \dots$). By virtue of the above properties, the number of such sets in Π is denumerable. Figures 4.45–4.48 show only those domains Π_k^m ($m = 1, 2, 3, 4$; $k = 0, 1, 2, \dots$) which are relatively large in Π. The sets Π_k^1 ($k = 0, 1, 2, \dots$) in Fig. 4.45 correspond to the domains of parameter values $P = (\omega; C_f)$ where a locally stable 1-cycle exists. Let us take Π_0^1 as the basic set and denote it by Π_0.

We have found that the sets Π_k^m ($m = 1, 2, 3, 4, 5$; $k = 0, 1, 2, \dots$) have the following properties: $\Pi_k^1 \bigcap \Pi_{k+1}^1 \neq \emptyset$, $k = 0, 1, 2, \dots$, $\Pi_k^m \bigcap \Pi_{k+1}^m = \emptyset$, $m > 1$, $k = 0, 1, 2, \dots$ ($\Pi_0^m = \emptyset$). Furthermore, from Fig. 4.44 it follows that the domains Π_k^m for $m = 2, 3, 4, 5$; $k = 0, 1, 2, \dots$ are symmetrical with respect to the section $\omega = \omega_0$.

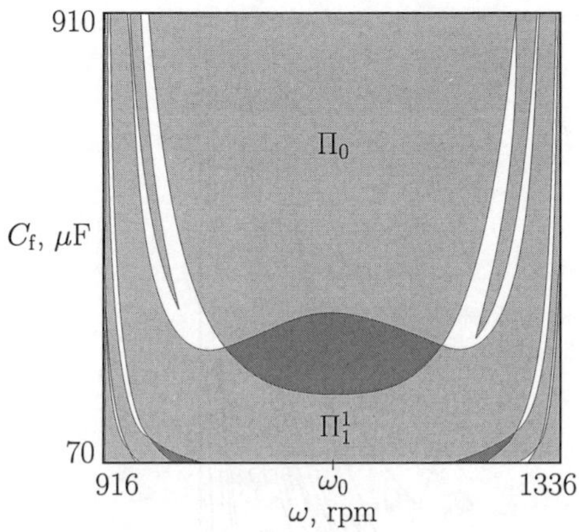

Figure 4.45. Domains of existence for different 1-cycles.

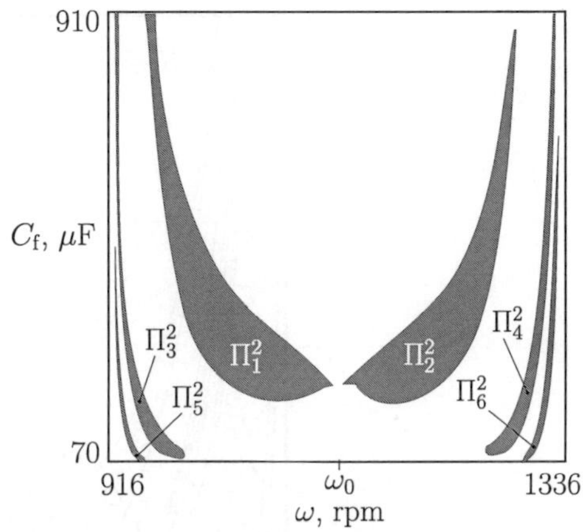

Figure 4.46. Domains of existence for different 2-cycles.

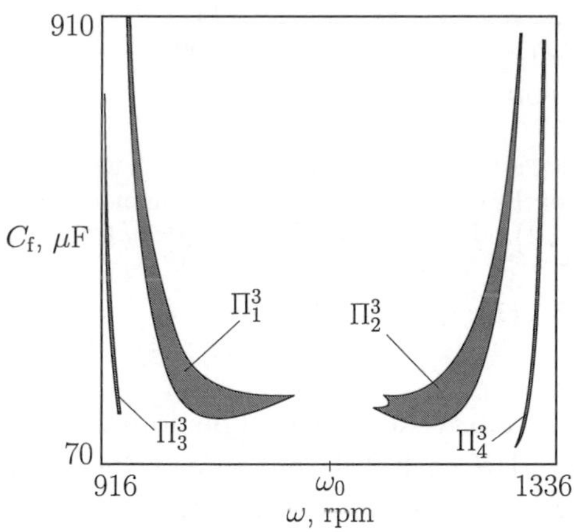

Figure 4.47. Domains of existence for different 3-cycles.

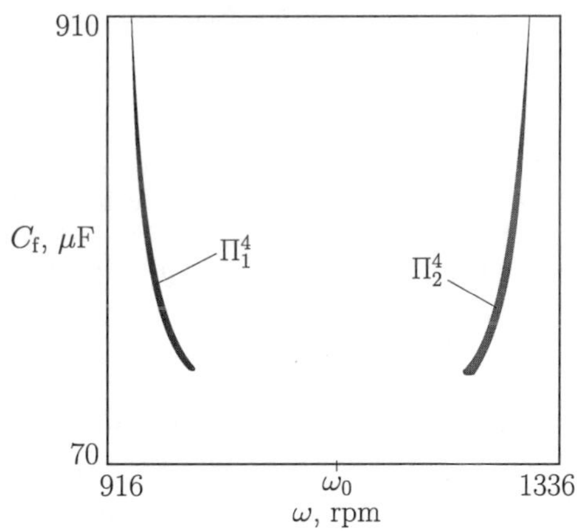

Figure 4.48. Domains of existence for different 4-cycles.

In Fig. 4.45 the areas where the sets Π_k^1, $k = 0, 1, 2, \ldots$ intersect are dark-shaded. Figures 4.46–4.48 show the sets Π_k^2, $k = 0, 1, 2, \ldots$; Π_k^3, $k = 0, 1, 2, \ldots$; Π_k^4, $k = 0, 1, 2, \ldots$ ($\Pi_0^m = \emptyset, m = 2, 3, 4, 5$), respectively. Inspection of the figures shows that the sets Π_k^1, $k = 0, 1, 2, \ldots$ intersect Π_k^m, $m = 2, 3, 4, 5$; $k = 0, 1, 2, \ldots$ $\Pi_k^1 \bigcap \Pi_k^m \neq \emptyset$, $m = 2, 3, 4, 5$; $k = 0, 1, 2, \ldots$

Figure 4.49 provides a more detailed overview of the oscillation modes in the

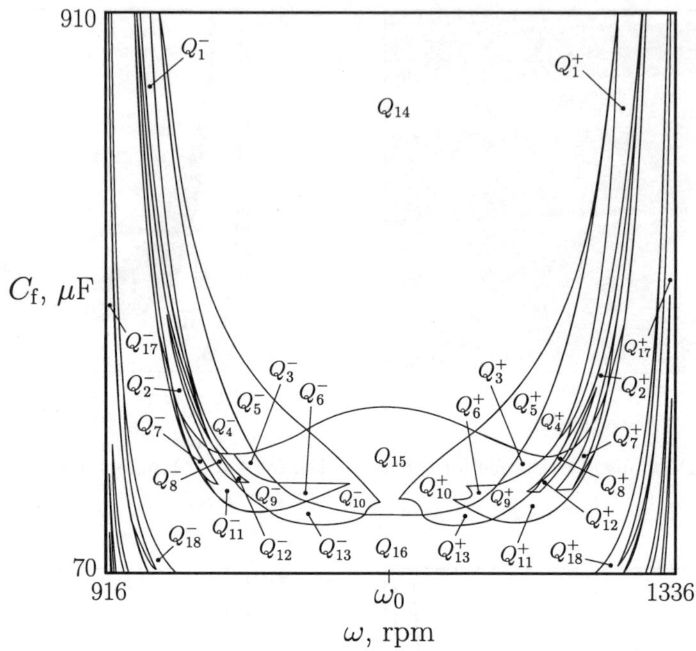

Figure 4.49. Intersections of the domains of existence of different periodic modes. The regions $Q_k^{\pm}$, $k = 1, 2, 3, \ldots$ correspond to the intersections of the domains of existence of the modes with the following periods: $Q_1^{\pm} : \{1\}$; $Q_2^{\pm} : \{3\}$; $Q_3^{\pm} : \{1, 2\}$; $Q_4^{\pm} : \{2\}$; $Q_5^{\pm} : \{1, 2\}$; $Q_6^{\pm} : \{1, 1, 2, 3\}$; $Q_7^{\pm} : \{1, 3, 4\}$; $Q_8^{\pm} : \{1, 3, 5\}$; $Q_9^{\pm} : \{1, 2, 3\}$; $Q_{10}^{\pm} : \{1, 1, 2\}$; $Q_{11}^{\pm} : \{1, 3\}$; $Q_{12}^{\pm} : \{1, 2, 3, 5\}$; $Q_{13}^{\pm} : \{1, 2\}$; $Q_{14} : \{1\}$; $Q_{15} : \{1, 1\}$; $Q_{16} : \{1\}$; $Q_{17}^{\pm} : \{1\}$; $Q_{18}^{\pm} : \{1, 2\}$.

domains where the non-uniqueness of the motions takes place. The domains $Q_i^{\pm}$, $i = 1, 2, 3, \ldots$ correspond to the areas where Π_k^m with different m or k intersect. The periods of cycles coexisting within $Q_i^{\pm}$ are indicated in the branches (see figure caption). When increasing the value of C_f the sets Π_k^m, $m = 1, 2, 3, 4, 5$; $k = 1, 2, 3, \ldots$ tend to become narrower, while Π_0 increases.

To examine regularities of the appearance of m-cycle and the character of the intersections of Π_k^m, $m = 1, 2, 3, 4, 5$; $k = 0, 1, 2, \ldots$ we have calculated one-parameter bifurcation diagrams and the branching patterns in the sections $C_f = 280 + 70 \cdot (k - 1)$ μF, $k = \overline{1, 12}$. This has shown that the properties of the branches in these sections are similar to those shown in Fig. 4.40. But some differences have been observed. These differences relate to branches that belong to Π_k^1, $k = 0, 1, 2, \ldots$ and Π_k^2, $k = 1, 2, 3, \ldots$. Let us consider the differences in

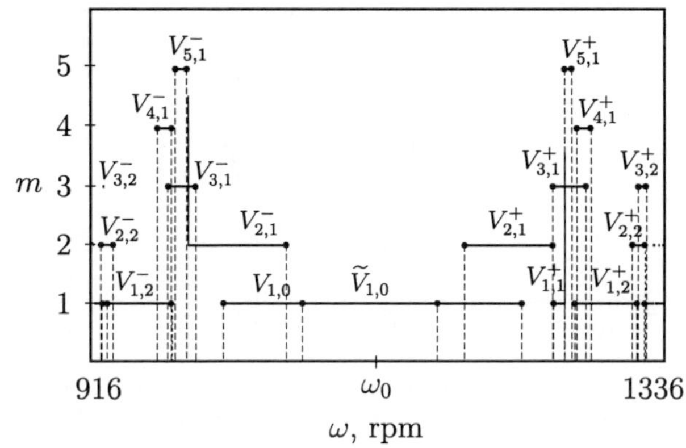

Figure 4.50. Branching pattern calculated in the section $C_f = 350\ \mu$F.

detail.

Figure 4.50 shows the branching pattern calculated in the section $C_f = 350\ \mu$F. This figure demonstrates that there are two branches $V_{1,0} \in \Pi_0$, $\widetilde{V}_{1,0} \in \Pi_1^1$, with $V_{1,0} \bigcap \widetilde{V}_{1,0} \neq \emptyset$. The range of existence for the branch $\widetilde{V}_{1,0}$ ($\widetilde{V}_{1,0} \in \Pi_1^1$) by the parameter ω is much wider than that of $V_{1,0}$. The boundaries of $\widetilde{V}_{1,0}$ correspond to the parameters of hard appearance of a 1-cycle.

The boundaries of the branch $V_{1,0} \in \Pi_0$ correspond to the bifurcation values of the parameter ω where subcritical period-doubling bifurcations are observed. The branch $V_{1,2}^{+} \in \Pi_2^2$ begins with the hard appearance of a 2-cycle. When ω continues to increase, two solutions branch off the 2-cycle, a stable and an unstable 2-cycle. Then, at the value of ω where the 1-cycle loses its stability (the right boundary of $V_{1,0} \in \Pi_0$), the unstable 2-cycle merges with the 1-cycle. By virtue of the branching pattern symmetry with respect to $\omega = \omega_0$, the same bifurcation transition from 1-cycle to a 2-cycle takes place close to the left boundary of $V_{1,0} \in \Pi_0$. Figures 4.51 and 4.52 show bifurcation diagrams corresponding to the left and right parts, respectively, of the branching pattern presented in Fig. 4.50. Here unstable 1- and 2-cycles are schematically shown by the dashed lines. The left boundary of $V_{1,2}^{-} \in \Pi_1^2$ and the right boundary of $V_{1,2}^{+} \in \Pi_2^2$ correspond to such values of ω where a quasiperiodic motion branches off the 2-cycle.

The branch $V_{1,2}^{+}$ consists of the intervals, where a 1-cycle and a 2-cycle exists, with the 2-cycle arising smoothly from the 1-cycle via a period-doubling bifurcation. When moving along the parameter ω, three types of bifurcations take place

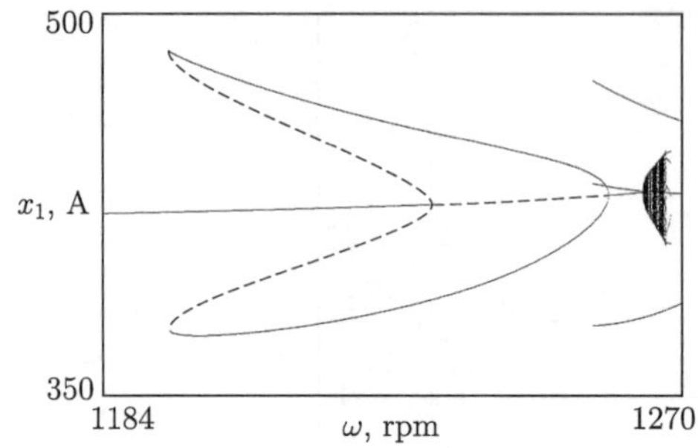

Figure 4.51. Super- and subcritical period-doublings of a 1-cycle and appearance of a quasiperiodic motion.

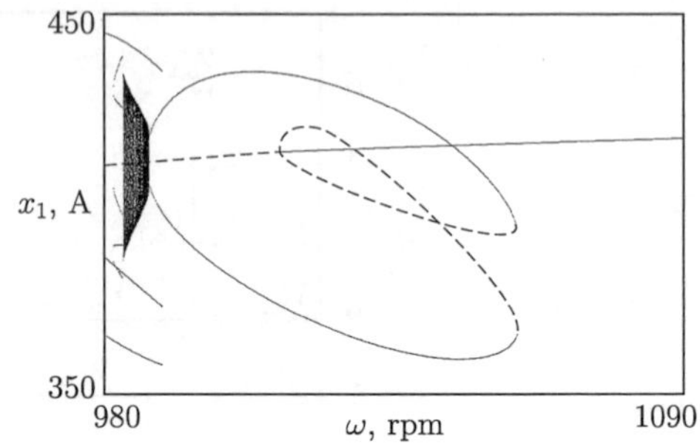

Figure 4.52. Appearance of a quasiperiodic motion from a 2-cycle.

within the ranges of the branches $V_{1,0}$ and $V_{1,2}^+$: a subcritical period-doubling bifurcation, a supercritical period-doubling bifurcation and the bifurcation of the appearance (disappearing) of a quasiperiodic motion. The first type has been analyzed before. The other two are illustrated in Figs. 4.51 and 4.52. Figure 4.51 show the supercritical period-doubling and the appearance of a quasiperiodic motion from a period-1 cycle. The appearance of a quasiperiodic motion from a 2-cycle is given in Fig. 4.52.

The domains where the motions have an aperiodic nature are rather small. The boundaries of $V_{1,2}^- \in \Pi_1^2$ and $V_{1,2}^+ \in \Pi_2^2$, corresponding to the loss of stability of the 2-cycle and the appearance of the quasiperiodic motion, are shown in the branching pattern in Fig. 4.50 as vertical lines.

Because of the non-uniqueness of the motions, when ω increases, an abrupt change in the oscillation modes will take place and, under the influence of external disturbances, there is the possibility of a jump from one oscillation mode to another. Figure 4.53 shows the diagrams that illustrate the possible scenarios of the change in the dynamics under external disturbances. These calculations have been made for $C_f = 350$ μF and $\omega = 1170$ rpm. At these parameter values one can observe three oscillation modes (Fig. 4.50).

The diagrams shown in Fig. 4.53 illustrate the transitions from one stable motion to another under the disturbance effected by the mains supply. In the numerical experiments the external disturbance was applied to the input voltage in the form of a short-time pulse with a duration about one control period with an amplitude of 300 V relative to the supply voltage rating of $E_0 = 3000$ V.

Figure 4.53 shows that up to a certain moment the system operates in a

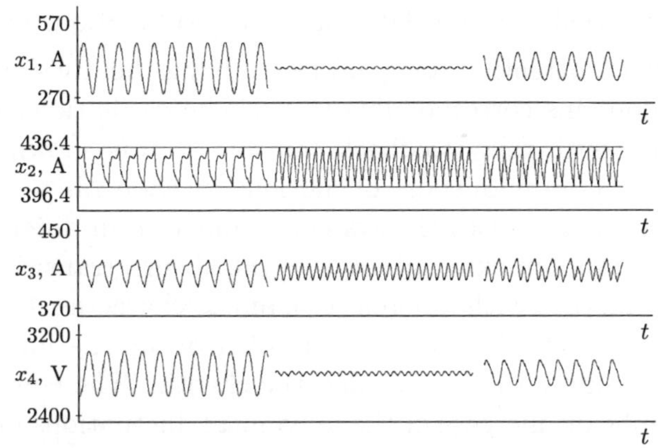

Figure 4.53. Transitions among different stable limit cycles under external disturbance.

1-cycle mode. This motion corresponds to the branch $\widetilde{V}_{1,0}$ (see Fig. 4.51). As a reaction to the disturbing action of the power supply a non-stationary dynamics arises which terminates in the establishment of the periodic motion corresponding to the second 1-cycle that corresponds to $V_{1,0}$. The remaining part of the diagram represents transition from a 1-cycle to a 2-cycle.

As we can see from the above diagrams, the amplitudes of voltage and current pulses of the input filter at a 1-cycle mode exceed the oscillation amplitudes at the 1-cycle mode that corresponds to $V_{1,0}$ by more than 10 times, and at a 2-cycle mode approximately 8 times more. Because of large differences in the motion characteristics such transitions will have a catastrophic character.

Figures 4.44, 4.45 and 4.50 demonstrate that the decrease in the wave resistance of the input filter produces a widening of the domain of convergence. Domains with motions different from a 1-cycle occupy smaller ranges of ω.

4.6 On the Concept of Normal Structures

We have presented results of an investigation of the oscillating properties for relay control systems of two different types, both described by a four-dimensional set of autonomous piecewise-smooth differential equations. Both cases described realistic technical situations, and in both cases the system were found to exhibit an abnormal structure, implying the occurrence of hard (or dangerous) transitions between different solutions under variation of the conditions of operation. Let us briefly recapitulate the main results of our analysis.

In the section 4.2 attention was paid to studying the division of the parameter plane into domains corresponding to periodic oscillations and chaos for a DC/DC converter with relay feedback control. It was shown, that for relay systems there is a non-uniqueness of motions in wide ranges of the realistic parameters space. This causes the system to function in different modes, and in the presence of random external impacts the abrupt transition from one dynamical state to another with different dynamical characteristics is possible, in particular, from one periodic motion to another or alternation of two extreme types of motion, namely the regular and the chaotic. The consequence of this is not only a significant increase of the dynamical fluctuations in the regulation and a worsening of the qualitative measures of performance, but also the possibility of sudden breakdowns. We showed that catastrophical phenomena are typical for abnormal structures. This is connected with the non-uniqueness of motion and the possibility of non-empty intersections of the sets Π_h^m, $\hat{\Pi}$ with the base set Π_0 (see (4.3)).

Let us comment on some theses in the works by Postnikov [16, 17]. The problem discussed in the second of Postnikov's papers [17] is rather different from the problem that we considered in the present chapter. The basic result of Postnikov's work was the design of a control law providing modes of chaotic oscillations, in which the best control quality performance is achieved. However, this problem may not be so important in practice, if the structure of the dynamical system is abnormal. It is clear, that the control algorithms designed by Postnikov [17] will provide the working modes of oscillations meeting the quality criteria only if $\Pi = \Pi_{\text{conv}}$ or if the domain of convergence is large enough to provide the desired motion while the external parameters vary in an uncontrollable manner during the operation. It is natural to suppose that the extent of abnormality will vary. This extent is determined by the magnitude of the difference between characteristics of the motions that exist with the same parameters.

The problem of distinguishing between different structures takes special significance when designing technical systems. Thus for example Neimark [26], referring to works by Gal'perin (see Refs. in [26]) on the development of control systems in which a single stable motion is realized at all parameter values, notice that "control systems which display a low sensitivity to variations of their parameters or are robust by the terminology of Andronov and Pontryagin are sought nowadays all over the world. Hence, it is most important to design systems that are robust by virtue of their structure".

The conception of normal structures [13, 14, 19] is here only the first step of the understanding of the problem of quality design of technical system. Further development of this concept, as it is indicated in [13], is above all connected with the investigation of sets Π_h^m and their completeness properties in the aggregate. Set Π is complete with respect to $\bigcup_{m,h} \Pi_h^m$ if $\Pi = \bigcup_{m,h} \Pi_h^m$ and for any point $P \in \Pi$ with exception, perhaps, of points at the boundaries of the domains Π_h^m every point in the phase space of dynamical system (4.1) either belongs to the basin of attraction of some motion $X_c(t, P_{\text{int}}^{\text{opt}}, P)$ or lies on a trajectory of unstable stationary motion. Here $P_{\text{int}}^{\text{opt}} \in \Pi_{\text{int}}$ is the point of set of internal parameters which is defined from the solution of some optimization problem with respect to prescribed quality criterion within the limits of the fixed structure of regulation and with nominal values of the external parameters P_0.

In Sec. 4.3 we also studied bifurcation transitions in detail. Systematized information about regularities in the occurrence of various m-cycles and about the nature of oscillation complication is represented as branching patterns. Analysis of such branching patterns allows us to extend the statements concerning causes of occurrence of non-determinate modes and catastrophic chaotization of oscillations, which was given earlier in works of Baushev and Zhusubaliyev [14, 18]. The basic branch $V_{1,1}$ intersects with some $V_{i,j}$ as it was seen in Fig. 4.11. Here the system may choose different motions depending on initial conditions and, if a certain ratio between the radii of the basins of attractions and the magnitude of the random disturbances is realized, then either non-determinate dynamics occurs, or it becomes possible for the system to catastrophically jump from determinate to non-determinate dynamics and backwards.

In Sec. 4.4 we showed that non-determinate modes in the dynamics of a relay system can occur even if the single stable periodical motion exists. This happens when the phase point jumps from one part of the trajectory of a stable periodical motion to another part of the same motion owing to smallness of radius of the inner basin of attraction (see Figs. 4.31, 4.32). The results of these investigations, based on earlier works by Baushev *et al.* [13, 14, 18] and by Neimark and Landa [15], and supported by the data of numerical experiments allow us to clearly describe the causes of non-determinate behavior and chaotization of oscillation in the dynamics of relay control systems.

Finally, in Sec. 4.5 we performed a detailed analysis of the complex behavior of a relay system with hysteresis and time deadzone. A DC electric drive, in which the motor torque was stabilized by a relay control system, was considered as an example.

Bibliography

[1] Yu. K. Rozanov, Semiconductor Transformers with a Link of High Frequency (Energoatomizdat, Moscow, 1987), in Russian.

[2] P. R. K. Chetty, Switch-Mode Power Supply Design (TAB Books Inc., 1986).

[3] B. K. Bose (Editor), Modern Power Electronics: Evolution, Technology, and Applications (IEEE Press, New York, 1992).

[4] N. Mohan, T. M. Undeland, and W. P. Robbins, Power Electronics: Converters, Applications and Design (Wiley, New York, 1995).

[5] A. Magauer and S. Banerjee, *Bifurcations and Chaos in the Tolerance Band PWM Technique*, IEEE Trans. Circuits Syst.-I: Fund. Theory Appl. **47**, 254–259 (2000).

[6] S. Banerjee and G. C. Verghese (Editors), Nonlinear Phenomena in Power Electronics (IEEE Press, New York, 2001).

[7] O. Wasynczuk, S. D. Sudhoff, T. D. Tran, D. H. Clayton, and H. J. Hegner, *A Voltage Control Strategy for Current-Regulated PWM Inverters*, IEEE Trans. Power Electron. **11**, 7–15 (1996).

[8] V. S. Baushev, Zh. T. Zhusubaliyev, and Yu. V. Kolokolov, *On Analysis of Relay Control Systems for Regulation of Electric Current in Modes of Electrodynamic Braking of High-Speed Electric Trains*, Electrichestvo **7**, 66–70 (1989), in Russian.

[9] Yu. V. Kolokolov, Zh. T. Zhusubaliyev, L. Yu. Veitsman, and R. M. Berzin, *Automatic System Controlling the Traction Electric Drives of the Second High-Speed Electric Train ER200*, Elektrotekhnika **9**, 49–52 (1990), in Russian.

[10] P. S. Landa, Nonlinear Oscillations and Waves in Dynamical Systems (Kluwer Academic Publ., Dordrecht, 1996).

[11] Zh. T. Zhusubaliyev, E. A. Soukhoterin, V. N. Rudakov, Yu. V. Kolokolov, and E. Mosekilde, *Bifurcations and Chaotic Oscillations in an Automatic Control Relay System with Hysteresis*, Int. J. Bifurcation and Chaos **11**, 1193–1231 (2001).

[12] M. di Bernardo and Ch. K. Tse, *Chaos in Power Electronics: An Overview*, in Chaos in Circuits and Systems, edited by G. Chen and T. Ueta (World Scientific, Singapore, 2002), pp. 317–340.

[13] V. S. Baushev, A. V. Kobzev, and Yu. N. Tanovitski, *Normal Structures of Dynamical Objects*, in Hardware and Software Means for Automation of Technological Processes (Tomsk State University Publ., Tomsk, 1997), pp. 146–152, in Russian.

[14] V. S. Baushev, Zh. T. Zhusubaliyev, and S. G. Michal'chenko, *Stochastic Features in the Dynamic Characteristics of a Pulse-Width Controlled Voltage Stabilizer*, Electrical Technology (1), 137–150 (1996), Elsevier Science.

[15] Yu. I. Neimark and P. S. Landa, Stochastic and Chaotic Oscillations (Nauka Publ., Moscow, 1987).

[16] N. S. Postnikov, *Dynamic Chaos in Relay Systems with Hysteresis*, Computational Mathematics and Modeling **8**, 62–72 (1997).

[17] N. S. Postnikov, *Stochasticity of Relay Systems with Hysteresis*, Avtomatika i Telemekhanika **3**, 57–68 (1998), in Russian.

[18] V. S. Baushev and Zh. T. Zhusubaliyev, *Indeterminable States of a Voltage Regulator with Pulse-Width Control*, Electrical Technology (3), 85–98 (1992), Elsevier Science.

[19] Zh. T. Zhusubaliyev, *Investigation of Chaotic Regimes of a Voltage Converter with Pulse-Width Modulation*, Elektrichestvo **6**, 40–46 (1997), in Russian.

[20] V. I. Arnol'd, Catastrophe Theory (Springer–Verlag, 1992).

[21] A. A. Krasovski, The Reference Book on Automatic Control Theory (Nauka Publ., Moscow, 1987), in Russian.

[22] V. S. Anishchenko, Dynamical Chaos — Models and Experiments. Appearance Routes and Structure of Chaos in Simple Dynamical Systems (World Scientific, Singapore, 1995).

[23] V. S. Anishchenko and G. I. Strelkova, *Irregular Attractors*, Discrete Dynamics in Nature and Society **2**, 53–72 (1998).

[24] V. S. Anishchenko, T. E. Vadivasova, G. I. Strelkova, and A. S. Kopeikin, *Chaotic Attractors of Two-dimensional Invertible Maps*, Discrete Dynamics in Nature and Society **2**, 249–256 (1998).

[25] N. V. Butenin, Yu. I. Neimark, and N. A. Fufayev, Introduction to the Theory of Nonlinear Oscillations (Nauka Publ., Moscow, 1987), in Russian.

[26] Yu. I. Neimark, *Dynamical System as a Fundamental Model of Modern Science*, Avtomatika i Telemekhanika **3**, 196–201 (1999).

[27] Zh. T. Zhusubaliyev and E. A. Soukhoterin, *Oscillations in a Relay Control System with Hysteresis and Time Dead Zone*, Mathematics and Computers in Simulation **58**, 329–350 (2002).

Chapter 5

CHAOTIC OSCILLATIONS IN PULSE-WIDTH MODULATED SYSTEMS

5.1 Application of Pulse-Width Modulation in Power Electronics

Pulse modulated systems have a wide range of applications in control engineering, communication, and information technology. Similar types of regulation is also used in many biological control systems. It is well-known, for instance, that the insulin producing pancreatic β-cells in intact islets of Langerhans display a bursting behavior with alternations between an active (spiking) phase and a silent phase [1–4]. It is also established [5, 6] that the secretion of insulin depends on the fraction of time that the cells spend in the active state, and that this fraction increases with the concentration of glucose in the extracellular environment. The bursting dynamics controls the influx of Ca^{+2} ions into the cell, and calcium is an essential trigger for the release of insulin. In this way the bursting dynamics takes part in the control of the glucose concentration in the blood. At glucose concentration below 5 mM, the cells do not burst at all. For high glucose concentrations ($>$ 22 mM), on the other hand, the cells spike continuously, and the secretion of insulin saturates [7].

Pulse modulation is also applied in many areas of science where the technique is used, for instance, to stabilize the experimental conditions and minimize the significance of exogenous noise or gradual drift. An example is the X-ray scattering experiments performed by Carlson *et al.* [8, 9] at IBM Research Center in the early 1970's on one of the first fully-computerized experimental set-ups.

The purpose was to study the spectral and angular distribution of piezoelectrically amplified sound flux in highly doped GaAs where the dominant acoustic frequencies typically fall in the 50–100 GHz. The experiments were performed by means of a differential photon counting technique with one counter operating during a relatively short period (1 μs) of acoustic amplification and another counter operating during a subsequent period of vanishing acoustic intensity. The measurements were performed with a repetition frequency of 50–100 Hz over an 8 h period and with an IBM 1800 system on-line for controlling the diffractometer, rotating the crystal and the detectors, and logging the data. With these introductory remarks, let us restrict the following considerations to problems related to power engineering.

In the 1960's highly reliable, effective and light-weight power systems for applications in spacecrafts appeared. Strict limitations to the on-board power consumption had led to the development of pulse-modulated methods of electrical energy conversion. This marked the start of the present epoch of modern power electronics. Methods developed in the space industry were soon applied for power conversion in other industrial fields, and today limitations on the power consumption have become a major issue in the design of most electrotechnical equipment [10–15].

Power electronics is used to convert electrical energy from one form to another. To ensure a high efficiency in the conversion, the active semiconductor devices (thyristors, power transistors, etc.) always operate in a switching mode in combination with passive components (power diodes, inductors, capacitors, and transformers). By using high *switching frequencies* one can in general reduce the sizes and masses of capacitors, inductors and transformers. Furthermore, increasing the frequency often makes it possible to raise the efficiency. However, the higher frequency also produces higher switching losses as well as other losses and limitations associated with high-frequency operation of the various components. Switching frequencies in the range 15–300 kHz are common at the present time [10–14].

Power electronics technology is widely used in the home as well as in the industry [10–13]. Examples are the domestic light dimmer, switched-mode power supplies in all electronic appliances including personal computers, heating and lighting controls, electronic ballasts for fluorescent lamps, drives for industrial motion control, induction heating, battery chargers, traction regulators in locomotives, solid-state relays and circuit breakers, offline DC power supplies, uninterruptible power supplies, conditioning for alternative energy sources, automobile electronics, and electric vehicles [10–14, 16].

Since electrical sources can be either DC (direct current) or AC (alternating current) there are four basic types of converters: AC/DC, DC/AC, AC/AC, and DC/DC. DC/DC converters are among the most widely used circuits in power electronics.

In DC/DC converters the constant output voltage is obtained by rectifying and filtering of an AC voltage. There are two methods of DC to AC voltage conversion. According to the first method the input constant voltage is broken at high frequency to produce a rectangular waveform. This voltage is subsequently brought to higher or lower levels by means of a transformer. The transformer output voltage (displaying the pulse waveform) is rectified and then filtered with capacitors and/or inductors that are configured to respond to its average value (i.e., its DC component). Changing the pulse ratio we can control the value of the DC component. This control approach is referred to as *pulse-width modulation.* DC/DC converters using such control are also referred to as *switching regulators*, *switched-mode converters* or *high-frequency PWM DC/DC converters* [10–16].

With another method, referred to as the *resonance method*, the constant input voltage is applied to a resonance LC circuit via control switches. In this case the frequency of conversion is determined by the resonance frequency of the circuit rather than by the clock frequency as for the switch mode approach [10–16].

The control of DC/DC converters usually applies one of two approaches, namely *voltage-mode* control or *current-mode* control [17, 18]. In voltage-mode control, the output voltage is compared with a reference to generate a control signal that drives the pulse width modulator via some typical feedback compensation configuration. For current-mode control, an inner current loop is used in addition to the voltage feedback loop, the aim of which is to force the peak inductor current to follow a reference signal which is derived from the output voltage feedback loop. The result of current-mode control is often a faster response [17, 18].

Figure 5.1 shows the typical functional scheme of a *switched-mode power supply* with pulse-width modulated feedback control that is a particular example of high-frequency PWM DC/DC converters.

The supply includes the input rectifier with filter. The rectifier may be

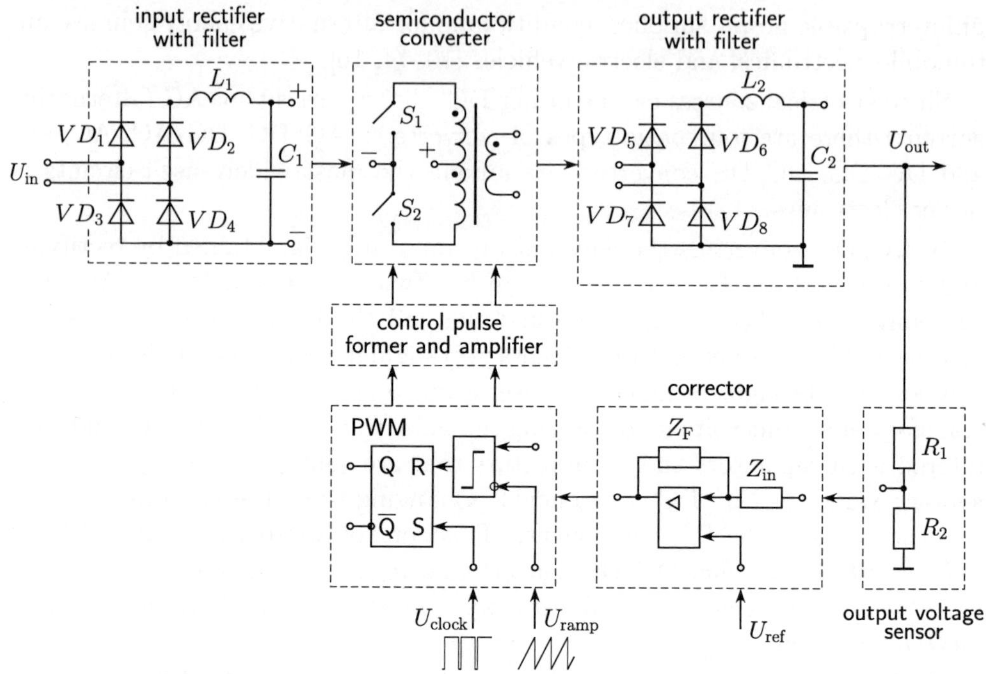

Figure 5.1. Functional scheme of switched-mode power supply with pulse-width modulated feedback control.

single-phased or multi-phased, depending on the available mains supply and the magnitude of power being converted. The filter, as a rule, is intended to suppress disturbances and compensate the recuperation currents caused by the peculiarities of electromagnetic elements operating at high frequencies. The ripple of the mains is usually not suppressed by the input filter; this function is entrusted to the control system. The output voltage of the input filter is supplied to the power converter. The transformer of the power converter scales voltages and currents, and also provides electrical isolation between source and load. The converter transforms the non-stabilized constant voltage to high-frequency square pulses with variable polarity. Connected to the secondary winding of the transformer, the demodulator rectifies this pulse sequence of variable polarity. The rectified voltage is supplied to the power LC-filter, and the stabilized constant voltage U_{out} is taken from the output of this filter.

The divider (output voltage sensor) is connected to the output of the filter.

The output signal U_{FB} of the divider is supplied to one of the inputs of the corrector, while the reference voltage U_{ref} is supplied to another input of the corrector. The error signal amplifier compares the signal of the output voltage sensor with the reference voltage U_{ref}, then corrects and amplifies the obtained difference and supplies the resulting signal to the pulse-width modulator. The modulator transforms the continuous signal to a pulse-width modulated signal with variable pulse ratio, which drives the power switches of the converter.

There are several different kinds of pulse-width modulation and of modifications with different circuitries and functioning principles, which can be chosen depending on the particular operating conditions and modes of the converters. There is no generally accepted classification of the different types of pulse-width modulation in systems theory. However, many specialists in the field of power electronics and control theory follow the classification given by Tsypkin [19] (see also [15, 20]). We will also follow this classification. The main and practically most important of these classes are denoted PWM-1 and PWM-2. For both of these methods, pulse-width modulation may be performed on the trailing edge, the leading-edge, or both edges of the output pulse. This is referred to as *trailing-edge*, *leading-edge*, and *double-edge modulation*, respectively.

Figure 5.2(a) shows the circuitry for pulse-width modulation of the second kind (PWM-2). The circuit compares the output signal of the corrector with the sawtooth voltage $U_{ramp}(t)$ and produces square pulses. This example shows the trailing-edge modulation.

Figure 5.2(b) displays time plots that explain the principle of control pulse generation for pulse-width modulation of the first and the second kind. The algorithm of PWM-1, in contrast to PWM-2, sets the data points that determine the values of U_{corr} to be compared with the sawtooth voltage $U_{ramp}(t)$ in discrete time instants with a constant interval that is given by the period of the clock pulses a.

Power electronic circuits can display a variety of nonlinear phenomena, including local, homoclinic and border-collision bifurcations, chaos, multistability, quasiperiodic behavior, etc. The wide use of power electronic circuits in engineering obviously motivates the quest for a deeper understanding of the nature of their bifurcations and the conditions for the appearance of complex behavior.

Since first reported by Hamill [21], bifurcation phenomena and chaotic oscillations in the simplest power converters with pulse-width modulation have been studied in numerous works [18, 22–32]. di Bernardo and Tse [18] provide

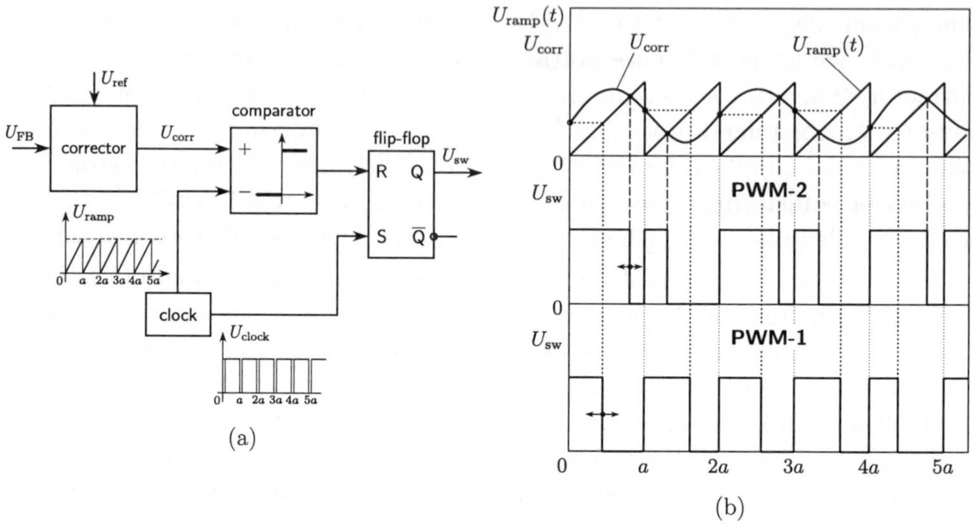

Figure 5.2. (a) Circuitry for a control feedback with pulse-width modulation of the second kind. This control configuration is routinely and widely used in single-chip implementations. (b) Time plots that explain the principle of control pulse generation for pulse-width modulation of the first (PWM-1) and the second (PWM-2) kind.

a collection of important results in the fields of numerical and experimental investigation of chaotic behavior in power electronic circuits. A chronological survey of results obtained in the analysis and modeling of nonlinear phenomena in power electronic circuits and systems is also presented in the recent collection of papers edited by Banerjee and Verghese [14]. Among the earliest works providing a detailed numerical and analytical investigation of bifurcations and transitions to chaos in DC/DC converters we should recall the paper by Baushev and Zhusubaliyev [33] (see also [34]). These authors showed that such systems can display a situation where there is a wide range of parameters with many locally stable limit cycles with different dynamic characteristics, including regions of coexistence. The cycles typically arise in hard transitions, for example through a saddle-node or border-collision bifurcation [35, 37–40], and with changing parameters each of them can undergo either a finite or an infinite sequence of period-doubling bifurcations, resulting in the transition to chaos. Similar results were obtained by Fossas and Olivar [36] using an approach similar to one described in the work by Baushev and Zhusubaliyev [33].

El Aroudi and Leyva [31] studied quasiperiodic behavior in a pulse-width modulated DC/DC boost converter. The observed phenomena include a subcritical torus birth (Neimark-Sacker, or Andronov-Hopf) bifurcation, the associated hysteretic transitions, and the onset of chaos via torus destruction. The authors described two different ways for the destruction of the torus: via period-doubling of resonance cycles and through loss of torus smoothness.

The complex behavior of switching power converters connected with border-collision and grazing bifurcations [35, 37–40] have been described by numerous researchers [18, 22–29].

Banerjee and Grebogi [23] (see also [25]) considered the general class of two-dimensional piecewise-smooth maps that depend on a single parameter. They present results from analyses of border-collision bifurcations by means of a so-called normal form, i.e., a piecewise linear approximation in the neighborhood of the border. According to the *normal form theory* of border collisions, the kind of bifurcation can be determined in terms of the eigenvalues of the Jacobian matrix of the border crossing fixed point before and after the border collision. A classification of border-collision bifurcations for two-dimensional piecewise-smooth maps is given, and it is shown that a complete classification includes eleven qualitatively different types of border-collision bifurcations depending on the parameters of the normal form. The obtained results are used to explain the bifurcation behavior of an electronic DC/DC converters. Banerjee *et al.* [24] also presented a classification of all possible border-collision bifurcations for one-dimensional piecewise-smooth maps in terms of the parameters of the normal form. These authors applied their results to explain the dynamics of two power electronic switching circuits. Banerjee *et al.* [22] showed that robust chaotic dynamics can occur in two-dimensional piecewise-smooth maps and obtained the conditions for such a dynamics to arise. The transition to chaos was illustrated for a two-dimensional model of a current mode controlled boost converter.

Dutta *et al.* [41] demonstrated that piecewise smooth systems can exhibit border-collision bifurcations in which multiple coexisting attractors are created simultaneously. These phenomena were described for a simple electronic switching circuit. Dutta *et al.* refer to these bifurcations as *multiple attractor bifurcations*, while Kapitaniak and Maistrenko [42] call them *multiple choice bifurcations.* Similar results were independently obtained by Zhusubaliev *et al.* [26] when analyzing a multi-dimensional set of piecewise-smooth differential equations describing the behavior of DC/DC converters.

Zhusubaliev *et al.* [26] first presented a detailed analysis of various possible types of multiple attractor bifurcations in multi-dimensional piecewise-smooth systems. They showed that the typical case is a border-collision bifurcation where a chaotic attractor arises together with a stable cycle. It was also observed that a denumerable set of unstable cycles with periods that are multiple to the period of the basic cycles provides a skeleton for the chaotic attractor.

di Bernardo *et al.* [43] studied dynamical peculiarities of piecewise-smooth systems connected with the grazing, skipping and sliding phenomena in DC/DC converters. A particular case of border-collision bifurcation, referred to as a corner-collision bifurcation, was emphasized.

Zhusubaliev *et al.* [27, 44] investigated scenarios for the transition to chaos via quasiperiodicity, resonance cycles, and torus destruction for a DC/DC converter with pulse-width modulation. These authors first showed that the resonance solutions arise via border-collision bifurcations on a two-dimensional torus. The arrangement of the resonance domains within the parameter plane was shown to be related to the Farey series and the internal structure of the domains was described. It was shown that transitions to chaos mainly occur through finite sequences of border-collision bifurcations. Some other possible causes of complex dynamics were considered, including a subcritical Neimark-Sacker (Andronov-Hopf) bifurcation and the associated hysteretic phenomena.

In the present chapter we study the bifurcations and main routes to chaos for pulse-modulated control systems. As a specific example we consider a constant voltage converter with pulse-width modulation of the second kind and with proportional-plus-integrating feedback corrector. The behavior of such a converter is described by a three-dimensional system of piecewise-smooth differential equations with an external periodic action (see Sec. 5.2). Investigation of this system is reduced to studying the properties of a three-dimensional mapping (*Baushev-Zhusubaliyev map*). Section 5.3 is devoted to the methods for determining the periodic motions and investigating their local stability, a necessary starting point for the bifurcation analysis.

We first perform a more detailed analysis of some of the peculiarities of the complex behavior of multidimensional systems with piecewise-smooth differential equations. We show that border-collision bifurcations can be accompanied not only by period-doubling, but also by period-tripling, -quadrupling, -quintupling etc. of the oscillations, as well as by the soft emergence of a number of cycles with multiple periods. More complex bifurcational transitions

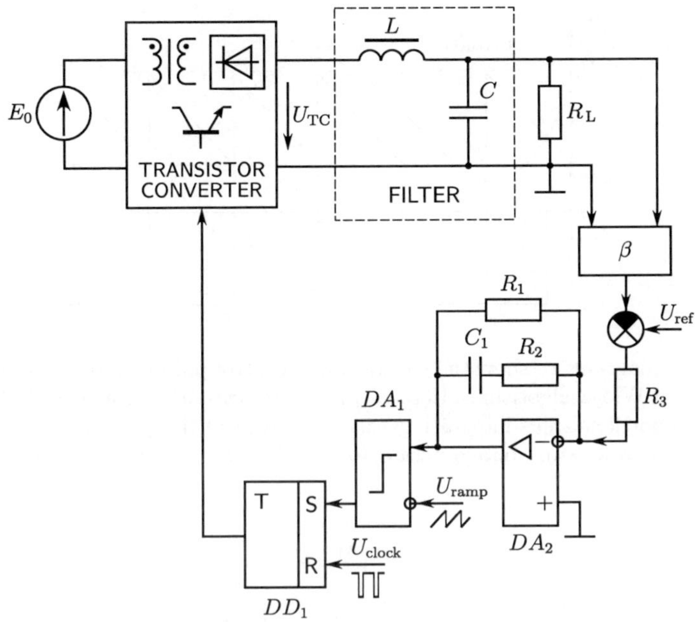

Figure 5.3. Block diagram of a constant voltage converter with pulse-width modulation of the second kind (PWM-2). Here U_{ramp} is a generator that produces sawtooth and square voltage pulses of the same period; DA_1 is the comparator; DA_2 is the error signal amplifier; U_{ref} is the reference voltage; U_{TC} is the transistor converter output voltage.

with fundamentally new properties are cited as examples, for instance, finite sequence of period-tripling bifurcations, soft transition from one family of cycles (the number of which is finite) to another family of cycles (the number of which can be finite or infinite) with multiple periods.

5.2 Voltage Converter with Pulse-Width Modulation

The block diagram of the constant voltage converter, a particular instance of a control system with pulse-width modulation, is given in Fig. 5.3. The output voltage is controlled according to the method of pulse-width modulation of the second kind. The time diagrams shown in Fig. 5.4 illustrate the principles of producing the control pulses by the switch elements of the transistor converter (see also Fig. 5.2).

The control diagram of the above control system is given in Fig. 5.5. Here

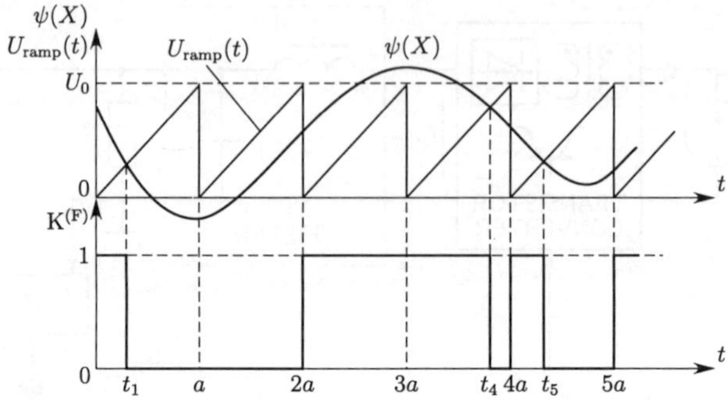

Figure 5.4. Time diagrams to show the generation of control pulses by the transistor converter according to the PWM-2 algorithm. Here $U_{\rm ramp}$ is the sawtooth voltage; $\psi(X)$ is the output signal of the proportional-plus-integrating corrector; $K^{(F)}$ is the pulse-width modulator output signal; a is the sawtooth voltage period; t_k are the switching instants of the pulse-width modulator.

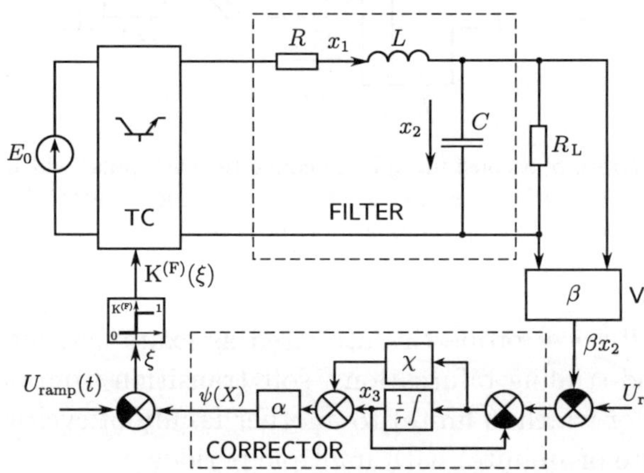

Figure 5.5. Control diagram of the constant voltage converter with pulse-width modulation of the second kind (PWM-2). Here TC is the transistor converter; CORRECTOR is the proportional-plus-integrating corrector; VS is the voltage sensor. In the following analysis the amplification constant α of the corrector will be considered as the main bifurcation parameter. In our two-dimensional bifurcation diagrams, the transfer constant χ serves as the other bifurcation parameter.

CORRECTOR is the proportional-plus-integrating correcting link; TC is the transistor converter; and VS is the voltage sensor. In accordance with the diagram, the behavior of the considered system is described by the three-dimensional piecewise-linear nonautonomous system of ordinary differential equations:

$$\frac{\mathrm{d}X}{\mathrm{d}t} = G(t, X); \tag{5.1}$$

$$X = (x_1, x_2, x_3)^{\mathrm{T}};\ G(t, X) = AX + B(t, X);$$

$$A = \begin{pmatrix} -\dfrac{R}{L} & -\dfrac{1}{L} & 0 \\ \dfrac{1}{C} & -\dfrac{1}{CR_{\mathrm{L}}} & 0 \\ 0 & -\dfrac{\beta}{\tau} & -\dfrac{1}{\tau} \end{pmatrix};\ B(t, X) = \begin{pmatrix} \dfrac{E_0}{L}\,\mathrm{K}^{(\mathrm{F})}(\xi) \\ 0 \\ \dfrac{U_{\mathrm{ref}}}{\tau} \end{pmatrix};\ \xi = \xi(t, X).$$

Here X is the state vector. x_1 and x_2 are the current in the ripple LC-filter reactor's inductance coil and the output voltage, respectively, and x_3 is the voltage of the corrector's integrator. L and C are the inductance and the capacity of the LC-filter; R is the resistor characterizing the dissipation in the filter inductance coil and in the converter. R_{L} is the load resistor; E_0 is the input voltage; $\tau = C_1(R_1 + R_2)$ is the integration time constant of the corrector and ξ is the control signal. $\mathrm{K}^{(\mathrm{F})}(\xi)$ is the output signal of the pulse-width modulator; U_{ref} is the prescribed value of voltage, and β is the sensitivity of the voltage sensor.

Within the range $(k-1)\,a < t < ka$ the output signal of the pulse-width modulator $\mathrm{K}^{(\mathrm{F})}(\xi)$ is determined as

$$\mathrm{K}^{(\mathrm{F})}(\xi) = \begin{cases} 1, & (k-1)\,a < t < t_k; \\ 0, & t_k < t < ka, \end{cases} \qquad k = 1, 2, 3, \ldots$$

where a is the external action period (modulation period). Depending on ξ the value of t_k is determined by the algorithm [33, 34]:

$$t_k = \begin{cases} (k-1)\,a, & \xi\big((k-1)\,a, X_{k-1}\big) \le 0; \\ \to \xi\big(t_k, X(t_k)\big) = 0, & \xi\big((k-1)\,a, X_{k-1}\big) > 0 \text{ and } \xi(ka, X_k) < 0; \\ ka, & \xi\big(t, X(t)\big) > 0 \text{ for } (k-1)\,a \le t \le ka, \end{cases} \tag{5.2}$$

where

$$\xi(t, X) = \psi(X) - U_{\mathrm{ramp}}(t);$$
$$\psi(X) = \alpha\big(\chi(U_{\mathrm{ref}} - \beta x_2) + (1-\chi)\,x_3\big);$$

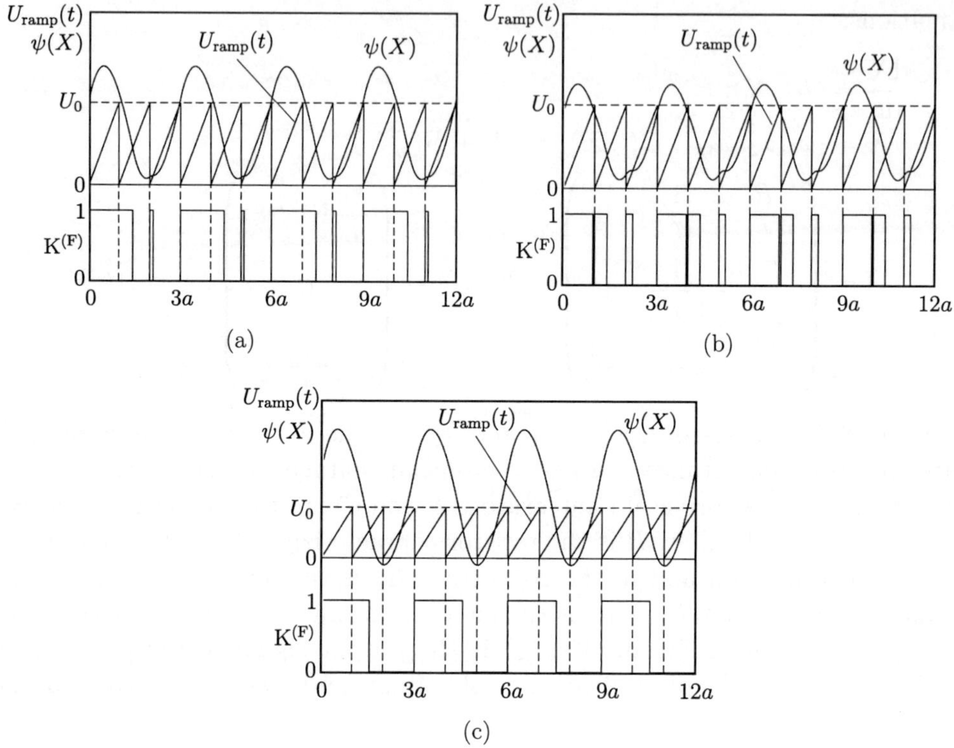

Figure 5.6. Forming of the control pulses $\mathrm{K}^{(\mathrm{F})}$ for different relationships between the feedback signal $\psi(X)$ and the sawtooth signal $U_{\mathrm{ramp}}(t)$.

$$\begin{aligned} U_{\mathrm{ramp}}(t) &= U_0\big(t/a - E_1(t/a)\big); \\ X_k &= X(ka), \end{aligned}$$

or

$$\xi(t, X) = \alpha\big(\chi(U_{\mathrm{ref}} - \beta x_2) + (1-\chi)\,x_3\big) - U_0\big(t/a - E_1(t/a)\big).$$

Here $\chi = R_2/(R_1 + R_2)$ and $\alpha = R_1/R_3$ are the transfer and amplification constants of the corrector, respectively. These two parameters will be considered as the main bifurcation parameters in the analyses to be presented below. $E_1(t/a)$ is a function defined as the greatest integer that does not exceed t/a.

In the first of the above cases the pulse duration is equal to zero, in the second case it is determined as the smallest root of the equation $\xi\big(t_k, X(t_k)\big) = 0$, and in the third case it is equal to the magnitude of the clock interval [33, 34].

Figure 5.6 illustrates some peculiarities of the formation of control pulses in accordance with Eqs. (5.2). Control systems with PWM are designed in such a way that no more than one switching of the converter can take place during one period of the external action, i.e. during each interval $(k-1)\,a < t < ka$, $k = 1, 2, 3, \ldots$ This is an important requirement which is ensured by using a latch that locks as soon as the signal $\psi(X)$ becomes equal to the external periodic action $U_{\text{ramp}}(t)$. In the simplest case this function is implemented by a flip-flop (the element DD_1 in Fig. 5.3).

The dynamical system (5.1) was investigated for the following parameters values: $E_0 = 104$ V, $R = 10.6\ \Omega$, $L = 0.1$ H, $C = 10^{-6}$ F, $R_L = 100\ \Omega$, $U_{\text{ref}} = 5$ V, $U_0 = 10$ V, $a = 10^{-4}$ s, $\tau = 4a$, $\beta = 0.1$, $0 < \chi \leq 1$, $\alpha > 0$. These values correspond to the parameters of an actual DC/DC converter used in the power supply for radio equipment [15].

With the chosen parameter values, the eigenvalues λ_i, $i = \overline{1,3}$ of the matrix A are real and negative:

$$\lambda_1 = -\frac{1}{2}\left(\frac{R}{L} + \frac{1}{CR_{\text{L}}}\right) - \sqrt{\frac{1}{4}\left(\frac{R}{L} + \frac{1}{CR_{\text{L}}}\right)^2 - \frac{1}{LC}\left(1 + \frac{R}{R_{\text{L}}}\right)};$$

$$\lambda_2 = -\frac{1}{2}\left(\frac{R}{L} + \frac{1}{CR_{\text{L}}}\right) + \sqrt{\frac{1}{4}\left(\frac{R}{L} + \frac{1}{CR_{\text{L}}}\right)^2 - \frac{1}{LC}\left(1 + \frac{R}{R_{\text{L}}}\right)};$$

$$\lambda_3 = -\frac{1}{\tau}.$$

Let us pass from the variables x_1, x_2, x_3 to the variables w_1, w_2, w_3 via the expressions

$$\begin{aligned} x_1 &= \frac{E_0/L}{\lambda_1 - \lambda_2}\left(\frac{\lambda_2 + R/L}{\lambda_1} w_1 - \frac{\lambda_1 + R/L}{\lambda_2} w_2\right); \\ x_2 &= \frac{E_0/(LC)}{\lambda_2 - \lambda_1}\left(\frac{w_1}{\lambda_1} - \frac{w_2}{\lambda_2}\right); \\ x_3 &= -\lambda_3 \sum_{i=1}^{3} \mu_i \frac{w_i}{\lambda_i} + U_{\text{ref}} \end{aligned} \tag{5.3}$$

where

$$\mu_1 = -\frac{\beta E_0}{LC(\lambda_2 - \lambda_1)(\lambda_1 - \lambda_3)};$$

$$\mu_2 = \frac{\beta E_0}{LC(\lambda_2 - \lambda_1)(\lambda_2 - \lambda_3)};$$

$$\mu_3 = -(\mu_1 + \mu_2).$$

Substituting these expressions into Eqs. (5.1) and solving with respect to the derivatives we obtain

$$\frac{\mathrm{d}w_i}{\mathrm{d}t} = \lambda_i\big(w_i - \mathrm{K}^{(\mathrm{F})}(\xi)\big),\ i = \overline{1,3}; \tag{5.4}$$

$$\xi(t, w_1, w_2, w_3) = \sum_{i=1}^{3} \gamma_i \left(\chi - \frac{\lambda_3}{\lambda_i}\right) w_i - \frac{Q}{\alpha}\left[\frac{t}{a} - E_1\left(\frac{t}{a}\right)\right] + q;$$

$$\gamma_1 = \frac{\lambda_1}{\lambda_1 - \lambda_3};\ \gamma_2 = \frac{\lambda_1}{\lambda_3 - \lambda_2};\ \gamma_3 = -(\gamma_1 + \gamma_2);$$

$$Q = \frac{U_0(\lambda_1 - \lambda_2)(R_\mathrm{L} + R)}{\beta E_0 \lambda_2 R_\mathrm{L}};$$

$$q = \frac{U_\mathrm{ref}}{U_0} Q.$$

The purpose of the substitution (5.3) is to pass from Eqs. (5.1) to the simpler system (5.4). Having the results expressed via the variables w_1, w_2, w_3 one can always return to the variables x_1, x_2, x_3 by means of Eqs. (5.3).

Within the domain $(k-1)\,a < t < t_k$ the system (5.4) has the form

$$\frac{\mathrm{d}w_i}{\mathrm{d}t} = \lambda_i(w_i - 1),\ w_i\big((k-1)a\big) = w_{i(k-1)}.$$

The solution of this system is

$$w_i(t) = 1 + (w_{i(k-1)} - 1)\mathrm{e}^{\lambda_i(t-(k-1)a)},$$

or, at the time of switching,

$$w_i(t_k) = 1 + (w_{i(k-1)} - 1)\mathrm{e}^{\lambda_i(t_k-(k-1)a)}.$$

Within the domain $t_k < t < ka$ the system (5.4) will take the form

$$\frac{\mathrm{d}w_i}{\mathrm{d}t} = \lambda_i w_i,$$

with the solution

$$w_i(t) = w_i(t_k)\mathrm{e}^{\lambda_i(t-t_k)}.$$

Therefore for $t = ka$

$$w_i(ka) = w_i(t_k)\mathrm{e}^{\lambda_i(ka-t_k)},$$

or

$$w_{ik} = (w_{i(k-1)} - 1)\mathrm{e}^{a\lambda_i} + \mathrm{e}^{\lambda_i(ka-t_k)}.$$

In this way, the investigation of the dynamical system (5.1) can be reduced to the study of the properties of the three-dimensional mapping (*Baushev-Zhusubaliyev map*) [33, 34, 45]

$$\begin{cases} w_{1k} = \mathrm{e}^{a\lambda_1} w_{1(k-1)} + \mathrm{e}^{a\lambda_1(1-z_k)} - \mathrm{e}^{a\lambda_1}; \\ w_{2k} = \mathrm{e}^{a\lambda_2} w_{2(k-1)} + \mathrm{e}^{a\lambda_2(1-z_k)} - \mathrm{e}^{a\lambda_2}; \\ w_{3k} = \mathrm{e}^{a\lambda_3} w_{3(k-1)} + \mathrm{e}^{a\lambda_3(1-z_k)} - \mathrm{e}^{a\lambda_3}. \end{cases} \tag{5.5}$$

or

$$w_{ik} = \mathrm{e}^{a\lambda_i} w_{i(k-1)} + \mathrm{e}^{a\lambda_i(1-z_k)} - \mathrm{e}^{a\lambda_i}, \; i = 1, 2, 3; \; k = 1, 2, \ldots$$

Here $w_{ik} = w_i(ka)$; $z_k = t_k/a - (k-1)$, $0 \leq z_k \leq 1$ is the relative pulse duration (pulse ratio).

The recurrence relations (5.5) must be supplemented by an algorithm for calculating z_k [33, 34]:

$$z_k = \begin{cases} 0, & \varphi(0) \leq 0; \\ z^*, & \varphi(0) > 0 \text{ and } \varphi(1) < 0; \\ 1, & \varphi(z) > 0 \text{ for } 0 \leq z \leq 1. \end{cases} \tag{5.6}$$

Here z^* is the smallest root of the equation $\varphi(z) = 0$,

$$\varphi(z) = -\sum_{i=1}^{3} \gamma_i \left(\chi - \frac{\lambda_3}{\lambda_i}\right) \left(1 - w_{i(k-1)}\right) \mathrm{e}^{a\lambda_i z} - \frac{Q}{\alpha} z + q - \frac{\lambda_1}{\lambda_2} + 1.$$

The combination of the recurrence relations (5.5) and the algorithm (5.6) determines a shift map in the space of variables (w_1, w_2, w_3, t). The algorithm for solving the Cauchy problem by means of the mapping (5.5), (5.6) is: first to evaluate z_k using Eq. (5.6) for $w_{i(k-1)}$ and then to substitute z_k into Eq. (5.5) to obtain w_{ik}.

It is not difficult to show, that when $\chi = 1$ the map (5.5), (5.6) becomes two-dimensional

$$\begin{cases} w_{1k} = \mathrm{e}^{a\lambda_1} w_{1(k-1)} + \mathrm{e}^{a\lambda_1(1-z_k)} - \mathrm{e}^{a\lambda_1}; \\ w_{2k} = \mathrm{e}^{a\lambda_2} w_{2(k-1)} + \mathrm{e}^{a\lambda_2(1-z_k)} - \mathrm{e}^{a\lambda_2}. \end{cases}$$

Here z_k is calculated by the algorithm (5.6), where

$$\varphi(z) = \frac{\lambda_1}{\lambda_2}\left(1 - w_{2(k-1)}\right) e^{a\lambda_2 z} - \left(1 - w_{1(k-1)}\right) e^{a\lambda_1 z} - \frac{Q}{\alpha} z + q - \frac{\lambda_1}{\lambda_2} + 1.$$

Such a map describes the behavior of a constant voltage converter with proportional feedback corrector. We have already considered this map in Chapter 2 where different bifurcation scenarios were discussed. A more detailed analysis may be found in the original literature [33, 34].

5.3 Periodic Solutions and Their Local Stability

A. The period T of the periodic motion of the dynamical system (5.1) is multiple to the external action period, i.e. $T = ma$, $m = 1, 2, \ldots$. As before we denote the motion with such a period as an m-cycle [33, 34].

Let us consider the general form of recurrence relations (5.5):

$$w_{ik} = e^{a\lambda_i} w_{i(k-1)} + e^{a\lambda_i(1-z_k)} - e^{a\lambda_i}, \; i = 1, 2, 3; k = 1, 2, \ldots$$

From this expression we can successively obtain

$$\begin{aligned}
w_{i1} &= e^{a\lambda_i}(w_{i0} - 1) + e^{a\lambda_i(1-z_1)};\\
w_{i2} &= e^{a\lambda_i}(w_{i1} - 1) + e^{a\lambda_i(1-z_2)} = e^{2a\lambda_i} w_{i0} + e^{a\lambda_i(2-z_1)} + e^{a\lambda_i(1-z_2)} - e^{a\lambda_i} - e^{2a\lambda_i};\\
&\ldots\\
w_{ik} &= e^{ka\lambda_i} w_{i0} + e^{a\lambda_i(k-z_1)} + \cdots + e^{a\lambda_i(1-z_k)} - e^{a\lambda_i}(1 + e^{a\lambda_i} + \cdots + e^{\lambda_i(k-1)a}).
\end{aligned}$$

Taking into consideration, that the expression within the brackets is the sum of a geometric series, we write

$$w_{ik} = e^{ka\lambda_i} w_{i0} + \sum_{j=1}^{k} e^{a\lambda_i(k+1-j-z_j)} - e^{a\lambda_i}\frac{1 - e^{ka\lambda_i}}{1 - e^{a\lambda_i}}. \tag{5.7}$$

Using the periodicity condition $w_{i0} = w_{im}$, from (5.7) we find for $k = m$:

$$w_{i0} = e^{ma\lambda_i} w_{i0} + \sum_{j=1}^{m} e^{a\lambda_i(m+1-j-z_j)} - e^{a\lambda_i}\frac{1 - e^{ma\lambda_i}}{1 - e^{a\lambda_i}}.$$

Hence

$$w_{i0} = \frac{1}{1 - e^{ma\lambda_i}} \sum_{j=1}^{m} e^{a\lambda_i(m+1-j-z_j)} - \frac{e^{a\lambda_i}}{1 - e^{a\lambda_i}}. \tag{5.8}$$

Finally, the expression for w_{ik} takes the form

$$w_{ik} = \frac{1}{1-\mathrm{e}^{ma\lambda_i}} \sum_{j=1}^{m} \mathrm{e}^{a\lambda_i(m+k+1-j-z_j)} + \sum_{j=1}^{k} \mathrm{e}^{a\lambda_i(k+1-j-z_j)} - \frac{\mathrm{e}^{a\lambda_i}}{1-\mathrm{e}^{a\lambda_i}}. \tag{5.9}$$

Let us now find the expression for $1 - w_{i(k-1)}$:

$$\begin{aligned}
1 - w_{i(k-1)} &= 1 - \frac{1}{1-\mathrm{e}^{ma\lambda_i}} \sum_{j=1}^{m} \mathrm{e}^{a\lambda_i(m+k-j-z_j)} - \sum_{j=1}^{k-1} \mathrm{e}^{a\lambda_i(k-j-z_j)} + \frac{\mathrm{e}^{a\lambda_i}}{1-\mathrm{e}^{a\lambda_i}} \\
&= \frac{1}{1-\mathrm{e}^{a\lambda_i}} - \frac{1}{1-\mathrm{e}^{ma\lambda_i}} \sum_{j=1}^{m} \mathrm{e}^{a\lambda_i(m+k-j-z_j)} - \sum_{j=1}^{k} \mathrm{e}^{a\lambda_i(k-j-z_j)} + \mathrm{e}^{-a\lambda_i z_k} \\
&= (1-\sigma_{ik})\,\mathrm{e}^{-a\lambda_i z_k},
\end{aligned}$$

where

$$\sigma_{ik} = \mathrm{e}^{a\lambda_i z_k} \left[\frac{1}{1-\mathrm{e}^{ma\lambda_i}} \sum_{j=1}^{m} \mathrm{e}^{a\lambda_i(m+k-j-z_j)} + \sum_{j=1}^{k} \mathrm{e}^{a\lambda_i(k-j-z_j)} - \frac{1}{1-\mathrm{e}^{a\lambda_i}} \right], i = 1,2,3.$$

Substituting $1 - w_{i(k-1)}$ into the expression (5.6) for $\varphi(z)$ we find

$$\varphi_k(z_1, z_2, \ldots, z_m) = \sum_{i=1}^{3} \gamma_i \left(\chi - \frac{\lambda_3}{\lambda_i} \right) \sigma_{ik} - \frac{Q}{\alpha} z_k + q,\ k = \overline{1,m}.$$

Finally, using (5.6) we obtain the system of transcendental equations with respect to z_k, $k = \overline{1,m}$ [33]:

$$\varphi_k(z_1, z_2, \ldots, z_m) = \mu_k(z_k),\ k = \overline{1,m}. \tag{5.10}$$

$$\mu_k(z_k) = \begin{cases} 0, & 0 < z_k < 1; \\ > 0, & z_k = 1; \\ < 0, & z_k = 0. \end{cases}$$

The above algorithm allows us to evaluate not only the stable m-cycles but also the unstable cycles. If z_k, $k = \overline{1,m}$ are the solutions to the Eq. (5.10) then the m-cycle w_{ik}, $k = \overline{1,m}$ can be evaluated according to Eq. (5.9).

We solve the system (5.10) by means of Newton's method:

$$\begin{cases} \dfrac{\partial \Phi(Z^{(s)})}{\partial Z} \Delta Z^{(s)} = -\Phi(Z^{(s)}); \\ Z^{(s+1)} = Z^{(s)} + \Delta Z^{(s)}; \end{cases} \quad s = 0, 1, 2, \ldots,$$

where, provided that $0 < z_k < 1$ for all $k = \overline{1, m}$

$$Z = \begin{pmatrix} z_1 \\ z_2 \\ \vdots \\ z_m \end{pmatrix}; \; \Phi(Z) = \begin{pmatrix} \varphi_1(z_1, z_2, \dots, z_m) \\ \varphi_2(z_1, z_2, \dots, z_m) \\ \vdots \\ \varphi_m(z_1, z_2, \dots, z_m) \end{pmatrix}; \; \mu_k(z_k) = 0;$$

$$\frac{\partial \Phi(Z)}{\partial Z} = \begin{pmatrix} \dfrac{\partial \varphi_1}{\partial z_1} & \dfrac{\partial \varphi_1}{\partial z_2} & \cdots & \dfrac{\partial \varphi_1}{\partial z_m} \\ \dfrac{\partial \varphi_2}{\partial z_1} & \dfrac{\partial \varphi_2}{\partial z_2} & \cdots & \dfrac{\partial \varphi_2}{\partial z_m} \\ \cdots & \cdots & \cdots & \cdots \\ \dfrac{\partial \varphi_m}{\partial z_1} & \dfrac{\partial \varphi_m}{\partial z_2} & \cdots & \dfrac{\partial \varphi_m}{\partial z_m} \end{pmatrix};$$

$$\frac{\partial \varphi_k}{\partial z_j} = \sum_{i=1}^{3} \gamma_i \left(\chi - \frac{\lambda_3}{\lambda_i} \right) \frac{\partial \sigma_{ik}}{\partial z_j} - \frac{Q}{\alpha} \delta_{k,j}, \quad k, j = \overline{1, m}; \quad \delta_{k,j} = \begin{cases} 0, & k \neq j; \\ 1, & k = j; \end{cases}$$

$$\frac{\partial \sigma_{ik}}{\partial z_j} = \begin{cases} \dfrac{a\lambda_i \mathrm{e}^{a\lambda_i z_k}}{1 - \mathrm{e}^{ma\lambda_i}} \left(\sum\limits_{j=1}^{k-1} \mathrm{e}^{a\lambda_i(k-j-z_j)} + \sum\limits_{j=k+1}^{m} \mathrm{e}^{a\lambda_i(m+k-j-z_j)} - \dfrac{1 - \mathrm{e}^{ma\lambda_i}}{1 - \mathrm{e}^{a\lambda_i}} \right), & j = k; \\ -\dfrac{a\lambda_i}{1 - \mathrm{e}^{ma\lambda_i}} \mathrm{e}^{a\lambda_i(k-j+z_k-z_j)}, & j < k; \\ -\dfrac{a\lambda_i}{1 - \mathrm{e}^{ma\lambda_i}} \mathrm{e}^{a\lambda_i(m+k-j+z_k-z_j)}, & j > k. \end{cases}$$

B. Investigation of the local stability of periodic solutions can be performed in accordance with the theory outlined in Chapter 3.

The equation with respect to $F(t)$, which corresponds to (3.20), has the form

$$\frac{\mathrm{d}F}{\mathrm{d}t} = \Big[A + A_k(\vartheta_1)\delta(t - t_k) \Big] F, \; F(0) = E. \tag{5.11}$$

Here

$$A_k(\vartheta_1) = \left. \frac{1}{\dfrac{\partial \xi}{\partial t} + \left(\dfrac{\partial \xi}{\partial X} \right)^{\mathrm{T}} \left(G_k^- + \vartheta_1 \Delta G_k \right)} \Delta G_k \left(\frac{\partial \xi}{\partial X} \right)^{\mathrm{T}} \right|_{\substack{X = X_c \\ t = t_k}},$$

where

$$\xi(t,X) = \chi(U_{\rm ref} - \beta x_2) + (1-\chi)\,x_3 - \frac{U_0}{\alpha}\big(t/a - E_1(t/a)\big);$$

$$\left(\frac{\partial \xi}{\partial X}\right)^{\rm T} = \left(\frac{\partial \xi}{\partial x_1}, \frac{\partial \xi}{\partial x_2}, \frac{\partial \xi}{\partial x_3}\right) = \big(0, -\chi\beta, 1-\chi\big);$$

$$\frac{\partial \xi}{\partial t} = \begin{cases} -\dfrac{U_0}{\alpha a}, & t_k \neq ka; \\ \dfrac{U_0}{\alpha a}\,\delta(t-ka), & t_k = ka; \end{cases}$$

$$\Delta G_k = G_k^+ - G_k^-,\ G_k^{\pm} = \lim_{t\to t_k \pm 0} G_{\pm}(t, X_{\rm c}),$$

$$G_-(t, X_{\rm c}) = AX_{\rm c} + B(1),\ G_+(t, X_{\rm c}) = AX_{\rm c} + B(0),$$

$$B(1) = \begin{pmatrix} E_0/L \\ 0 \\ U_{\rm ref}/\tau \end{pmatrix},\ B(0) = \begin{pmatrix} 0 \\ 0 \\ U_{\rm ref}/\tau \end{pmatrix},\ \Delta G_k = \begin{pmatrix} -E_0/L \\ 0 \\ 0 \end{pmatrix}.$$

Let us make two important remarks before proceeding to determine the basic matrix $F(T)$:

First, when $t = ka$

$$\frac{\partial \xi}{\partial t} = \frac{U_0}{\alpha a}\,\delta(t-ka).$$

Hence $F(t)$ is continuous at the points $t = ka$ [46].

Second, it is not difficult to show [46] that the condition

$$\left(\frac{\partial \xi}{\partial X}\right)^{\rm T} \Delta G_k \Bigg|_{\substack{X=X_{\rm c}\\ t=t_k}} = 0$$

is fulfilled for the dynamical system (5.1). Hence, the updating matrix M_k is determined by the expression

$$M_k = \big(E - A_k^+\big)^{-1} = E + A_k^-,\ A_k^+ = A_k(1),\ A_k^- = A_k(0).$$

In order to simplify the following calculations we will obtain the basic matrix using Eq. (5.4). The equation for $F(t)$ has the form

$$\frac{{\rm d}F}{{\rm d}t} = \Big[\Lambda + \hat{A}_k(\vartheta_1)\delta(t-t_k)\Big]\,F,\ F(0) = E. \tag{5.12}$$

Here

$$\Lambda = \begin{pmatrix} \lambda_1 & 0 & 0 \\ 0 & \lambda_2 & 0 \\ 0 & 0 & \lambda_3 \end{pmatrix};$$

$$\hat{A}_k = \left. \frac{1}{\dfrac{\partial \xi}{\partial t} + \sum\limits_{i=1}^{3} \dfrac{\partial \xi}{\partial w_i}\left[\lambda_i(w_i - 1 + \vartheta_1)\right]} \Delta \hat{G}_k \left(\frac{\partial \xi}{\partial W}\right)^{\mathrm{T}} \right|_{\substack{W = W_c \\ t = t_k}};$$

$$W = \begin{pmatrix} w_1 \\ w_2 \\ w_3 \end{pmatrix}; \ \Delta \hat{G}_k = \begin{pmatrix} \lambda_1 \\ \lambda_2 \\ \lambda_3 \end{pmatrix}; \ \frac{\partial \xi}{\partial W} = \left(\frac{\partial \xi}{\partial w_1}, \frac{\partial \xi}{\partial w_2}, \frac{\partial \xi}{\partial w_3}\right)^{\mathrm{T}},$$

$$\frac{\partial \xi}{\partial w_i} = \gamma_i \left(\chi - \frac{\lambda_3}{\lambda_i}\right), \ i = 1, 2, 3;$$

$$\frac{\partial \xi}{\partial t} = \begin{cases} -\dfrac{Q}{\alpha a}, & t_k \neq ka; \\ \dfrac{Q}{\alpha a}\,\delta(t - ka), & t_k = ka; \end{cases}$$

Within the domain $(k-1)\,a < t_k < t_k$ the system (5.12) has the form

$$\frac{\mathrm{d}F}{\mathrm{d}t} = \Lambda F, \ F\big((k-1)a\big) = F_{k-1}$$

with the solution

$$F(t) = \mathrm{e}^{\Lambda(t-(k-1)a)} F_{k-1},$$

or, evaluated at the switching time,

$$F(t_k - 0) = F_k^- = \mathrm{e}^{\Lambda(t_k-(k-1)a)} F_{k-1}.$$

Within the domain $t_k < t_k < ka$ the system (5.12) has the form

$$\frac{\mathrm{d}F}{\mathrm{d}t} = \Lambda F, \ F(t_k + 0) = M_k F_k^-$$

and the solution is

$$F(t) = \mathrm{e}^{\Lambda(t-t_k)} M_k \mathrm{e}^{\Lambda(t_k-(k-1)a)} F_{k-1}.$$

When $t = ka$

$$\frac{\partial \zeta}{\partial t} = \frac{Q}{\alpha a}\,\delta(t - ka).$$

hence

$$F(ka) = F_k = \mathrm{e}^{\Lambda(ka-t_k)} M_k \mathrm{e}^{\Lambda(t_k-(k-1)a)} F_{k-1},$$

or

$$F_k = \mathrm{e}^{\Lambda(1-z_k)a} M_k \mathrm{e}^{\Lambda z_k a} F_{k-1},$$
$$M_k = E + \hat{A}_k^-,$$

where

$$\hat{A}_k^- = \frac{1}{s_k} V.$$

V is a matrix with the elements $v_{ij} = \lambda_i \gamma_j \left(\chi - \lambda_3/\lambda_j\right)$, $i, j = 1, 2, 3$;

$$\begin{aligned} s_k &= -\frac{Q}{\alpha a} + \sum_{i=1}^{3} \gamma_i (\chi \lambda_i - \lambda_3)\big(w_i(t_k) - 1\big) \\ &= -\frac{Q}{\alpha a} + \sum_{i=1}^{3} \gamma_i (\chi \lambda_i - \lambda_3)(\sigma_{ik} - 1). \end{aligned}$$

Then the basic matrix can be determined by the recurrent formula

$$F_k = \mathrm{e}^{a\Lambda(1-z_k)} D_k \mathrm{e}^{a\Lambda z_k} F_{k-1},\ F_0 = E,\ k = \overline{1, m};\ F(T) = F_m.$$

Here

$$F_k = F(ka);$$
$$D_k = \begin{cases} E + A_k^-, & 0 < z_k < 1; \\ E, & z_k = 1 \text{ or } z_k = 0. \end{cases}$$

5.4 Transitions to Chaos via Local Bifurcations

In the following investigations we will consider the qualitative changes in behavior that occur when the parameters of the corrector α and χ are varied within the range $\Pi = \big\{(\alpha, \chi) : 1.0 \leq \alpha \leq 250.0; 0.0 < \chi \leq 1.0\big\}$. The remaining parameters are supposed to be constant. A chart of the distribution of dynamical modes within the parameter plane for the map (5.5), (5.6) is depicted in Fig. 5.7. Here we use the same designations as introduced in connection with our analysis of the Hénon map in Chapter 2. The domains where one can observe aperiodic oscillations (including chaotic and quasiperiodic) are denoted $\Pi_{\mathrm{Chaos},j}$,

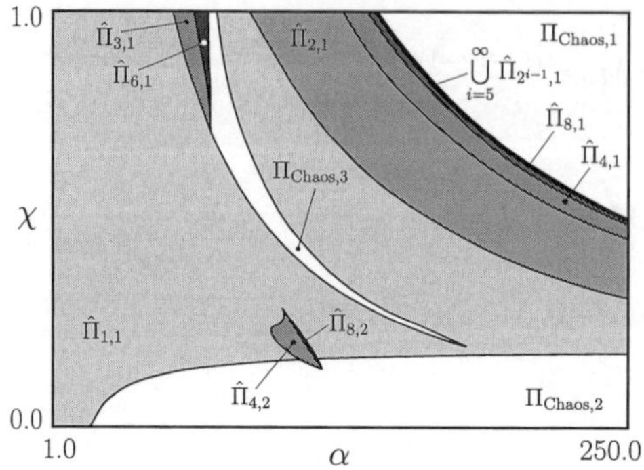

Figure 5.7. Two-parameter chart of dynamical modes in the parameter plane $\{(\alpha, \chi) : 1.0 \leq \alpha \leq 250.0; 0.0 \leq \chi \leq 1.0\}$. As previously defined α and χ are the amplification and transfer constants of the corrector, respectively.

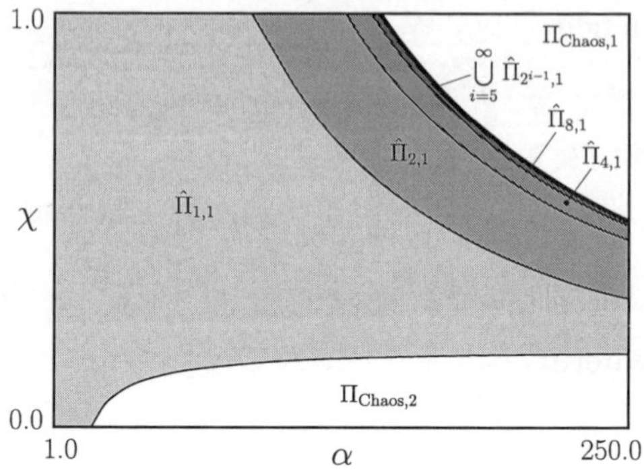

Figure 5.8. The collection of the sets $\hat{\Pi}_{2^{i-1},1}$, $i = 1, 2, \ldots$ that forms $\Pi_{1,1}$. $\Pi_{\text{Chaos},1}$ and $\Pi_{\text{Chaos},2}$ denote domains of chaos and quasiperiodicity.

$j = 1, 2, 3$. There are many narrow periodic windows $\Pi_{k,j}$ within $\Pi_{\text{Chaos},j}$. Figure 5.7 shows only those $\Pi_{k,j}$ that are relatively large in Π. The whole map of dynamical modes is much more complicate than we can present in Fig. 5.7. In Fig. 5.8 we separately show the aggregate $\Pi_{1,1} = \bigcup_{i=0}^{\infty} \hat{\Pi}_{2^i,1}$ and $\Pi_{\text{Chaos},j}$, $j = 1, 2$. To the left the domain $\hat{\Pi}_{1,1}$ is bounded by the bifurcation curve $\Gamma_{1,1}$ in the points of which the cycle period doubles. As an example, Fig. 5.9 shows the results of a linear stability analysis for the 1-cycle in dependence of α with $\chi = 0.25$. It is easily seen from Fig. 5.9(a) that the maximum multiplier (in absolute value) of the 1-cycle turns into -1 at the point $P = (304.81; 0.25)$ situated on the curve $\Gamma_{1,1}$. When the trajectory of deformation intersects the curve $\Gamma_{1,1}$, the 1-cycle loses stability, a 2-cycle softly arises and the 1-cycle becomes saddle.

From below the domain $\hat{\Pi}_{1,1}$ is bounded by the curve N_φ, in the points of which two-frequency quasiperiodic oscillations softly arise from the period-1 cycle via an Andronov-Hopf bifurcation (see Fig. 5.9(b)). The period-doubling curve $\Gamma_{1,1}$ is supported by the curve N_φ in the point of codimension two.

The bifurcation curves $\Gamma_{1,2^i}$, $i = 1, 2, 3, \ldots$ accumulate in the same way as for the Hénon map and there exist transversal curves along which period-doubling sequences occur. The domains $\Pi_{\text{Chaos},1}$ and $\Pi_{1,1}$ are separated by the curve which is an aggregate of accumulation points, where the motions are aperiodic,

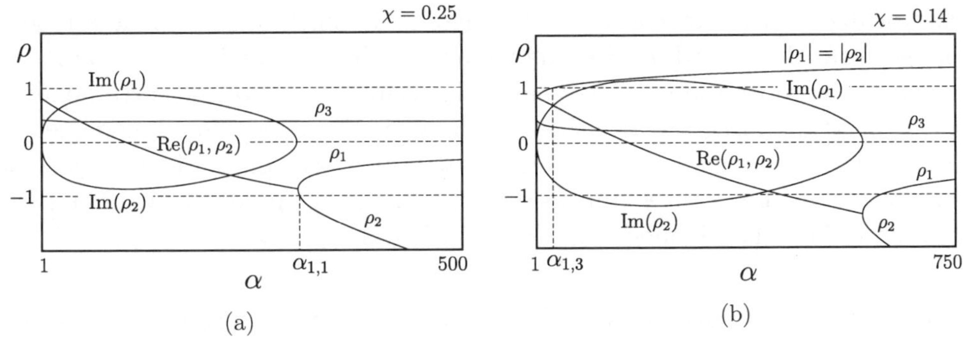

Figure 5.9. Variation of the multipliers with the amplification constant α. The figures illustrate: (a) a period-doubling bifurcation; (b) an Andronov–Hopf bifurcation.

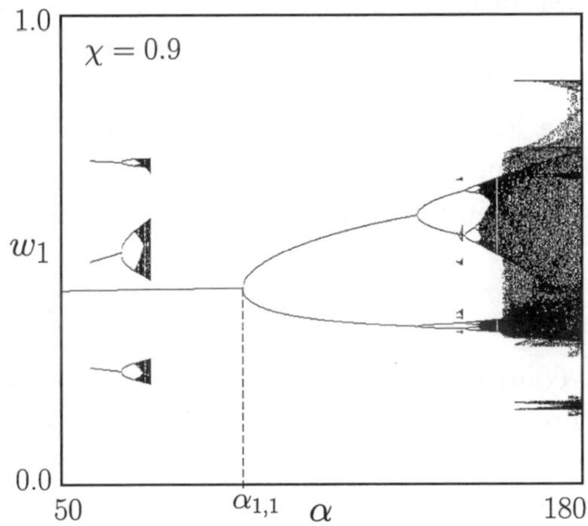

Figure 5.10. Bifurcation diagram showing the transition to chaos through a period-doubling sequence for the main branch.

but not chaotic since the relevant Lyapunov exponent vanishes. This type of behavior is sometimes referred to as a Feigenbaum attractor. The period-doubling curves $\Gamma_{1,2^i}$, $i = 1, 2, 3, \ldots$ are supported by the curve N_φ in the points of codimension two. The set of these points along the curve N_φ is denumerable.

As an example in Fig. 5.10 we provide a bifurcation diagram built in the section $\Omega = \{(\alpha, \chi) : 50.0 < \alpha < 180.0; \chi = 0.9\}$, $\Omega \in \Pi_{1,1}$. Besides the period-doubling route to chaos for the main cycle, the diagram also illustrates the formation of a period-3 subharmonic oscillation near $\alpha = 63$. This mode follows its own route to chaos, to be discussed in Sec. 5.5, and finally disappears in a crisis near $\alpha = 76$. Around $\alpha = 165$ there is a similar interval where modes derived from the period-6 subharmonic are stable. Figure 5.11 illustrates the change of the waveform for the dynamical variable x_2 and of the corresponding power spectrum during the transition to chaos via the period-doubling sequence. The oscillation shape for the variable x_2, the form of the phase space trajectory in a projection onto a three-dimensional space, and the power spectrum for the variable x_2 in a chaotic mode are shown in Fig. 5.12.

5.5 Border-Collision Bifurcations and Transitions to Chaos

For the dynamical system under consideration, we have found that transitions to chaos via border-collision bifurcations only occur for those domains $\Pi_{k,j}$ which intersect $\hat{\Pi}_{1,1}$, i.e. $\hat{\Pi}_{1,1} \cap \Pi_{k,j} \neq \emptyset$, $k > 1$. Only two of them ($\Pi_{4,2}$ and $\Pi_{3,1}$) are shown in the Fig. 5.7. More detailed numerical calculations for the higher resolution of the map of dynamical regimes show that the number of such domains in Π can be infinite, each of them being an aggregate of the sets:

$$\Pi_{m,j} = \bigcup_{k=m}^{s} \hat{\Pi}_{k,j}, \ \hat{\Pi}_{k,j} \subset \Pi, \ m \geq 3.$$

Here $\hat{\Pi}_{k,j}$ are simply connected sets of parameters such that for any points $P \in \hat{\Pi}_{k,j}$, $P = (\alpha, \chi)$, there exists a locally stable k-cycle that varies continuously with the parameters in $\hat{\Pi}_{k,j}$, where k are multiplies of m. The value s can be either finite or infinite, and m is a minimum, that is, there are no k-cycles in $\Pi_{m,j}$ with $k < m$. As before, the index j is introduced to distinguish sets that are related to motions with the same m. Detailing of the two-parameter diagram reveals a more subtle picture of the inner organization of $\Pi_{m,j}$ and an extraordinary great variety of bifurcational transitions some of which, as it will be demonstrated below, have unusual properties.

To be more specific we shall study the bifurcational transitions at the boundaries of the parameter domain where a 5-cycle and motions with periods multiple to 5 exist, i.e.

$$\Pi_{5,1} = \bigcup_{k=5}^{s} \hat{\Pi}_{k,1}.$$

This restriction of the problem is consistent with our goal. We do not intend to study the behavior of the dynamical system (5.1) for all parameter values.

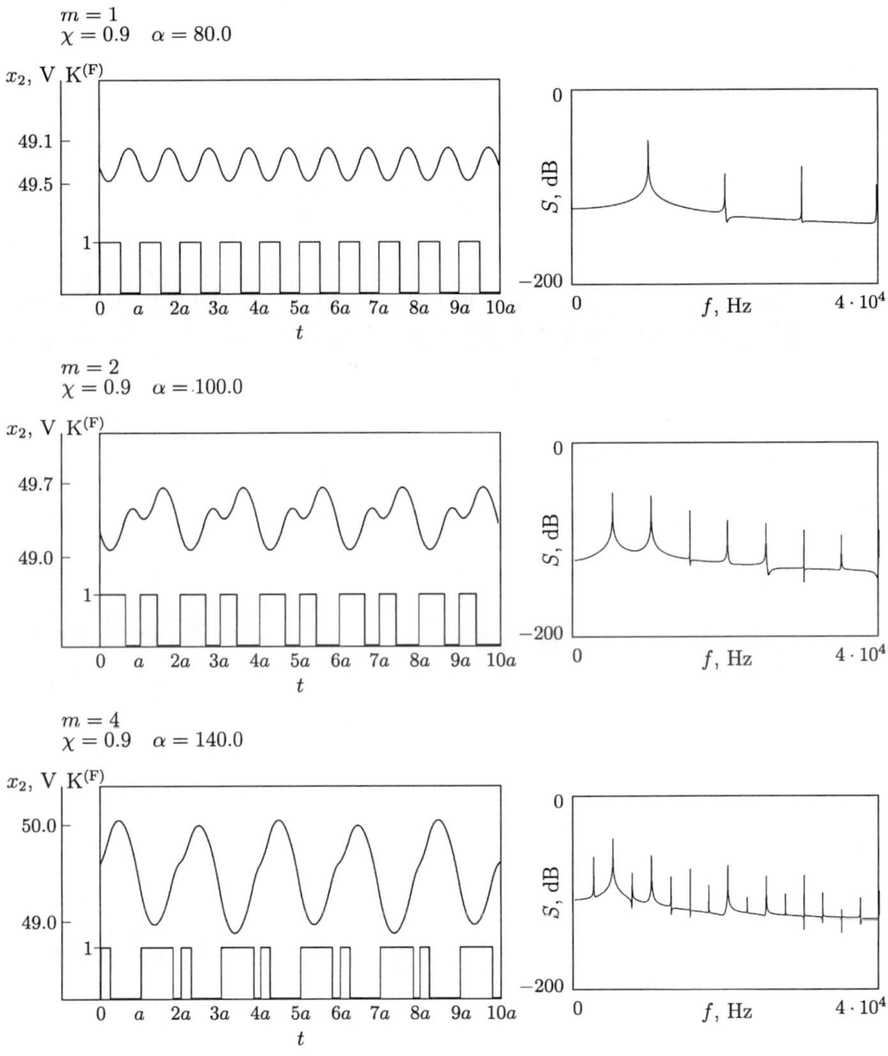

Figure 5.11. Time series for the variable x_2 and output signal of the pulse-width modulator $K^{(F)}$ (left panels) and power spectra (right panels) for cycles that arise during the period-doubling sequence $1 \to 2 \to 4 \to 8 \to 16 \to 32$ shown in Fig. 5.10.

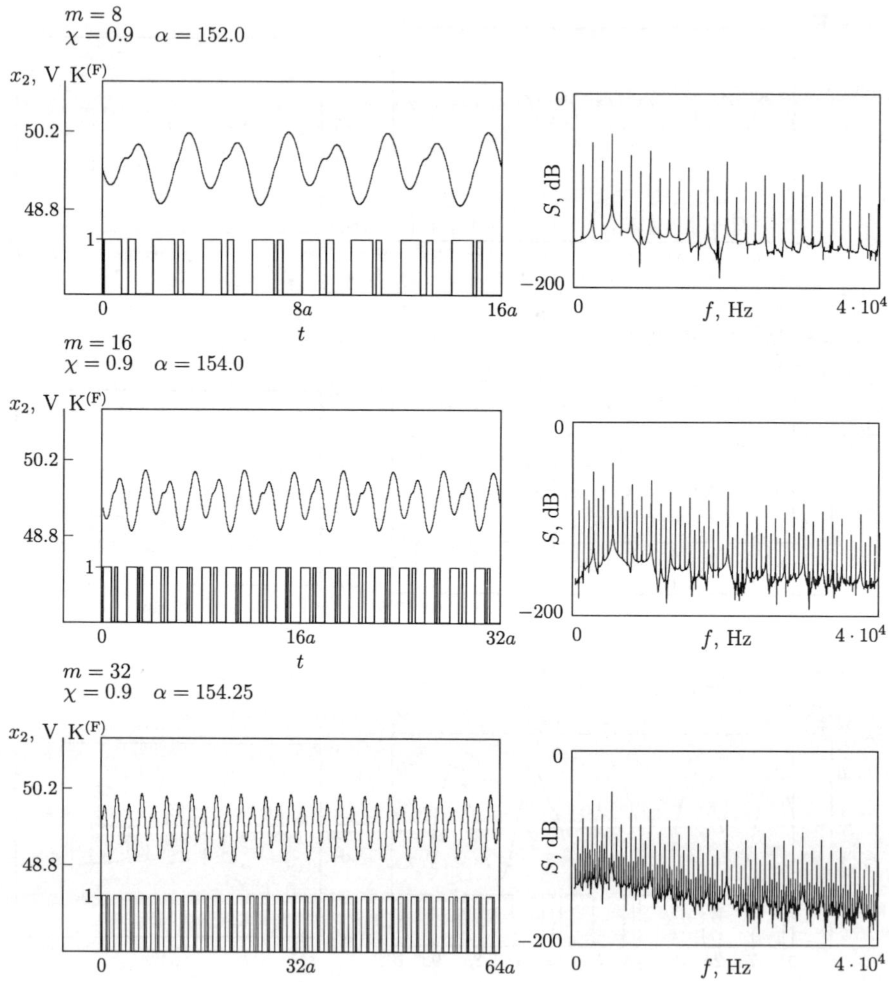

Figure 5.11. (continued)

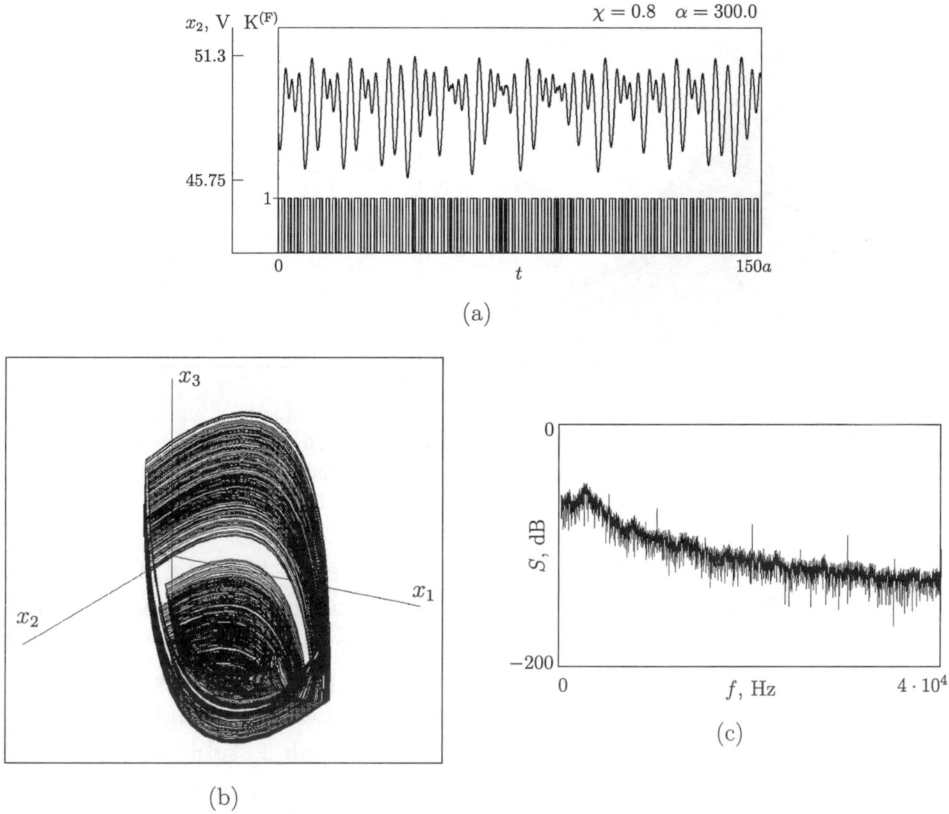

Figure 5.12. (a) Time series for the variable x_2 and output signal of the pulse-width modulator $K^{(F)}$, (b) three-dimensional projection of the phase space trajectory, and (c) power spectrum for a chaotic oscillation.

Rather, the aim is to analyze the possible scenarios for the development of complex behavior that are related to border-collision (BC) bifurcations.

The results of our numerical calculations of the domains of existence of various stable cycles are presented in Fig. 5.13. In this figure the sets $\hat{\Pi}_{k,j}$, $k = 5$, 15, 20, 25, 30, 40, 45, 60, $j = 1, 2, 3$ correspond to the domains of parameter values where locally stable 5-, 15-, 20-, 25-, 30-, 40-, 45-, 60-cycles, respectively, exist. This diagram contains only those $\hat{\Pi}_{k,j}$ which are relatively large in Π. In reality the diagram is much more complicated than that of Fig. 5.13. The results of a two-parameter bifurcation analysis are depicted in Fig. 5.14 [47]. This diagram does not show the domains $\Pi_{30,1}$, $\Pi_{30,2}$, $\Pi_{40,1}$, $\Pi_{45,1}$, $\Pi_{50,1}$, $\Pi_{60,1}$

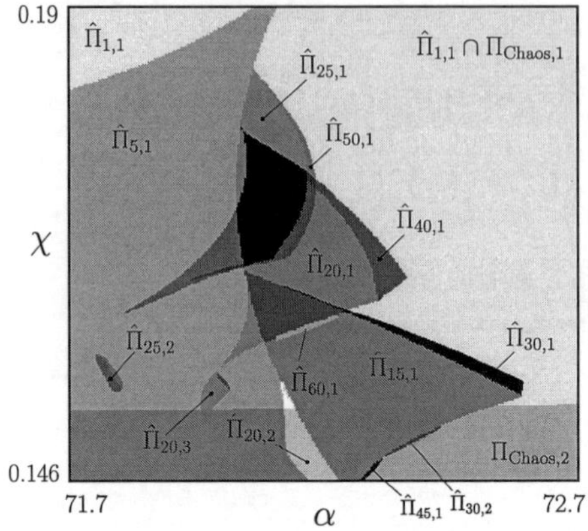

Figure 5.13. Domains of existence for the different oscillatory modes in the system parameter space.

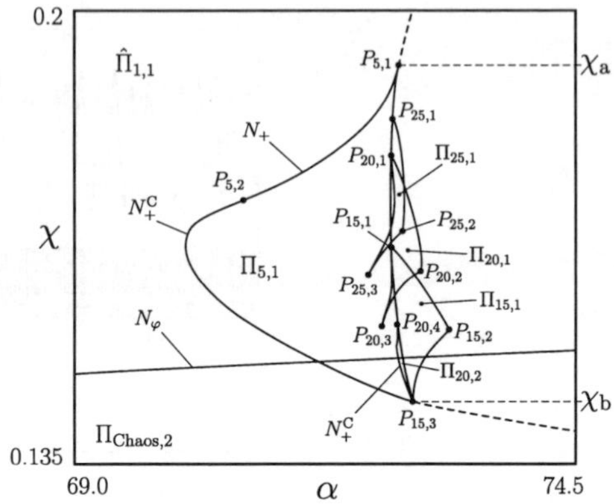

Figure 5.14. Boundary structure of the domains of existence for the different oscillatory modes shown in Fig. 5.13.

where locally stable 30-, 40-, 45-, 50-, 60-cycles exist because of their small sizes, nor does it show the domains of chaos adjoining the boundaries of $\Pi_{5,1}$.

Let us now consider the two-parameter bifurcation diagram presented in Fig. 5.14 in more detail. The left boundary of $\hat{\Pi}_{5,1}$ consists of two bifurcation curves. The curve $P_{5,1}$–$P_{5,2}$ in the upper left hand side corresponds to the curve N_+ of a saddle-node bifurcation at the points of which a stable 5-cycle merges with an unstable cycle of the same periodicity and disappears (or arises). The line $P_{5,2}$–$P_{15,3}$ corresponds to the BC-bifurcation boundary N_+^C where a stable 5-cycle of one type merges with an unstable cycle of another type and disappears (or arises). The right boundary of $\hat{\Pi}_{5,1}$ is a combination of BC-bifurcation curves of various types.

By analyzing the diagram (Figs. 5.13 and 5.14) it is possible to single out a number of interesting behavioral cases of the system when it passes through the points of BC-boundaries.

First let us consider such bifurcational transitions where the trajectory of deformation transverses the BC-boundary of the domain $\hat{\Pi}_{5,1}$ in the points located on the line $P_{15,1}$–$P_{20,1}$. The results of a calculation performed in the section $\Omega = \{(\alpha, \chi) : 71.95 < \alpha < 72.35; \chi = 0.167\}$ show that, with α increasing, the value of one of the cyclic points $z_j^c(\alpha)$, $j = \overline{1,5}$ diminishes to vanish at the bifurcation point $\alpha_{5,1}$ ($\alpha_{5,1} \approx 72.03580$). When passing through $\alpha_{5,1}$ a 5-cycle sewed from 8 parts loses its stability and transforms into a saddle 5-cycle

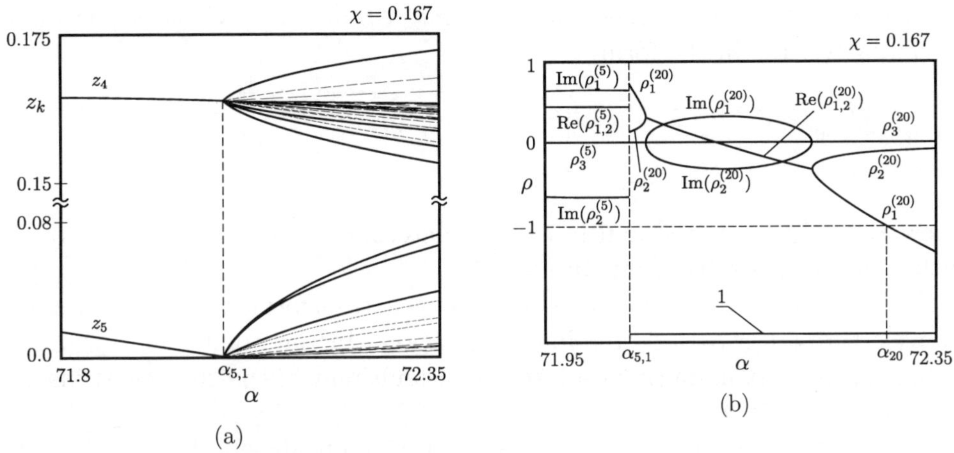

Figure 5.15. Border-collision bifurcation of the 5-cycle. The solution type changes with the loss of stability which is accompanied with the appearance of a stable 20-cycle and a set of unstable cycles with multiple-to-5 periods: (a) Bifurcation diagram, where z_4, z_5 denote two of the original 5-cycle pulse ratios. (b) The cycle multiplier dependence on the parameter α. $\rho_j^{(k)}$, $j = 1, 2, 3$ denotes the jth multiplier of the k-cycle.

of another type, sewed from 6 parts. It is seen from Fig. 5.15(a) that the loss of stability of the 5-cycle is accompanied with the soft appearance of a family of cycles with periods that are multiple to 5. One of the 20-cycles is locally stable while the other cycles are unstable. The number of unstable cycles may be finite or infinite. The diagram in Fig. 5.15(a) represents nine cycles that we have succeeded to obtain numerically, namely:

- one unstable 10-cycle;
- two unstable 15-cycles of different types;
- three 20-cycles of different types, one of which is locally stable, while the others are unstable;
- two unstable 25-cycles of different types; and
- one unstable 30-cycle.

The variations of $z_k(\alpha)$ for the stable 5-cycle and the stable 20-cycle are marked by thick lines.

The asymptotic dependence of z_k around a bifurcation value of parameter α_* can be written in the form

$$\lim_{|\alpha^+-\alpha^-|\to 0} |z_k(\alpha^-) - z_{k+m(r-1)}(\alpha^+)| = 0,\ k = \overline{1,m};\ r = \overline{1,N};\ N = 1,2,\ldots$$
$$\alpha^+ > \alpha_*,\ \alpha^- < \alpha_*,$$

where the value N is the multiplier that determines the period $N \times m$ of the cycle which arises softly from an m-cycle.

With further changes of α the stable 20-cycle transforms into a locally stable 40-cycle through a period-doubling bifurcation. In the point $\alpha = \alpha_{20}$ ($\alpha_{20} \approx 72.29597$) the maximum (in absolute value) multiplier of the 20-cycle turns into -1 (see Fig. 5.15(b)).

Figure 5.15(b) illustrates the variations of the multipliers $\rho_i^{(5)}$, $\rho_i^{(20)}$, $i = 1,2,3$ for the stable 5-cycle and the stable 20-cycle with the amplification parameter α. Also shown is the dependence (designated by the number 1) of the maximum multiplier (in absolute value) of the saddle 5-cycle on α. Figure. 5.15(b) clearly shows that $\rho_i^{(5)}(\alpha)$, $i = 1,2,3$ smoothly depend on α, with the only exception of the point $\alpha_{5,1}$, in the neighborhood of which $\rho_i^{(5)}(\alpha)$, $i = 1,2,3$ undergo simple discontinuities:

$$\rho_i(\alpha) = \rho_i^-(\alpha) + \left(\rho_i^+(\alpha) - \rho_i^-(\alpha)\right) \eta(\alpha - \alpha_{5,1}),\ i = 1,2,3$$

whereas $z_j^c(\alpha)$ remain continuous functions of the parameter α. Here $\rho_i^-(\alpha)$, $\rho_i^+(\alpha)$, $i = 1,2,3$ are sufficiently smooth functions such, that

$$\begin{aligned} \rho_i^-(\alpha) &= \rho_i(\alpha),\ \alpha < \alpha_{5,1};\\ \rho_i^+(\alpha) &= \rho_i(\alpha),\ \alpha > \alpha_{5,1};\\ \lim_{\alpha\to\alpha_{5,1}-0} |\,\rho_i(\alpha)| &< 1,\ \max_{1\le i\le 3}\ \lim_{\alpha\to\alpha_{5,1}+0} |\,\rho_i(\alpha)| > 1. \end{aligned}$$

This property is easy to prove and is typical for border-collision bifurcations. Another important property is that, unlike what holds for smooth dynamical systems, the multipliers of the stable m-cycle do not cross the boundary of the unit circle while passing through the bifurcational value of a parameter.

In addition to the results of the bifurcation analysis presented in Fig. 5.15, Fig. 5.16(a) displays the bifurcation diagram obtained at $\chi = 0.168$ by direct calculation. Inspection of this figure shows that a chaotic attractor [37–39] softly branches off the 5-cycle together with a stable 20-cycle. Such a picture

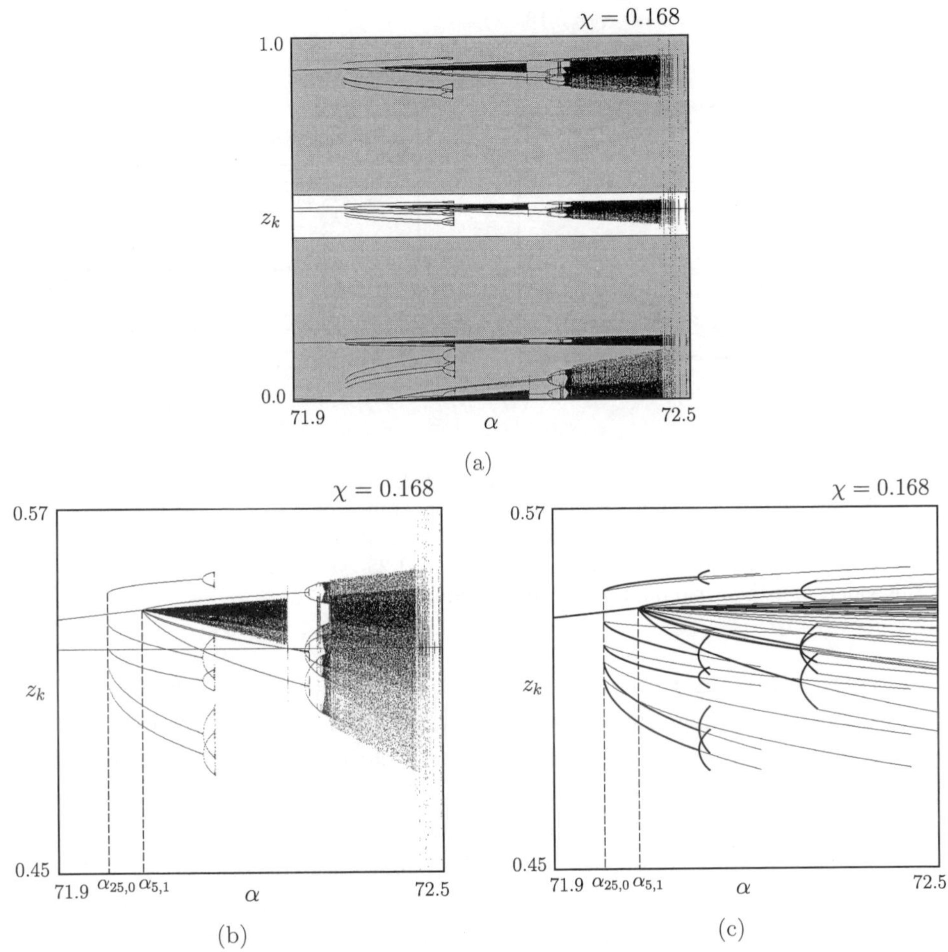

Figure 5.16. Border-collision bifurcation where a chaotic attractor arises together with a stable 20-cycle: (a) Bifurcation diagram. (b) The magnified light portion of Fig. 5.16(a). (c) Results of a detailed bifurcation analysis of the diagram shown in Figs. 5.16(a) and 5.16(b). The stable cycles are marked in the diagram by thick lines.

is observed in numerical experiments at all the points of the line $P_{15,1}$–$P_{20,1}$. A magnified portion of this diagram is presented in Fig. 5.16(b) to better illustrate the bifurcational transition. The results of the detailed bifurcation analysis are given in Fig. 5.16(c). The figure contains the evolution of stable and unstable cycles with periods, that are multiple to 5, up to 50. It is easy to see, that a pair

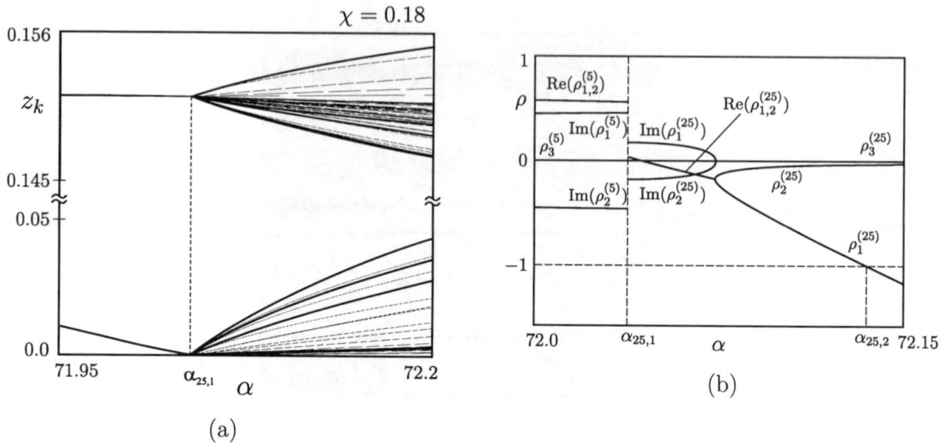

Figure 5.17. Border-collision bifurcation of the 5-cycle. The solution type changes with the loss of stability which is accompanied with the appearance of a stable 25-cycle and a set of unstable cycles with multiple-to-5 periods: (a) The bifurcation diagram. The stable cycles that take part in the border-collision bifurcation (the 5- and 25-cycles) are marked with thick lines. (b) Dependence of the cycle multipliers on the parameter α.

of 25-cycles (stable and unstable) arises in the points of the line $P_{20,1} - P_{25,3}$. This branch is not shown in Fig. 5.15(a). While changing the parameter the stable 25-cycle undergoes a period-doubling bifurcation. When further increasing the transfer parameter χ the conditions of existence of the 50-cycle are broken, and this cycle disappears. The stable 20-cycle evolution is similar to that of the 25-cycle with the only difference, that when changing the parameter the stable 40-cycle, which is born from the 20-cycle via a period-doubling, transforms into an unstable 40-cycle of another type via a border-collision bifurcation.

Figures 5.17 and 5.18 contain the results of a couple of bifurcation analyses performed in the sections $\Omega_1 = \{(\alpha, \chi) : 72.0 < \alpha < 72.15; \chi = 0.18\}$ and $\Omega_2 = \{(\alpha, \chi) : 71.75 < \alpha < 72.25; \chi = 0.183\}$, respectively. These sections cross the curve $P_{20,1}$–$P_{25,1}$. The diagrams illustrate the bifurcational transition when the stable 25-cycle arises via 5-cycle period-quintupling. As in the previous case (see Fig. 5.15), the stability loss of the 5-cycle is accompanied by the appearance of a set of cycles with periods that are multiple to 5 (Fig. 5.17). This 5-cycle arises in the points of the saddle-node bifurcation curve (line $P_{5,1}$–$P_{5,2}$ in Fig. 5.14), which is shown in Fig. 5.18(a). It loses stability at the point $\alpha = \alpha_{25,1}$, $\alpha_{25,1} \approx 72.05856$ (see Fig. 5.18(b)). Here, beside the stable

m	j	z_1–z_5	z_6–z_{10}	z_{11}–z_{15}	z_{16}–z_{20}	z_{21}–z_{25}	z_{26}–z_{30}	N
5	1	$\mathcal{L}$						6
10	1	$\mathcal{D}$	$\mathcal{L}$					14
15	1	$\mathcal{D}$	$\mathcal{D}$	$\mathcal{L}$				22
	2	$\mathcal{D}$	$\mathcal{L}$	$\mathcal{L}$				20
20	1	$\mathcal{D}$	$\mathcal{D}$	$\mathcal{D}$	$\mathcal{L}$			30
	2	$\mathcal{D}$	$\mathcal{D}$	$\mathcal{L}$	$\mathcal{L}$			28
	3	$\mathcal{D}$	$\mathcal{L}$	$\mathcal{L}$	$\mathcal{L}$			26
25	1	$\mathcal{D}$	$\mathcal{D}$	$\mathcal{D}$	$\mathcal{D}$	$\mathcal{L}$		38
	2	$\mathcal{D}$	$\mathcal{D}$	$\mathcal{D}$	$\mathcal{L}$	$\mathcal{L}$		36
	3	$\mathcal{D}$	$\mathcal{L}$	$\mathcal{D}$	$\mathcal{L}$	$\mathcal{L}$		34
	4	$\mathcal{D}$	$\mathcal{L}$	$\mathcal{L}$	$\mathcal{L}$	$\mathcal{L}$		32
30	1	$\mathcal{D}$	$\mathcal{L}$	$\mathcal{L}$	$\mathcal{L}$	$\mathcal{L}$	$\mathcal{L}$	38

Table 5.1. Symbolic characteristics of cycles with periods that are multiple to 5 at $\alpha = 72.0586$ and $\chi = 0.184$. Here $\mathcal{L}$ denotes the string "1***0" and $\mathcal{D}$ denotes "1****" in our symbolic notation. The asterisk "*" means that the point z_k falls between 0 and 1.

25-cycle, a set of unstable cycles arise. The 25-cycle loses its stability at the point $\alpha_{25,2}$ through a period-doubling bifurcation. Table 5.1 gives the results of numerical calculations for 10 of the cycles that we have succeeded to determine numerically, namely:

- one 10-cycle;
- two 15-cycles of different types;
- three 20-cycles of different types;
- three 25-cycles of different types;
- one 30-cycle.

The first column of this table contains the period of the cycle, and the second column provides an index to distinguish cycles with the same period but of different type. The third column contains the symbolic characteristics of the cycle. The symbolic characteristics of an m-cycle consists of m symbols "0", "1", and "*", each related to the corresponding z_k-values, $k = \overline{1,m}$: "0" for $z_k = 0$, "1" for $z_k = 1$, and "*" for $0 < z_k < 1$. The turning of one of the points z_k into 0 or 1 (see Figs. 5.15 and 5.17) results in a reduction of the number of portions from which the phase trajectory of the m-cycle is sewed up by 2. Thus this number is determined as $2\big(m - (N_0 + N_1)\big)$, where $N_{0,1}$ are the

m	j	ρ_1	ρ_2	ρ_3
5	1	−0.17384	−2.23026	0.00472
10	1	−0.05140	−3.72987	0.00002
15	1	−0.02872	−3.33046	$6.42 \cdot 10^{-8}$
	2	0.00863	8.61442	$8.25 \cdot 10^{-8}$
20	1	−0.02229	−2.14074	$2.35 \cdot 10^{-10}$
	2	0.00470	7.89266	$3.03 \cdot 10^{-10}$
	3	−0.00150	19.26318	$3.90 \cdot 10^{-10}$
25	1	−0.02631	−0.90532	$8.64 \cdot 10^{-13}$
	2	0.00355	5.21486	$1.11 \cdot 10^{-12}$
	3	−0.00044	−32.14474	$1.38 \cdot 10^{-12}$
	4	0.00026	42.97079	$1.84 \cdot 10^{-12}$

Table 5.2. Values of multipliers for different cycles at the bifurcation point.

number of symbols “0” and “1” in symbolic characteristics, respectively. The value of N is indicated in the last column of Table 5.1.

The multipliers of the m-cycles at the bifurcational value of the parameter α are given in Table 5.2. A view on the table shows that all cycles except one of the 25-cycles are unstable. Besides the cycles indicated in Table 5.1, the unstable 35-, 40-, 45-, 50-cycles also arise at the point $\alpha_{5,1}$. This is depicted separately in Fig. 5.18(c).

The described bifurcational transition is schematically shown in Fig. 5.19, where the numbers indicate the periodicity of the cycles. Equal numbers correspond to cycles with the same period but of different type. The number M can be finite or infinite. The solid and dashed lines mark the dependence of stable and unstable cycles on α, respectively. This picture is typical for all points of the line $P_{5,1}$–$P_{15,3}$.

Let us look at one more property of the dynamical system (5.1). This property is related to the BC-bifurcational transitions whereat the loss of stability of an m-cycle softly generates a number of unstable cycles with periods that are multiple to m. Such transitions take place on the line $P_{5,1}$–$P_{25,1}$.

As an example, Fig. 5.20 presents the results of a bifurcation analysis performed in the section $\Omega = \{(\alpha, \chi) : 71.87 < \alpha < 72.25; \chi = 0.186\}$, crossing the lines $P_{5,1}$–$P_{5,2}$ and $P_{5,1}$–$P_{25,1}$. As it can be seen from the figure, in the point $P = (\alpha_{5,0}; 0.186)$, $\alpha_{5,0} \approx 71.91431$ lying on the line $P_{5,1}$–$P_{5,2}$, a couple of 5-cycles arise through a saddle-node bifurcation. When passing through the line $P_{5,1}$–$P_{25,1}$ the 5-cycle loses its stability which causes a set of unstable cycles

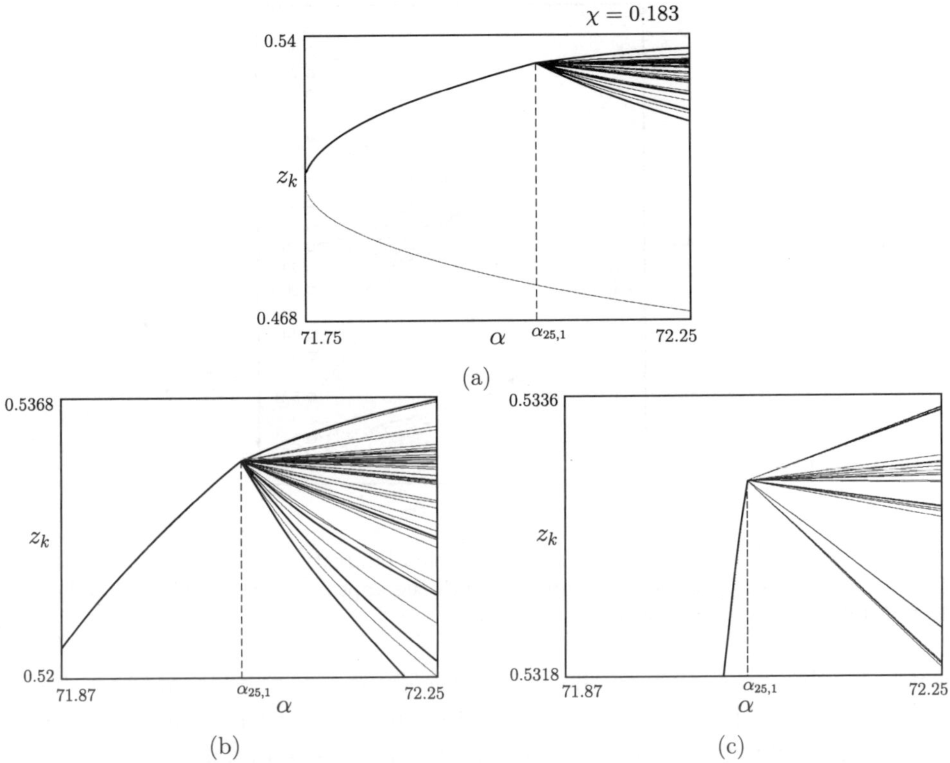

Figure 5.18. Border-collision bifurcation of the 5-cycle. The solution type changes with the loss of stability which is accompanied by the appearance of a stable 25-cycle and a set of unstable cycles with multiple-to-5 periods: (a) Bifurcation diagram. (b) Magnified part of the diagram (a) around the border-collision bifurcation. (c) Unstable cycles with high periods (35, 40, 45, 50) that take part in the border-collision bifurcation.

with periods that are multiples of 5 to softly branch off from the 5-cycle. The 5-cycle continues to exist but turns into a saddle. Figure 5.20(c) shows the dependence of the cycle multipliers on the parameter α. The diagram contains all multipliers of the stable and unstable cycles arising at the point $\alpha_{5,0}$ ($\rho_j^{(5)\text{S}}$ and $\rho_j^{(5)\text{U}}$, $j = 1, 2, 3$, respectively) and several maximum multipliers (in absolute value) of unstable cycles ($\rho_1^{(15)}$, $\rho_1^{(20)}$, $\rho_1^{(25)}$) produced in the border-collision bifurcation.

In Fig. 5.14 the dashed lines designate BC-boundaries where a couple of unstable 5-cycles and a set of unstable $5k$-cycles, $k = 2, 3, \ldots$ merge and then

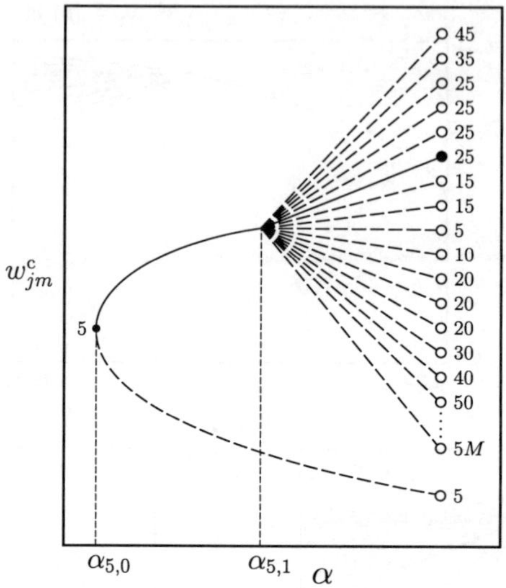

Figure 5.19. Schematic drawing of the border-collision bifurcation of the 5-cycle with solution type change. The bifurcation is accompanied with the appearance of a stable 25-cycle and a set of unstable cycles with multiple-to-5 periods.

disappear while moving into the parameter domain that is located to the left of these curves. Let us now consider Fig. 5.14 in order to understand the mechanism of this transition. The domain of existence of the 5-cycle becomes narrower when approaching the points $P_{5,1}$ and $P_{15,3}$, i.e. (see Figs. 5.14, 5.19)

$$\lim_{\chi\to\chi_a-0}|\alpha_{5,1}-\alpha_{5,0}|=0;$$
$$\lim_{\chi\to\chi_b+0}|\alpha_{5,1}-\alpha_{5,0}|=0.$$

A schematic drawing of the limiting case, i.e., the bifurcational transition that happens on the dashed lines of Fig. 5.14 is presented in Fig. 5.20(d).

Let us now consider the boundaries of the domains $\Pi_{15,1}$, $\Pi_{20,1}$, and $\Pi_{25,1}$. The right boundary of $\Pi_{15,1}$ consists of two bifurcation lines. The line beginning in $P_{15,1}$ and finishing in $P_{15,2}$, corresponds to the curve N_- of a 15-cycle period-doubling bifurcation. At the points of this curve the maximum multiplier (in absolute value) of a 15-cycle turns into -1. The second line, supported by the $P_{15,2}$ and $P_{15,3}$ points, corresponds to the BC-boundary, the structure of which is no less simple than of the BC-boundary for $\Pi_{5,1}$. When passing through the point of this curve the period of oscillation can double or triple.

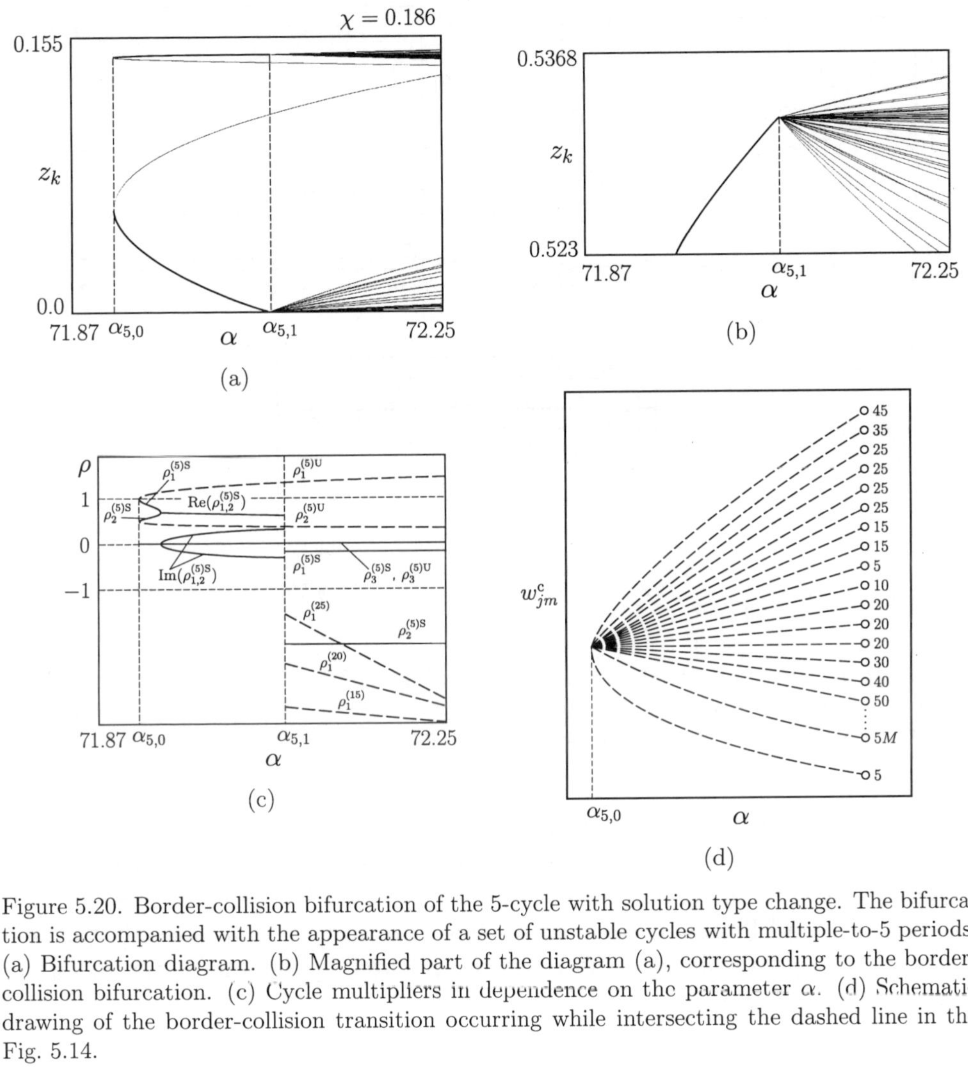

Figure 5.20. Border-collision bifurcation of the 5-cycle with solution type change. The bifurcation is accompanied with the appearance of a set of unstable cycles with multiple-to-5 periods. (a) Bifurcation diagram. (b) Magnified part of the diagram (a), corresponding to the border-collision bifurcation. (c) Cycle multipliers in dependence on the parameter α. (d) Schematic drawing of the border-collision transition occurring while intersecting the dashed line in the Fig. 5.14.

The boundary structures of the domains $\Pi_{20,1}$ and $\Pi_{25,1}$ are similar to that of $\Pi_{15,1}$. Hence, we shall not go into their detailed description but just add that the lines $P_{15,1}$–$P_{20,3}$ and $P_{20,1}$–$P_{25,3}$ correspond to the BC-bifurcation curves in the points of which a stable and an unstable cycle of similar periodicity arise (or disappear).

Our investigations of the changes in system behavior as we move along the

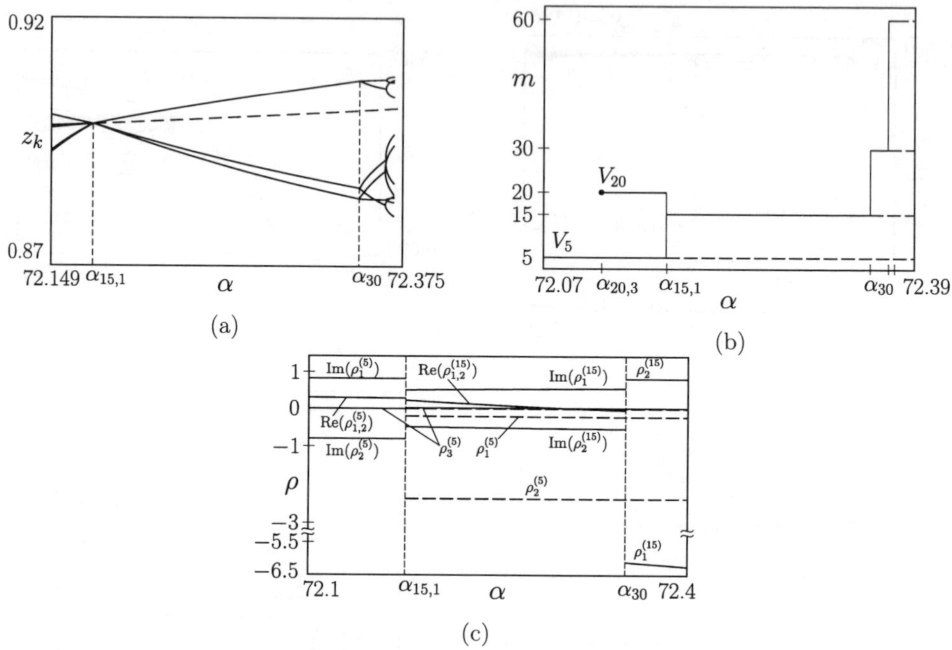

Figure 5.21. Results of a bifurcation analysis illustrating a soft transition of one group of stable motions into another with multiple periods: (a) The bifurcation diagram. (b) The branching pattern. (c) Cycle multipliers in dependence on the parameter α.

various sections of the parameter plane have revealed bifurcational transitions with extraordinary properties quite different from those studied above. The frames of the present work keep us from a more detailed analysis of such transitions. We shall limit ourselves to a demonstration of some possible variants when such atypical, yet not the most complicated, phenomena may occur in the bifurcational transitions.

Figures 5.21(a)–(c) present an example of a soft transition of one group of stable motions into another with multiple periods. This transition takes place in the points of the line $P_{15,3}$–$P_{20,4}$. Figure 5.21(a) is a bifurcation diagram made in the section $\Omega = \{(\alpha, \chi) : 71.149 < \alpha < 72.375; \chi = 0.1497\}$. On the vertical axis we have plotted the values of one of the cyclic points of period 5, 15, 20. We have used a dashed line to show the unstable 5-cycle. Figures 5.21(b) and 5.21(c) demonstrate the branching pattern, respectively the parameter dependence for the multipliers of the 5- and 15-cycles.

As shown in Fig. 5.21, with increasing α a stable 20-cycle and a stable 5-cycle at first monotonically approach each other and then merge at the point $\alpha = \alpha_{15,1}$ ($\alpha_{15,1} \approx 72.17629$)

$$\lim_{|\alpha^+ - \alpha^-| \to 0} |z_j(\alpha^-) - z_{j+5(r-1)}(\alpha^+)| = 0,\ j = \overline{1,5},\ r = \overline{1,4}.$$

On passing the point $\alpha = \alpha_{15,1}$, a stable 15-cycle branches off from the stable 5-cycle through a period-tripling BC-bifurcation

$$\lim_{|\alpha^+ - \alpha^-| \to 0} |z_j(\alpha^-) - z_{j+5(r-1)}(\alpha^+)| = 0,\ j = \overline{1,5},\ r = \overline{1,3},$$

while the 5-cycle turns into a saddle. We have managed to trace two more period-doubling bifurcations: $15 \to 30 \to 60$. The transition from a stable 15-cycle to the stable 30-cycle takes place through the period-doubling BC-bifurcation in the point α_{30}. Then with the parameter α increasing the maximum multiplier (in absolute value) of the 30-cycle leaves the unit circle through -1 and a 60-cycle arises softly, the 30-cycle turning into a saddle. Further the stable 60-cycle transforms into an unstable 60-cycle of another type.

At the reverse change of the parameter α from the values $\alpha > \alpha_{15,1}$ to the values $\alpha < \alpha_{15,1}$, as illustrated in Fig. 5.21, we observe a picture similar to the previous case: As α decreases the stable 15-cycle starts to monotonically approach the unstable 15-cycle and merges with it at the point $\alpha = \alpha_{15,1}$. Then, on the transition into the domain of the values $\alpha < \alpha_{15,1}$, a 20-cycle softly branches off the 5-cycle through the period-quadrupling BC-bifurcation, and the 5-cycle becomes stable. From Figs. 5.21(a) and 5.21(b) it follows that the transition from the stable 20-cycle to the stable 15-cycle and back occurs softly

$$\lim_{|\alpha^+ - \alpha^-| \to 0} \left| \frac{1}{4} \sum_{l=1}^{4} z_{j+m(l-1)}(\alpha^-) - \frac{1}{3} \sum_{r=1}^{3} z_{j+m(r-1)}(\alpha^+) \right| = 0,\ j = \overline{1,m}.$$

Here $m = 5$, $\alpha^+ > \alpha_{15,1}$, $\alpha^- < \alpha_{15,1}$.

Let us consider the bifurcational transition depicted in Figs. 5.21(a) and (b) in more detail. The results of our numerical calculations are presented in Fig. 5.22. It is seen from this figure, that a pair of unstable 20-cycles of different types (denoted by the numbers 1 and 2) arise at the point $\alpha = \alpha_{20,1}$, $\alpha_{20,1} \approx 71.99787$ via a BC-bifurcation. One of these cycles (denoted by the

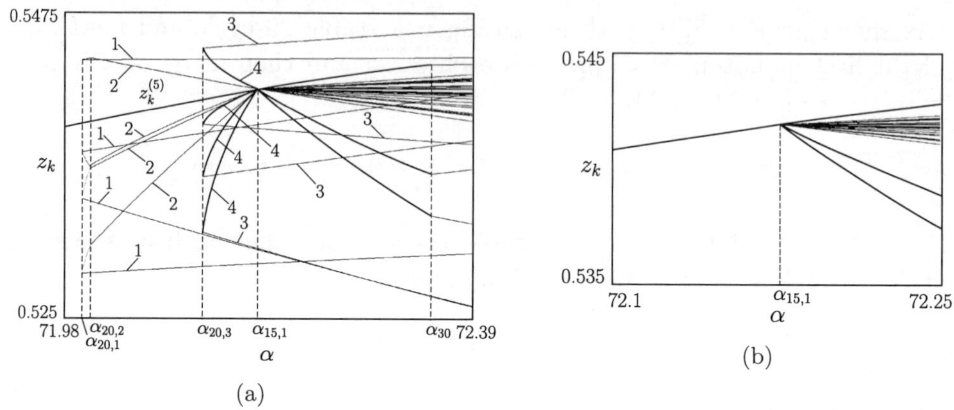

Figure 5.22. Magnified part of the diagram shown in Fig. 5.21(a), displaying the bifurcational structure near the point $\alpha_{15,1}$: (a) Bifurcation diagram. (b) Magnified part of the diagram (a) showing the cycles arising at the border-collision bifurcation, namely a stable 15-cycle and a set of unstable cycles with periods that are multiple to 5. The stable cycles are marked by thick lines in both diagrams.

number 2) undergoes a new BC-bifurcation of solution type change when parameter α passes the point $\alpha = \alpha_{20,2}$, $\alpha_{20,2} \approx 71.99657$ and then merges with a stable 5-cycle and a stable 20-cycle at the point $\alpha = \alpha_{15,1}$. The stable 20-cycle (denoted by the number 4) arises in a couple with an unstable cycle (denoted by the number 3) at the point $\alpha = \alpha_{20,3}$, $\alpha_{20,3} \approx 72.12043$ which is situated on the line N_+^{C} between nodes $P_{20,4}$ and $P_{15,3}$. As in the previous cases, a family of cycles with periods that are multiple to 5 branch off from the 5-cycle (Fig. 5.22(b)) while passing through the point $\alpha = \alpha_{15,1}$.

Let us consider one more example. Figure 5.23 contains the results of the bifurcation analysis performed in the section $\Omega = \{(\alpha, \chi) : 71.15 < \alpha < 72.30; \chi = 0.14723\}$, wherein the bifurcation pattern is qualitatively similar to that shown in Figs. 5.21 and 5.22. The difference is that in this section two period-tripling BC-bifurcations (at the points $\alpha_{15,1}$ and $\alpha_{15,2}$) take place: $5 \to 15 \to 45$ (see Fig. 5.23). As in the previous case, at the point $\alpha = \alpha_* - 0$ the 5-cycle is stable, while at the point $\alpha = \alpha_* + 0$ it becomes unstable, in the neighborhood of $\alpha = \alpha_*$ the stable 45-cycle transforms abruptly to a mode of chaotic oscillations. Figure 5.23 shows only the stable cycles taking part in the bifurcational transitions.

We shall not discuss the question about the nature of the BC-bifurcational transition connected with the loss of stability of the 60- and 45-cycles at the

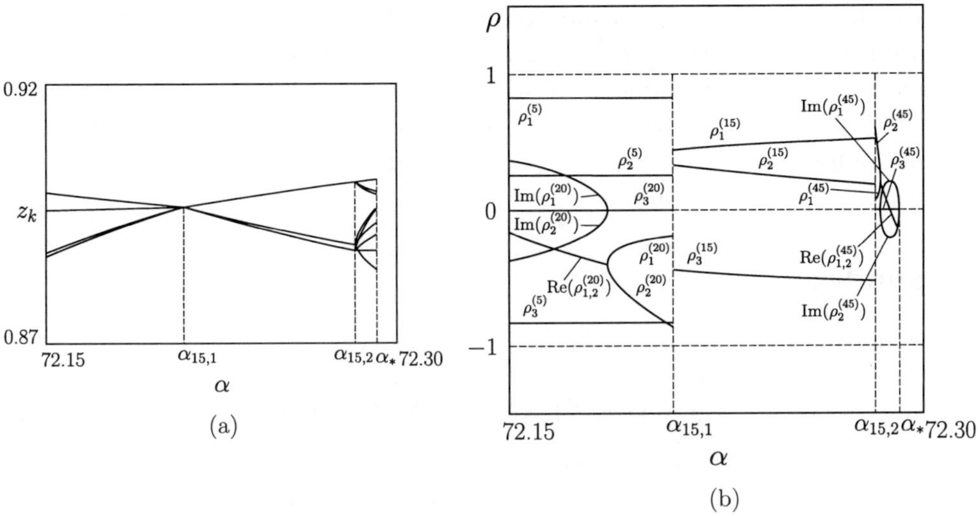

Figure 5.23. Two successive period-tripling bifurcations: (a) Bifurcation diagram depicting the stable cycles only. (b) Cycle multipliers in dependence on the parameter α.

points designated in Figs. 5.22 and 5.23 as α_*. The examples of Figs. 5.15, 5.17, 5.18, and 5.22 demonstrate that BC-bifurcations at which an m-cycle loses its stability can be accompanied with the doubling, tripling, etc., of the oscillation period or by the soft appearance of a set of cycles with periods that are multiple to m (see Figs. 5.21 and 5.22). With these results the question may arise: What happens in reality? This question may be too complicated to answer for the following reasons: The considered transitions can turn out to be extremely complex. Their profound description in the way we have done earlier for cycles with relatively small periods can be difficult because in the numerical studies we have observed a catastrophic chaotization of the oscillations when passing through $\alpha = \alpha_*$. One of the difficulties that may obtrude this analysis is that the limited precision of the computer calculations do not allow us to determine a stable cycle with a small radius of its inner basin of attraction, which is typical for the cycles with large periods. We have previously introduced the concept of an inner basin of attraction in connection with our discussion of the influence of noise (Chapter 4). A definition of the *inner basin of attraction* for flow systems is given in the Appendix below.

As to the bifurcational situation shown in Fig. 5.22 we can assume that the loss of stability of a 60-cycle is accompanied by the soft appearance of a stable cycle with a period that is a multiple to 60. Chaotization of the oscillations can occur because of the extreme smallness of the radius of the inner basin of attraction [34, 48] of a stable cycle when, under the influence of small random impacts connected, e.g., with truncation errors in the numerical calculations, a certain jump-over occurs from one part of the integral curve to another for one and the same stable motion.

Appendix

Let X_c be a stable m-cycle of the dynamical system

$$\frac{\mathrm{d}X}{\mathrm{d}t} = G(t, X);\ G(t + a, X) \equiv G(t, X).$$

Here X, G are n-dimensional column matrices, and a is the period of the external action. Let X be a perturbed motion: $X = X_c + \varepsilon$, $\varepsilon = (\varepsilon_1, \varepsilon_2, \ldots, \varepsilon_n)^{\mathrm{T}}$. The equation with respect to the disturbance has the form

$$\frac{\mathrm{d}\varepsilon}{\mathrm{d}t} = G(t, X_c + \varepsilon) - G(t, X_c),\ \varepsilon(t_0) = X_0 - X_c(t_0) = \varepsilon_0,\ X_0 = X(t_0). \quad (5.13)$$

Let D be the basin of attraction of an m-cycle in the ε-space with the radius r_D. The general solution of the system (5.13) depends on t_0 and ε_0

$$\varepsilon(t) = \varphi(t - t_0, \varepsilon_0).$$

Hence

$$\|\varepsilon(T)\| = \psi(t_0, \varepsilon_0)\, \|\varepsilon_0\|\,,\ \psi(t_0, \varepsilon_0) = \frac{\|\varphi(T, t_0, \varepsilon_0)\|}{\|\varepsilon_0\|}.$$

Definition 5.1. The aggregate of ε_0 for which

$$\max_{0<t_0<T} \psi(t_0, \varepsilon_0) < 1$$

is *the inner basin of attraction* D_{int}, $D_{\mathrm{int}} \subset D$.

Let us consider the function

$$\xi(r) = \sup_{\|\varepsilon_0\|=r} \big\{ \max_{0<t_0<T} \psi(t_0, \varepsilon_0) < 1 \big\}.$$

Definition 5.2. Let us call the smallest root of the equation

$$|\xi(r)| = 1,\ 0 < r < r_D \quad (5.14)$$

the radius of the inner basin of attraction and denote it as R_{int}.

When $\|\varepsilon_0\|$ is small

$$\varepsilon(t) = F(t - t_0)\,\varepsilon_0,$$

where $F(t - t_0)$ is the fundamental matrix which satisfies the equation

$$\frac{\mathrm{d}F(t)}{\mathrm{d}t} = \frac{\partial G(t, X_c)}{\partial X} F(t),\; F(t_0) = F_0.$$

Therefore when r is small

$$\xi(r) = \sup_{\|\varepsilon_0\|=r} \frac{\|F(T)\,\varepsilon_0\|}{\|\varepsilon_0\|} \le \sup_{\|\varepsilon_0\|\ne r} \frac{\|F(T)\|\,\|\varepsilon_0\|}{\|\varepsilon_0\|} = \|F(T)\|\,.$$

Remark 5.1. The value of R_{int} depends on the norm chosen in the vector space.

Remark 5.2. If $\|F(T)\| > 1$ then Eq. (5.14) has no roots. It means that any ball $\|\varepsilon_0\| \le r$ contains members of the set $D \setminus D_{\text{int}}$. But if $\|F(T)\| < 1$ then $0 < R_{\text{int}} < r_D$.

Remark 5.3. If $r > 0$ then the disturbances located within the area $D_{\text{int}} \cap \{\varepsilon_0 : \|\varepsilon_0\| \le R_{\text{int}}\}$ diminish monotonically.

Bibliography

[1] P. M. Dean and E. K. Matthews, *Glucose Induced Electrical Activity in Pancreatic Islet Cells*, J. Physiol. (Lond.) **210**, 255–264 (1970).

[2] I. Atwater and P. M. Beigelman, *Dynamic Characteristics of Electrical Activity in Pancreatic β-Cells*, J. Physiol. (Paris) **72**, 769–786 (1976).

[3] A. Sherman, J. Rinzel, and J. Keizer, *Emergence of Organized Bursting in Clusters of Pancreatic β-Cells by Channel Sharing*, Biophys. J. **54**, 411–425 (1988).

[4] E. Mosekilde, B. Lading, S. Yanchuk, and Yu. Maistrenko, *Bifurcation Structure of a Model of Bursting Pancreatic Cells*, BioSystems **63**, 3–13 (2001).

[5] S. Ozawa and O. Sand, *Electrophysiology of Endocrine Cells*, Physiol. Rev. **66**, 887–952 (1986).

[6] R. M. Miura and M. Pernarowski, *Correlations of Rate of Insulin Release from Islets and Plateau Functions for β-cells*, Bull. Math. Biol. **57**, 229–246 (1995).

[7] L. S. Satin and D. L. Cook, *Calcium Current Inactivation in Insulin-Secreting Cells is Mediated by Calcium Influx and Membrane Depolarization*, Pflügers Arch. **414**, 1–10 (1989).

[8] D. G. Carlson, A. Segmüller, E. Mosekilde, H. Cole, and J. A. Armstrong, *X-Ray Diffraction from Piezoelectrically Amplified Shear Waves in the 50 GHz Range*, Appl. Phys. Lett. **18**, 330 (1971).

[9] E. Mosekilde, D. G. Carlson, and A. Segmüller, *Quantum Anomaly in Acoustic Parametric Interaction*, J. Phys. C: Solid State Phys. **7**, 4281 (1974).

[10] B. K. Bose (Editor), Modern Power Electronics: Evolution, Technology, and Applications (IEEE Press, New York, 1992).

[11] R. E. Tarter, Solid-State Power Conversion Handbook (Wiley-Interscience, New York, 1993).

[12] N. Mohan, T. M. Undeland, and W. P. Robbins, Power Electronics: Converters, Applications and Design (Wiley, New York, 1995).

[13] J. G. Kasakian, M. F. Schlecht, and G. C. Verghese, Principles of Power Electronics (Addison-Wesley, Reading, MA, 1991).

[14] S. Banerjee and G. C. Verghese (Editors), Nonlinear Phenomena in Power Electronics (IEEE Press, New York, 2001).

[15] A. V. Kobzev, G. Ya. Mikhal'chenko, and N. M. Muzhitchenko, Modulating Power Supplies of Radioelectronic Devices (Radio i Svyaz, Tomsk, 1990), in Russian.

[16] P. R. K. Chetty, Switch-Mode Power Supply Design (TAB Books Inc., 1986).

[17] P. T. Krein, Elements of Power Electronics (Oxford University Press, New York, 1998).

[18] M. di Bernardo and Ch. K. Tse, *Chaos in Power Electronics: An Overview*, in Chaos in Circuits and Systems, edited by G. Chen and T. Ueta (World Scientific, Singapore, 2002), pp. 317–340.

[19] Ya. Z. Tsypkin, Sampling Systems Theory and Its Application (Macmillan, New York, 1964).

[20] A. Kh. Gelig and A. N. Churilov, Stability and Oscillations of Nonlinear Pulse-Modulated Systems (Birkhäuser, Boston, 1998).

[21] D. C. Hamill and D. J. Jefferies, *Subharmonics and Chaos in a Controlled Switched-Mode Power Converter*, IEEE Trans. Circuits Syst. **35**, 1059–1061 (1988).

[22] S. Banerjee, J. A. Yorke, and C. Grebogi, *Robust Chaos*, Phys. Rev. Lett. **80**, 3049–3052 (1998).

[23] S. Banerjee and C. Grebogi, *Border Collision Bifurcations in Two-Dimensional Piecewise Smooth Maps*, Phys. Rev. E **59**, 4052–4061 (1999).

[24] S. Banerjee, M. S. Karthik, G. Yuan, and J. A. Yorke, *Bifurcations in One-Dimensional Piecewise Smooth Maps — Theory and Applications in Switching Circuits*, IEEE Trans. Circuits Syst.-I: Fund. Theory Appl. **47**, 389–394 (2000).

[25] S. Banerjee, P. Ranjan, and C. Grebogi, *Bifurcations in Two-Dimensional Piecewise Smooth Maps — Theory and Applications in Switching Circuits*, IEEE Trans. Circuits Syst.-I: Fund. Theory Appl. **47**, 633–643 (2000).

[26] Zh. T. Zhusubaliyev, E. A. Soukhoterin, and E. Mosekilde, *Border-Collision Bifurcations and Chaotic Oscillations in a Piecewise-Smooth Dynamical System*, Int. J. Bifurcation and Chaos **11**, 2977–3001 (2001).

[27] Zh. T. Zhusubaliyev, E. A. Soukhoterin, and E. Mosekilde, *Border-Collision Bifurcations on a Two-Dimensional Torus*, Chaos, Solitons and Fractals **13**, 1889–1915 (2002).

[28] G. H. Yuan, S. Banerjee, E. Ott, and J. A. Yorke, *Border-Collision Bifurcations in the Buck Converter*, IEEE Trans. Circuits Syst.-I: Fund. Theory Appl. **45**, 707–716 (1998).

[29] M. di Bernardo and F. Vasca, *Discrete-Time Maps for the Analysis of Bifurcations and Chaos in DC–DC Converters*, IEEE Trans. Circuits Syst.-I: Fund. Theory Appl. **47**, 130–143 (2000).

[30] D. C. Hamill, J. H. B. Deane, and D. J. Jefferies, *Modeling of Chaotic DC–DC Converters by Iterated Nonlinear Mappings*, IEEE Trans. Power Electron. **7**, 25–36 (1992).

[31] A. El Aroudi and R. Leyva, *Quasi-Periodic Route to Chaos in a PWM Voltage-Controlled DC–DC Boost Converter*, IEEE Trans. Circuits Syst.-I: Fund. Theory Appl. **48**, 967–978 (2001).

[32] Zh. T. Zhusubaliyev, *Investigation of Chaotic Regimes of a Voltage Converter with Pulse-Width Modulation*, Elektrichestvo **6**, 40–46 (1997), in Russian.

[33] V. S. Baushev and Zh. T. Zhusubaliyev, *Indeterminable States of a Voltage Regulator with Pulse-Width Control*, Electrical Technology (3), 85–98 (1992), Elsevier Science.

[34] V. S. Baushev, Zh. T. Zhusubaliyev, and S. G. Michal'chenko, *Stochastic Features in the Dynamic Characteristics of a Pulse-Width Controlled Voltage Stabilizer*, Electrical Technology (1), 137–150 (1996), Elsevier Science.

[35] M. I. Feigin, Forced Oscillations in Systems with Discontinuous Nonlinearities (Nauka Publ., Moscow, 1994), in Russian.

[36] E. Fossas and G. Olivar, *Study of Chaos in the Buck Converter*, IEEE Trans. Circuits Syst.-I: Fund. Theory Appl. **43**, 13–25 (1996).

[37] H. E. Nusse and J. A. Yorke, *Border-Collision Bifurcations Including "Period Two to Period Three" for Piecewise Smooth Systems*, Physica D **57**, 39–57 (1992).

[38] H. E. Nusse, E. Ott, and J. A. Yorke, *Border-Collision Bifurcations: An Explanation for Observed Bifurcation Phenomena*, Phys. Rev. E **49**, 1073–1076 (1994).

[39] H. E. Nusse and J. A. Yorke, *Border-Collision Bifurcation for Piecewise Smooth One-Dimensional Maps*, Int. J. Bifurcation and Chaos **5**, 189–207 (1995).

[40] M. di Bernardo, M. I. Feigin, S. J. Hogan, and M. E. Homer, *Local Analysis of C-Bifurcations in n-Dimensional Piecewise-Smooth Dynamical Systems*, Chaos, Solitons and Fractals **10**, 1881–1908 (1999).

[41] M. Dutta, H. E. Nusse, E. Ott, J. A. Yorke, and G. Yuan, *Multiple Attractor Bifurcations: A Source of Unpredictability in Piecewise Smooth Systems*, Phys. Rev. Lett. **83**, 4281–4284 (1999).

[42] T. Kapitaniak and Yu. L. Maistrenko, *Multiple Choice Bifurcations as a Source of Unpredictability in Dynamical Systems*, Phys. Rev. E **58**, 5161–5163 (1998).

[43] M. di Bernardo, A. R. Champneys, and C. J. Budd, *Grazing, Skipping and Sliding: Analysis of the Nonsmooth Dynamics of the DC/DC Buck Converter*, Nonlinearity **11**, 858–890 (1998).

[44] Zh. T. Zhusubaliyev, E. A. Soukhoterin, and E. Mosekilde, *Quasiperiodicity and Border-Collision Bifurcations in a DC–DC Converter with Pulse-Width Modulation*, IEEE Trans. Circuits Syst.-I: Fund. Theory Appl., submitted.

[45] Zh. T. Zhusubaliyev, V. S. Titov, E. Yu. Emelyanova, and E. A. Soukhoterin, *C-Bifurcations in the Dynamics of Control System with Pulse-Width Modulation*, in Proc. of Second Int. Conf. "Control of Oscillations and Chaos" (COC'2000), vol. 1 (St.Petersburg, Russia, 2000), pp. 203–204.

[46] V. S. Baushev, Zh. T. Zhusubaliyev, Yu. V. Kolokolov, and I. V. Terekhin, *Local Stability of Periodic Solutions in Sample Data Control Systems*, Automation and Remote Control **53**, part 2, 865–871 (1992), Plenum Publishing.

[47] Zh. T. Zhusubaliyev and E. Yu. Emelyanova, *Analysis of the Structure of C-Bifurcation Boundaries in Piecewise Sewed Model of the Control System with Pulse-Width Modulation*, in Informational Technologies of Simulation and Control (Voronezh, Russia, 1999), pp. 40–46, in Russian.

[48] Zh. T. Zhusubaliyev, E. A. Soukhoterin, V. N. Rudakov, Yu. V. Kolokolov, and E. Mosekilde, *Bifurcations and Chaotic Oscillations in an Automatic Control Relay System with Hysteresis*, Int. J. Bifurcation and Chaos **11**, 1193–1231 (2001).

Chapter 6

BORDER-COLLISION BIFURCATIONS ON A TWO-DIMENSIONAL TORUS

6.1 Model of a Switching Power Converter

The purpose of this chapter is to discuss a number of new phenomena that can arise in piecewise-smooth dynamical systems whose motion involves two (or more) periodic components. When two self-oscillatory systems interact (or a self-oscillatory system is subjected to an external periodic forcing), the total motion can be viewed as occurring on the surface of a *two-dimensional torus* and, in the absence of resonances, the motion is said to be *quasiperiodic*. The problem that we are interested in is associated with the role that border-collision bifurcations play in the synchronization of the two modes and in the transition to chaos when the torus finally breaks down.

To examine these phenomena we shall consider a three-dimensional model of a DC/DC converter with pulse-width modulation. As described in Chapter 5, such converters are widely used in power electronics where they are applied in power supplies for industrial and aerospace equipment as well as in domestic appliances [1–5].

Synchronization is a universal phenomena in nonlinear systems [6]. The classical example is the synchronization of two mechanical clocks hanging side by side on the same wall. In his attempts to evaluate the accuracy of the newly developed pendulum clock, Huygens [7] observed that when two clocks could influence each others motion via "imperceptible vibrations" in the air or in the common support, the clocks tended to work in synchrony. Similar phenomena also occur in a wide range of different technical, biological, and economic systems [8, 9]. As discussed in Chapter 1, Togeby *et al.* [10], for instance, have described the synchronization of two thermostatically controlled radiators placed in the same room. Sturis *et al.* [11] have described how the production of insulin from the pancreas of healthy human subjects can synchronize to a sinusoidal variation in the rate of intravenous glucose infusion, and Schäfer *et al.* [12] have shown how the beating of the heart may synchronize with the respiratory cycle for young athletes.

In this connection it is worth emphasizing that synchronization can occur at many different locking ratios between the two periodicities. The heart, for instance, typically beats three to five times in each respiratory cycle. In nonlinear systems, a periodic motion usually contains *harmonic components* at frequencies of p times the fundamental frequency, where p is an integer. When two nonlinear oscillators interact, synchronization (*mode-locking* or *entrainment*) will occur whenever a harmonic component of one mode is sufficiently close to a harmonic component of the other. Hence, nonlinear oscillators tend to entrain so that one subsystem completes precisely p cycles each time the other subsystem completes q cycles, with p and q being integers.

The modern theory of synchronization was established primarily by Arnol'd [13, 14]. He reduced the problem to considering the simple one-dimensional sine-circle map

$$\vartheta_k = \vartheta_{k-1} + \Omega - \frac{K}{2\pi} \sin 2\pi\vartheta_{k-1} \mod 1 \tag{6.1}$$

that we discussed in Chapter 2 in connection with our analysis of the swallow tail structure. Here, Ω is the (normalized) phase advance of one of the oscillators per period of the other, and K is a measure of the coupling strength. The assumption involved in this reduction is that the disturbances that the coupling produces to the amplitudes of the two oscillators can be neglected to the first approximation. The amplitude of the individual oscillator is controlled by a balance between the inherent instability, the dissipation, and the nonlinear constraints.

Without coupling, however, the phases of the oscillators are *neutrally stable.* The phase difference can assume any value with equal probability, and there is no mechanism to correct for a relative shift in phase. Hence, the phase difference is particularly sensitive to the introduction of a coupling, and synchronization may occur for any finite value of the coupling parameter.

As illustrated in Fig. 2.29, the sine-circle map produces a dense set of resonance (or Arnol'd) tongues for $K < 1$, with each tongue corresponding to a particular synchronization ratio $p:q$. With increasing coupling parameter, the widths of the tongues grow monotonously, with most of the tongues displaying a cusp formed shape. Only the $1:1$ tongue (in Fig. 2.29) opens up with a finite angle between its edges.

The boundary of the resonance tongues in Fig. 2.29 is formed by saddle-node (or tangent) bifurcations. The boundary of the $1:1$ tongue, for instance, is associated with the mapping (6.1) loosing its intersections with the diagonal $\vartheta_k = \vartheta_{k-1}$ of the return map. Hence, a stable fixed point and a saddle point of the map are annihilated with one another. Within the resonance tongues, the two-dimensional torus (referred to as a *resonance torus*) is defined by the unstable manifolds of the saddle cycle.

The above scenario is generic in the sense that the same bifurcation structure is observed for a wide class of systems, independent of the precise equations underlying their dynamics. In particular, the edges of the resonance tongues are always associated with saddle-node bifurcations at small coupling strengths. For oscillators with more complicated, but still regular, dynamics (e.g., oscillators that have undergone a period-doubling bifurcation or bursting oscillators that alternate between a spiking state and a silent state), *phase multistability* may be observed [15, 16]. This implies that the same frequency locking ratio $p:q$ can be realized with two (or more) values of the relative phase. In this case the Arnol'd tongues display a substructure with two (or more) tongues embedded one into the other. At sufficiently high coupling strengths it is also possible that one of the oscillators will be strong enough to completely suppress the internal oscillations of the other. In this case, one may observe the phenomenon of *oscillator death* with its characteristic bifurcation structure [17–19].

For $K > 1$, the resonance tongues of the sine-circle map start to overlap, and quasiperiodic dynamics ceases to exist. At the same time, transitions to chaos start to occur both via period-doubling cascades (as described in connection with our discussion of the swallow tail structure in Chapter 2) and via torus destruction. It is well-known that resonance phenomena, that always precede the appearance of chaos, play an important role in the quasiperiodic scenario for the transition to chaos via torus destruction. For smooth dynamical systems, the basic theorem for the destruction of a two-dimensional torus was proved and the possible ways for the appearance of chaotic dynamics were described in the important work of Afraimovich and Shilnikov [20, 21]. Torus destruction can occur via three main mechanisms.

In all scenarios, the starting point is a smooth torus in a resonance region where a stable periodic orbit (node) coexists with an unstable orbit (saddle cycle) of the same periodicity. In one scenario, the unstable manifolds from the saddle cycle start to develop wrinkles near the points of the stable node. The first wrinkle is repeated indefinitely as the manifold approaches the node, and in this way the torus becomes nondifferentiable in the points of the stable cycle. As the system, under variation of a parameter, leaves the resonance zone, wrinkles and non-smoothness spread along the invariant manifolds, and the torus breaks up into a fractal structure.

In another scenario, the unstable manifolds from the saddle cycle start to intersect the nonleading manifolds of the node. This produces an infinitely folded structure accumulating at the node points, and again the torus loses its smoothness in these points. The torus is destructed in a *homoclinic bifurcation* when this folded structure makes contact with the stable manifold of the saddle cycle. This will typically occur for sufficiently strong coupling when the system approaches the edge of the resonance zone.

Finally, in the third scenario, the stable node is transformed into a stable focus, and the unstable manifolds from the saddle cycle start to spiral around the focus points. The torus is destructed as the stable cycle subsequently undergoes a Hopf bifurcation, and a new torus of higher dimension may be formed. Alternatively, the focus can again turn into a node, and the torus breaks down as this node undergoes a sequence of period-doubling bifurcations.

The generic character of the Afraimovich and Shilnikov analysis has been confirmed numerically as well as experimentally for a wide class of discrete time and flow systems [22–26], clearly demonstrating the possible routes to disintegration for a two-dimensional torus in smooth systems. One of the main ideas of the present chapter is to show that the transitions to chaos through two-frequency quasiperiodicity in piecewise smooth systems can differ fundamentally from the mechanisms described in the existing literature. At the same time we show that the boundaries of the Arnol'd tongues for non-smooth systems are formed by border-collisions bifurcations, rather than by the saddle-node bifurcations known from smooth systems. Moreover, the internal organi-

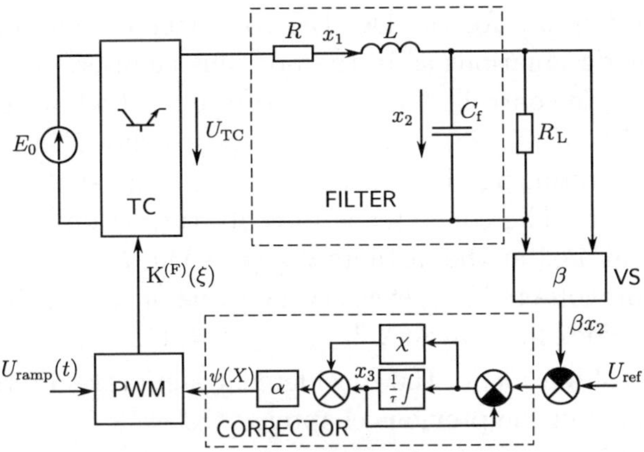

Figure 6.1. Coupling diagram for a constant voltage converter with pulse-width modulation of the first kind (PWM-1).

zation of the tongues is also quite different from the case of smooth systems. An introduction to some of these phenomena was given in Chapter 1 in connection with our discussion of the *rotating sawtooth map* (see Fig 1.29). In this chapter we will discuss a practical example from power electronics, and in Chapter 7 we shall show how similar phenomena arise in a managerial system.

We first present the coupling diagram and a mathematical model for a semiconductor converter made according to the same scheme as the converter considered in the previous chapter. The only change is that the output voltage is controlled in accordance with a slightly different method of pulse-width modulation. The behavior of such a system is described by a three-dimensional piecewise-smooth set of non-autonomous differential equations. The investigation of this dynamical system is reduced to studying a three-dimensional piecewise-smooth mapping, and methods of determining the cycles of such systems and analyzing their local stability are described. Section 6.2 is dedicated to the analysis of the two-parameter diagram of dynamical modes. Peculiarities in the structure of the bifurcation boundaries and in the transitions to chaos are studied in detail. In Sec. 6.3 we analyze the domains of synchronization on a two-dimensional torus, and the structure and inner organization of two different types of such domains are described. Typical border-collision bifurcations of cycles on a two-dimensional torus and transitions to chaos are investigated in Sec. 6.3.

The coupling diagram for the DC/DC converter is reproduced in Fig. 6.1. We use the same designations as in the previous chapter. The output voltage of the converter x_2 is controlled according to the method of *pulse-width modulation of the first kind* [27, 28]. The time diagrams shown in Fig. 6.2 illustrate the principles of producing the control pulses by the switch elements of the transistor converter. The converter is forcedly turned on in the beginning of each clock interval, i.e., at the instants $t = (k-1)a$, $k = 1, 2, \ldots$, and it turns off when the ramp voltage U_{ramp} exceeds the value of the feedback signal in the beginning of the clock interval $\psi(X)|_{t=(k-1)\,a}$, $k = 1, 2, \ldots$.

The equations of motion for the considered system take a form similar to the equations examined in the previous chapter,

$$\begin{aligned} \frac{\mathrm{d}X}{\mathrm{d}t} &= G(t, X); \\ X &= (x_1, x_2, x_3)^{\mathrm{T}}; \\ G(t, X) &= AX + B(t, X); \end{aligned} \tag{6.2}$$

with the matrices

$$A = \begin{pmatrix} -\dfrac{R}{L} & -\dfrac{1}{L} & 0 \\ \dfrac{1}{C_{\mathrm{f}}} & -\dfrac{1}{C_{\mathrm{f}} R_{\mathrm{L}}} & 0 \\ 0 & -\dfrac{\beta}{\tau} & -\dfrac{1}{\tau} \end{pmatrix}; \; B(t, X) = B\big(\mathrm{K}^{(\mathrm{F})}(\xi)\big) = \begin{pmatrix} \dfrac{E_0}{L}\,\mathrm{K}^{(\mathrm{F})}(\xi) \\ 0 \\ \dfrac{U_{\mathrm{ref}}}{\tau} \end{pmatrix}.$$

The only difference is associated with the algorithm used to determine the switching instants t_k. This algorithm reflects the kind of pulse-width modulation applied:

$$\mathrm{K}^{(\mathrm{F})}(\xi) = \begin{cases} 1, & (k-1)\,a < t < t_k; \\ 0, & t_k < t < ka, \end{cases} \quad k = 1, 2, \ldots, \tag{6.3}$$

$$t_k = \begin{cases} (k-1)\,a, & \xi\big((k-1)\,a, X_{k-1}\big) \le 0; \\ \to \xi(t_k, X_{k-1}) = 0, & \xi\big((k-1)\,a, X_{k-1}\big) > 0,\ \xi(ka, X_{k-1}) < 0; \\ ka, & \xi(t, X_{k-1}) > 0 \text{ for } (k-1)\,a \le t \le ka, \end{cases}$$

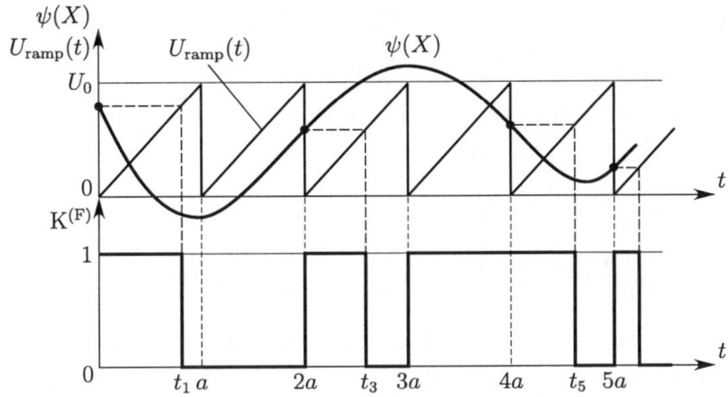

Figure 6.2. Time plots showing the principle of control pulse generation according to the algorithm for pulse-width modulation of the first kind (PWM-1). Note that the converter switches off when $U_{\text{ramp}}(t)$ exceeds the value of $\psi(X)$ in the beginning of the clock interval (when $t = (k-1)\,a$); $t_2 = a$, $t_4 = 4a$ according to the algorithm (6.3). Compare with Fig. 5.4 to see the difference between PWM-1 and PWM-2.

with

$$\begin{aligned} \xi(t, X) &= \psi(X) - U_{\text{ramp}}(t); \\ \psi(X) &= \alpha\big(\chi(U_{\text{ref}} - \beta x_2) + (1-\chi)\, x_3\big); \\ U_{\text{ramp}} &= U_0\big(t/a - E_1(t/a)\big); \\ X_k &= X(ka). \end{aligned}$$

When inserted one into the other, the last four equations may be expressed in the form

$$\xi(t, X) = \alpha\chi U_{\text{ref}} - UX - U_0\big(t/a - E_1(t/a)\big), \tag{6.4}$$

where $U = \big(0, \alpha\beta\chi, \alpha(\chi - 1)\big)$ is a row matrix that determines the control law. In order to make comparison of the results possible let us use the same parameter values as in Chapter 5: $R = 10.6\ \Omega$, $L = 0.1$ H, $R_{\text{L}} = 100\ \Omega$, $C_{\text{f}} = 10^{-6}$ F, $E_0 = 104$ V, $U_0 = 10$ V, $U_{\text{ref}} = 5$ V, $\beta = 0.1$, $a = 10^{-4}$ s, and $\tau = 4 \cdot 10^{-4}$ s, with $\alpha > 0$, $0 < \chi < 1$. The amplification constant α and the transfer constant χ of the corrector will again be used as the main bifurcation parameters.

By virtue of the periodic forcing, the dynamical system (6.2) may be reduced to the three-dimensional stroboscopic map

$$\begin{aligned} X_k &= CX_{k-1} + V_k; \\ C &= \mathrm{e}^{Aa}; \end{aligned} \tag{6.5}$$

$$V_k = (\mathrm{e}^{Aa} - \mathrm{e}^{Aa(1-z_k)})A^{-1}B(1) - (E - \mathrm{e}^{Aa(1-z_k)})A^{-1}B(0);$$

$$z_k = \begin{cases} 0, & U^* \le 0; \\ z^*, & 0 < U^* < U_0; \quad k = 1, 2, \ldots \\ 1, & U^* \ge U_0; \end{cases}$$

$$U^* = \alpha\chi U_{\text{ref}} - UX_{k-1};$$

$$z^* = \frac{\alpha\chi}{U_0} U_{\text{ref}} - \frac{1}{U_0} UX_{k-1}.$$

Periodic motions can be calculated numerically by solving the system of transcendental equations [3, 29, 30]

$$\varphi_k(z_1, z_2, \ldots, z_m) = \mu_k(z_k),\ k = \overline{1, m}; \tag{6.6}$$

$$\mu(z_k) = \begin{cases} < 0, & z_k = 0; \\ > 0, & z_k = 1; \\ 0, & 0 < z_k < 1; \end{cases}$$

$$\varphi_k(z_1, z_2, \ldots, z_m) = \alpha\chi U_{\text{ref}} - UX_{\text{c}(k-1)} - U_0 z_k;$$

$$X_{\text{c}(k-1)} = C^{k-1}(E - C^m)^{-1} \sum_{i=1}^{m} C^{m-i} V_i + \sum_{i=1}^{k-1} C^{k-1-i} V_i$$

with respect to z_1, z_2, ..., z_m. This system has been determined from Eq. (6.5) by applying the periodicity condition. The applied procedure was described in detail in the previous chapters.

Similarly to the procedure used in Chapter 5, we can pass from the variables x_1, x_2, x_3 to the variables w_1, w_2, w_3 via the linear transformation (5.3). The map (6.5) then takes the form (5.5):

$$w_{ik} = \mathrm{e}^{a\lambda_i} w_{i(k-1)} + \mathrm{e}^{a\lambda_i(1-z_k)} - \mathrm{e}^{a\lambda_i},\ i = 1, 2, 3;\ k = 1, 2, \ldots, \tag{6.7}$$

where

$$z_k = \begin{cases} 0, & \varphi(0) \le 0; \\ z^*, & \varphi(0) > 0 \text{ and } \varphi(1) < 0; \\ 1, & \varphi(z) > 0 \text{ for } 0 \le z \le 1, \end{cases}$$

and

$$\varphi(z) = \sum_{i=1}^{3} \gamma_i \left(\chi - \frac{\lambda_3}{\lambda_i}\right) w_{i(k-1)} - \frac{Q}{\alpha} z + q.$$

The value

$$z^* = \frac{\alpha}{Q}\left[\sum_{i=1}^{3}\gamma_i\left(\chi - \frac{\lambda_3}{\lambda_i}\right)w_{i(k-1)} + q\right]$$

now appears as the root of the equation $\varphi(z) = 0$.

Using the map (6.7), the expression (6.6) for $\varphi_k(z_1, z_2, \ldots, z_m)$ is reduced to the significantly simpler form:

$$\varphi_k(z_1, z_2, \ldots, z_m) = \sum_{i=1}^{3}\gamma_i\left(\chi - \frac{\lambda_3}{\lambda_i}\right)\hat{\sigma}_{ik} - \frac{Q}{\alpha}z_k + q,$$

$$\hat{\sigma}_{ik} = \frac{1}{1 - \mathrm{e}^{ma\lambda_i}}\sum_{j=1}^{m}\mathrm{e}^{a\lambda_i(m+k-j-z_j)} + \sum_{j=1}^{k-1}\mathrm{e}^{a\lambda_i(k-j-z_j)} - \frac{\mathrm{e}^{a\lambda_i}}{1 - \mathrm{e}^{a\lambda_i}},\ k = \overline{1, m}.$$

To define the type of periodic solution we have introduced a set of characteristics similar to those applied in the previous chapter [30]. These characteristics specify the order in which the phase point passes the various domains of piecewise continuity. Quite naturally we have chosen sequence of the cyclic points of the period m z_k^{c}, $z_k^{\mathrm{c}} = z_{k+m}^{\mathrm{c}}$, $k = \overline{1, m}$ as a characteristics. It is evident that if z_k^{c}, $k = \overline{1, m}$ denote the solutions of Eq. (6.6), then the sewing order for the trajectory of a periodic motion of a specific type will be determined by any of the characteristics

$$S_{m,1} = (z_1^{\mathrm{c}} z_2^{\mathrm{c}} \ldots z_{m-1}^{\mathrm{c}} z_m^{\mathrm{c}})$$
$$S_{m,2} = (z_2^{\mathrm{c}} z_3^{\mathrm{c}} \ldots z_m^{\mathrm{c}} z_1^{\mathrm{c}})$$
$$\ldots$$
$$S_{m,m} = (z_m^{\mathrm{c}} z_1^{\mathrm{c}} \ldots z_{m-2}^{\mathrm{c}} z_{m-1}^{\mathrm{c}}).$$

These characteristics transform one into the other under cyclic permutation of the components.

The sequence z_k^{c}, $k = \overline{1, m}$ can be replaced by the string of m symbols "0", "1", "*" such that "0" replaces $z_k = 0$, "1" replaces $z_k = 1$ and "*" replaces values $0 < z_k < 1$. Application of the symbolic characteristics obtained in this way significantly simplifies the description of the overall picture of possible motion types and bifurcations. In particular, the number of phase trajectory portions from which the trajectory of the m-cycle is sewed together equals

$$N = 2(m - r - l),$$

where r and l denote the number of zeros and ones, respectively, in the symbolic string. Changes in the number of portions from which a trajectory is sewed is reflected by the symbol "$*$" being transformed into zero or one, or vice versa.

When analyzing the local stability, the basic matrix $F(T)$ is calculated by the recurrence formula [29, 30]

$$F_k = \frac{\partial X_{\mathrm{c}k}}{\partial X_{\mathrm{c}(k-1)}} F_{k-1};\; k = 1, 2, \ldots, m;\; F_0 = E;\; F(T) = F_m;$$

$$\frac{\partial X_{\mathrm{c}k}}{\partial X_{\mathrm{c}(k-1)}} = C + \begin{cases} \dfrac{\partial X_{\mathrm{c}k}}{\partial z_k}\left(\dfrac{\partial z_k}{\partial X_{\mathrm{c}(k-1)}}\right)^{\mathrm{T}}, & 0 < z_k < 1; \\ O, & z_k = 1 \text{ or } z_k = 0, \end{cases}$$

where O is the zero matrix.

6.2 Two-Parameter Diagram of Dynamical Modes

The results of our numerical calculations of the two-parameter diagram of dynamical modes within the parameter plane $\Pi = \{(\alpha, \chi) : 9.0 < \alpha < 59.0; 0.0 < \chi < 1.0\}$ are presented in Fig. 6.3. Here, $\Pi_{m,k}$ denote simply connected sets of parameters such that for any point $P = (\alpha, \chi)$, $P \in \Pi_{m,k}$ there exist one or two locally stable period-m cycles. These cycles may not depend smoothly on the parameters because the domains $\Pi_{m,k}$ can contain fold singularities. Such singularities are formed by lines of border-collision bifurcations involving the merging of a stable and an unstable m-cycle, the fold lines being supported by points of codimension 2. The second index of the designation $\Pi_{m,k}$ is used to distinguish sets with the same m.

The domain $\Pi_{1,1}$ to the left is bounded by the curve N_φ of points at which quasiperiodic oscillations arise from a period-1 cycle, producing a two-dimensional torus in the phase space of our dynamical system. We shall also refer to quasiperiodic behavior as *ergodic beating*, referring to the characteristic temporal variation observed for the involved variables. Figure 6.4(a) shows the bifurcation diagram calculated for the section $\Omega_1 = \{(\alpha, \chi) : 32.5 < \alpha < 33.5; \chi = 0.55\}$ with the variation of the real and imaginary parts of the multipliers for the 1-cycle being displayed in Fig. 6.4(b). The third multiplier $\rho_3^{(1)}$ remains real and numerically less than unity. Inspection of Fig. 6.4(b) shows how a pair of complex conjugate multipliers for the 1-cycle leave the unit circle at the point $P = (\alpha_\varphi; 0.55)$, lying on the curve N_φ. The

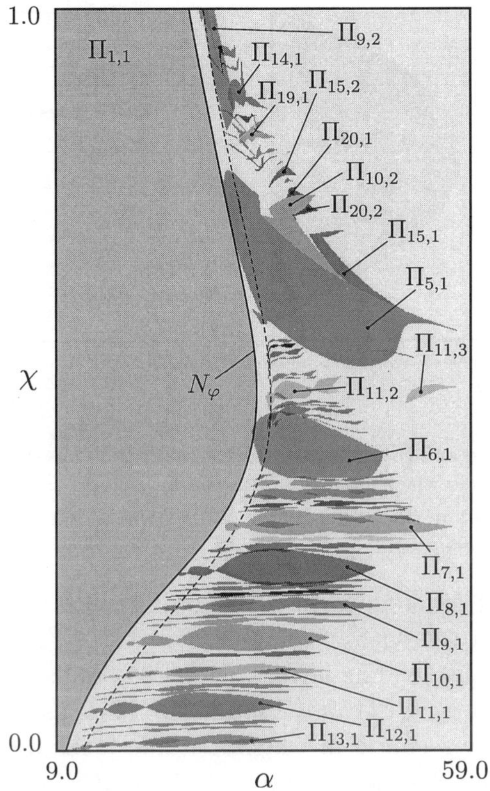

Figure 6.3. Domains of existence for the different oscillatory modes in the system parameter plane $\Pi = \{(\alpha, \chi) : 9.0 < \alpha < 59.0; 0.0 < \chi < 1.0\}$. $\Pi_{k,j}$ denotes the domains of existence for cycles of period k. The dashed line schematically shows the curve where a two-dimensional ergodic torus loses its smoothness. N_φ designates a torus bifurcation curve. To the left of this curve the converter displays a stable fixed point. α and χ are the amplification and the transfer constants of the corrector, respectively (see Fig. 6.1).

bifurcation diagram shows that a two-dimensional torus with irrational winding number softly arises from a stable period-1 cycle at this point.

In Fig. 6.3 the dashed line schematically shows the line where the two-dimensional ergodic torus loses its smoothness. In the following, the scenarios for the development of complex behavior and the transition to chaos with changing parameters will be described in terms of the bifurcations of the cycles defined within the domains $\Pi_{m,k}$, $m \neq 1$. Inspection of the Poincaré sections of the aperiodic oscillations to the right of the dotted curve shows that they

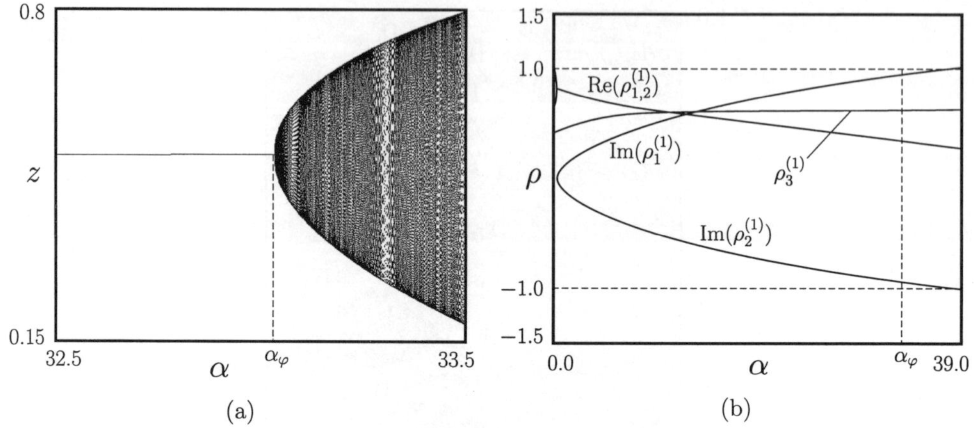

Figure 6.4. (a) Bifurcation and (b) multiplier diagrams for the appearance of a two-dimensional ergodic torus from a 1-cycle when intersecting the curve N_φ. $\alpha_\varphi \approx 33.03245$ denotes the point of torus formation (Andronov-Hopf bifurcation point). At this point $|\rho_{1,2}^{(1)}| = 1$.

can take a variety of different forms: The section can be a nonsmooth closed curve [Fig. 6.5(a)], a set with a complex structure [Fig. 6.5(b)], or disjunct sets of points [Fig. 6.5(c)]. The latter case corresponds to a so-called m-piece chaotic attractor, or a *cycle of chaotic intervals.* This case is associated with the Afraimovich-Shilnikov scenario in which a stable focus undergoes a secondary Hopf bifurcation. The chaotic attractor in Fig. 6.5(b), on the other hand, has obviously developed through the formation of wrinkles on the invariant manifolds close to a period-5 node. Figure 6.5 also displays the power spectra of the various oscillations. Here the marked discrete components $f_3 = f_0 - f_2$, $f_4 = f_0 + f_2$, $f_5 = f_1 - f_2$, $f_6 = f_1 + f_2$ correspond to the components that one can observe for the ergodic torus.

As the numerical analysis shows, the transitions to chaos via quasiperiodic oscillations are characterized by a significant variability. One of the possible scenarios of the transition to chaos is illustrated by the bifurcation diagram shown in Fig. 6.6. This diagram was calculated for the section $\Omega_2 = \{(\alpha, \chi) : 28.0 < \alpha < 37.0; \chi = 0.8\}$.

We shall postpone the more detailed discussion of the transition to chaos via torus breakdown to Sec. 6.3 and only make a few comments to the diagram here. We have tried to trace the evolution of the invariant curve in a section of the ergodic torus while changing the parameter α. To the left of the boundary

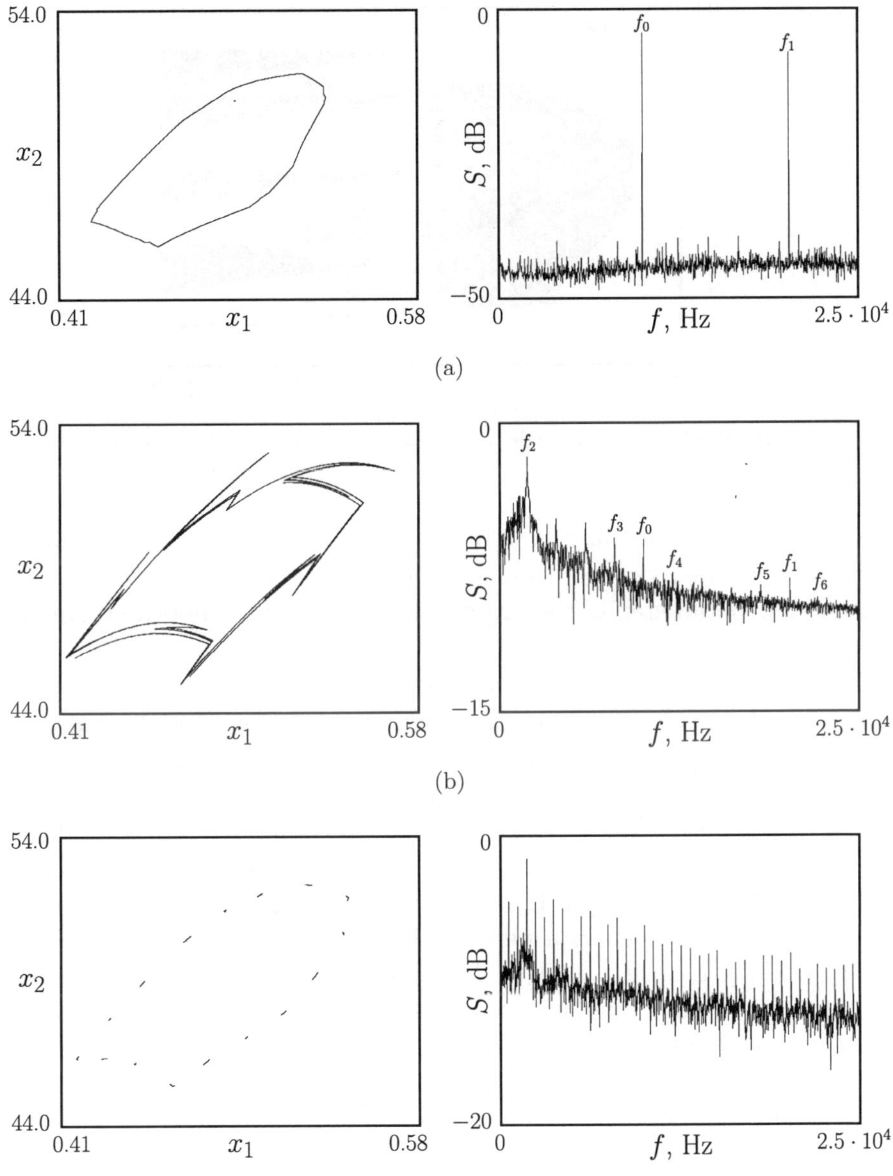

Figure 6.5. Two-dimensional projections of the Poincaré sections with the corresponding power spectra of the aperiodic motion for: (a) $\alpha = 35.5$ and $\chi = 0.55$; (b) $\alpha = 50.0$ and $\chi = 0.7$; (c) $\alpha = 40.0$ and $\chi = 0.53$.

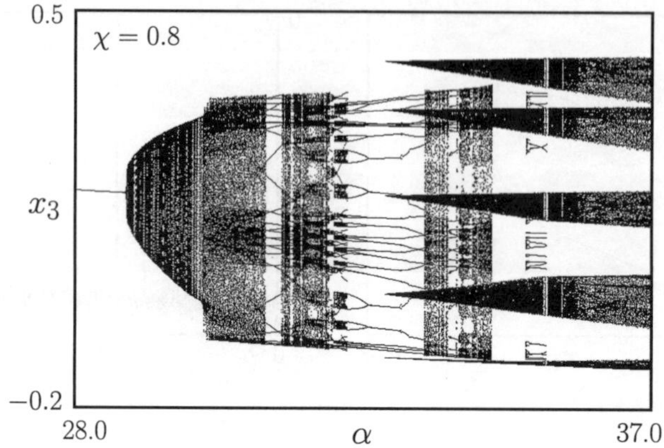

Figure 6.6. Transition to chaos through the appearance and destruction of an ergodic torus and bifurcations of resonance cycles.

(marked by a dashed line in Fig. 6.3) where the two-dimensional ergodic torus loses its smoothness, the Poincaré section is a smooth closed curve. When increasing the parameter α the ergodic torus loses its smoothness and then destroys.

In the quasiperiodic scenario of the transition to chaos, resonance phenomena play an important role [8]. Resonance dynamics always precedes the appearance of chaos. Some of the domains $\Pi_{m,k} \subset \Pi$ are domains of existence for stable cycles lying on the two-dimensional torus (domains of synchronization, or resonance tongues). The sets $\Pi_{m,k}$, $m \neq 1$ can be divided into several groups.

Let us first consider the domain of existence of the stable 5-cycle $\Pi_{5,1}$. Figure 6.7 is a magnified part of Π that contains the domains of existence of a 5-cycle $\Pi_{5,1}$ and of cycles with periods that are multiple to 5, namely $\Pi_{10,2}$, $\Pi_{15,1}$, $\Pi_{20,1}$, $\Pi_{20,2}$, $\Pi_{30,1}$, $\Pi_{30,2}$, $\Pi_{30,3}$, $\Pi_{30,4}$. The results of more precise numerical calculations for the boundaries of $\Pi_{5,1}$ are given in Fig. 6.8. The domain $\Pi_{5,1}$ has non-empty intersections with $\Pi_{1,1}$ as well as with the domain of existence of a two-dimensional ergodic torus. The area where the stable 5-cycle coexists with the ergodic two-dimensional torus is bounded by the lines N_φ and L_0 (the dashed line connecting the points 1 and 10) in Fig. 6.8. Let us illustrate how the 5-cycle and the two-dimensional ergodic torus arise in this part of the diagram. Consider a section $\Omega_1 = \{(\alpha, \chi) : 29.7 < \alpha < 32.7; \chi = 0.71\}$. Figure 6.9 shows the results of a bifurcation analysis for this section. Here, thick lines mark the stable cycles and dashed lines show the unstable cycles.

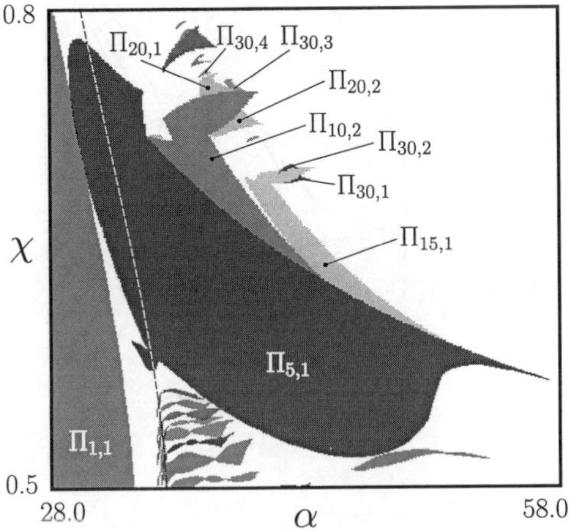

Figure 6.7. Domain of stability $\Pi_{5,1}$ for a 5-cycle with the associated domains of existence for cycles of periods 10, 15, 20, and 30.

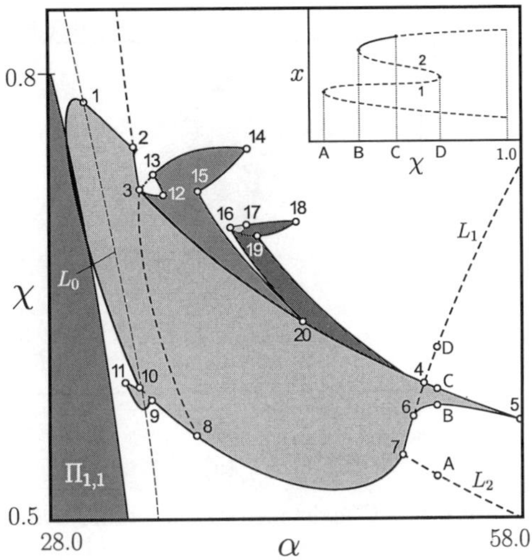

Figure 6.8. Bifurcation boundaries of the domain $\Pi_{5,1}$. The bifurcational transitions when intersecting the lines L_1 and L_2 are shown in the inset.

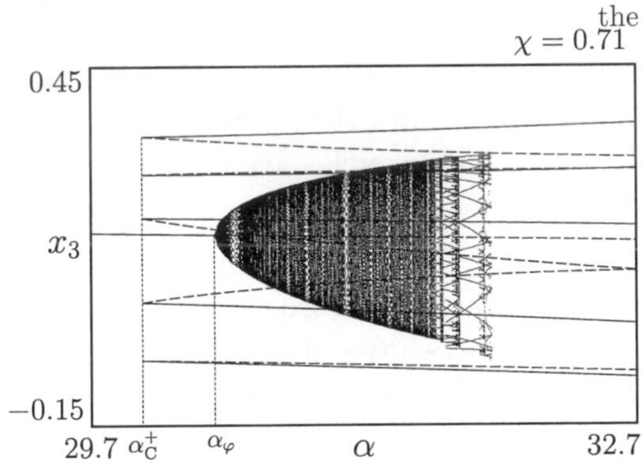

Figure 6.9. Bifurcation diagram for the region where the stable 5-cycle first coexists with the stable fixed point and then with a quasiperiodic motion. At the point α_C^+ two (one stable and one unstable) period-5 cycles arise in a border-collision bifurcation. α_φ is the Hopf bifurcation point.

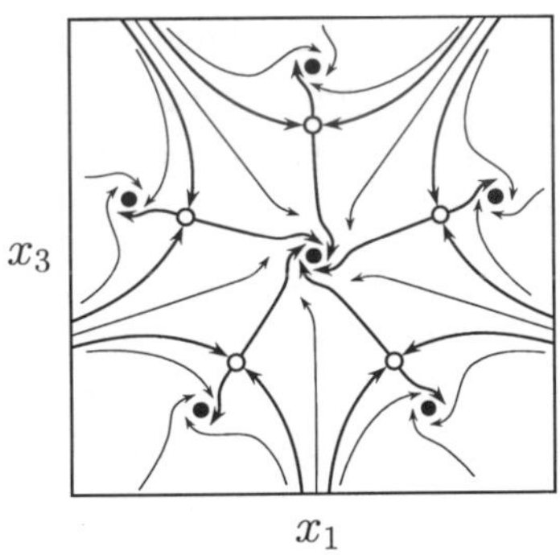

Figure 6.10. Sketch of phase portrait of the dynamical system for the case of coexistence of a stable fixed point and a stable period-5 cycle. The period-5 saddle cycle is indicated by a set of open circles.

Two (one stable and one unstable) period-5 cycles arise in a border-collision bifurcation at the point $\alpha_C^+ \approx 29.9824$ belonging to the curve 1–10 in Fig. 6.8.

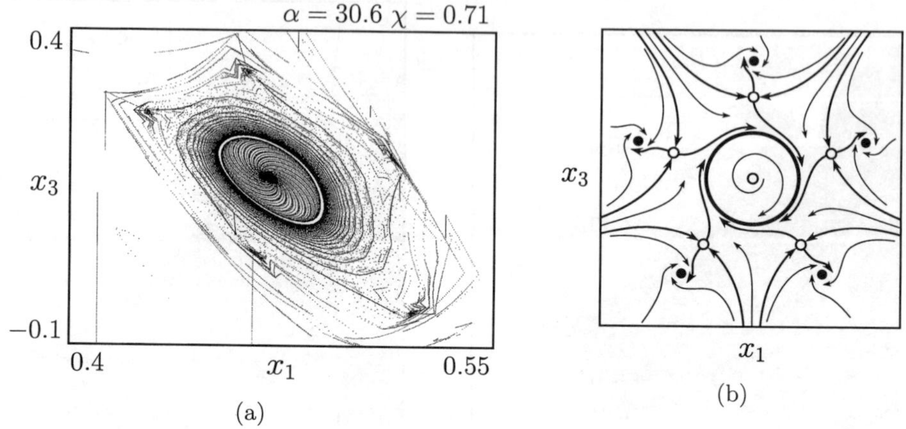

Figure 6.11. Coexistence of the stable two-dimensional torus and the stable period-5 cycle. (a) Two-dimensional projection of the Poincaré section. (b) Two-dimensional projection of the sketchy phase portrait.

Within the range $\alpha_C^+ < \alpha < \alpha_\varphi$, $\alpha_\varphi \approx 30.3946$, a stable fixed point coexists with a stable period-5 cycle. Figure 6.10 qualitatively shows the two-dimensional projection of the phase portrait of the dynamical system within this range. Black circles mark the points of the stable cycles and the open circles show the points of the period-5 saddle cycle. The two stable cycles are both represented as focuses.

In the point α_φ, $\alpha_\varphi \in N_\varphi$ the stable fixed point corresponding to period-1 cycle becomes unstable, and a two-dimensional ergodic torus arises from this focus via a Hopf bifurcation. Figure 6.11(a) displays a two-dimensional projection of the Poincaré section for $\alpha = 30.6$ and $\chi = 0.71$. The stable two-dimensional torus coexists with the stable period-5 cycle in this point of the parameter plane. In order to better illustrate this picture Fig. 6.11(b) schematically shows the two-dimensional projection of the phase portrait.

Many small domains of periodicity exist within the area of quasiperiodicity. These domains intersect each other producing areas of periodic multistability.

A nonsmooth two-dimensional torus arises when we cross the curve L_0 rightwards. Figure 6.12 shows a set of plots that illustrate the formation of a nonsmooth resonance torus. The stable ergodic torus coexists with the stable 5-cycle within the region between the curves N_φ and L_0 (Fig. 6.12(a), see also Fig. 6.11). On the curve L_0 the stable ergodic torus touches the saddle 5-cycle

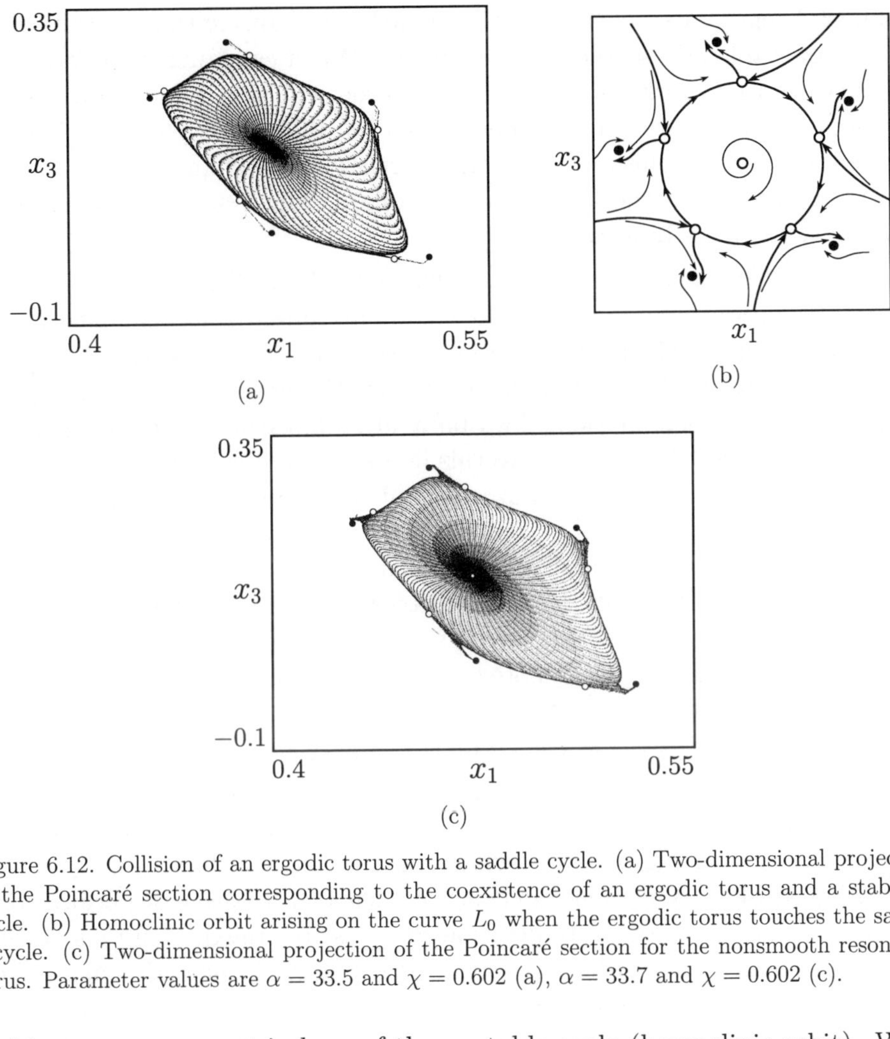

Figure 6.12. Collision of an ergodic torus with a saddle cycle. (a) Two-dimensional projection of the Poincaré section corresponding to the coexistence of an ergodic torus and a stable 5-cycle. (b) Homoclinic orbit arising on the curve L_0 when the ergodic torus touches the saddle 5-cycle. (c) Two-dimensional projection of the Poincaré section for the nonsmooth resonance torus. Parameter values are $\alpha = 33.5$ and $\chi = 0.602$ (a), $\alpha = 33.7$ and $\chi = 0.602$ (c).

and becomes a separatrix loop of the unstable cycle (homoclinic orbit). With further increase of α the stable ergodic torus disappears and the nonsmooth resonance torus shown in Fig. 6.12(c) arises. All trajectories are now attracted by the stable period-5 cycle. In Fig. 6.12 the black circles again mark the points of the stable cycles and the open circles show the points of the unstable cycle. On the BC-bifurcation curve 1–10 in Fig. 6.8, the 5-cycle abruptly transforms into a 1-cycle or into a mode of ergodic beating.

Let us return to the analysis of the diagram shown in Fig. 6.8. In this diagram the line 2–1 represents a curve of saddle-node bifurcations in the points of which a couple of stable and saddle 5-cycles appear. When changing the parameters, this curve continues into a curve 1–10 of border-collision bifurcations for the merging of a stable and an unstable 5-cycle of different types. The line 6–7 represents a border-collision bifurcation curve where a change of the solution type takes place with the loss of local stability. This line continues into the curve of saddle-node bifurcation 5–6.

In the previous chapter we showed that a border-collision bifurcation with change of solution type and loss of local stability often is accompanied by the appearance of a so-called attracting cycle of chaotic intervals, the skeleton of which is a denumerable set of unstable cycles with periods that are multiple to the period of the basic motion. Two bifurcation lines L_1 and L_2 are specially marked out in Fig. 6.8. The inset to this figure schematically presents a phase-parameter diagram illustrating bifurcations at the intersections with these lines. In the points of the line L_1 a pair of unstable 5-cycles of different types arises through a border-collision bifurcation. When the parameter χ is changed, one of these cycles, denoted 1 in the inset, merges with an other unstable cycle on the line L_2. If the transfer constant χ of the corrector is changed in the opposite direction, the unstable cycle denoted 2 disappears on the line of saddle-node bifurcation 5–6 as it merges with a stable 5-cycle.

The right boundary of the domain $\Pi_{5,1}$ consists of three border-collision bifurcation lines. Along the line 3–20 the period of a 5-cycle is doubled, and on the line 20–4 it is tripled. The line 4–5 corresponds to the border-collision bifurcation curve in the points of which a stable 5-cycle changes its type and becomes a saddle. A detailed analysis of such transitions was performed in Chapter 5. The results of our bifurcation analyses are presented in Figs. 6.13(a) and (b). They show that, when intersecting the line 20–4, the period of a 5-cycle is tripled while the 5-cycle itself changes its type and becomes a saddle. Here, besides a 15-cycle, a family of unstable cycles with periods that are multiple to 5 softly branch off from the period-5 cycle [see Fig. 6.13(b)]. We have succeeded in obtaining unstable cycles with periods up to 50.

Some of the results of our numerical calculations are summarized in Table 6.1. The first column of this table contains the periods of the cycles (the values of m), while the second column provides an index to distinguish the cycles with the same period but of a different type. The third column contains the symbolic characteristics of the individual cycle. The number of portions from which the

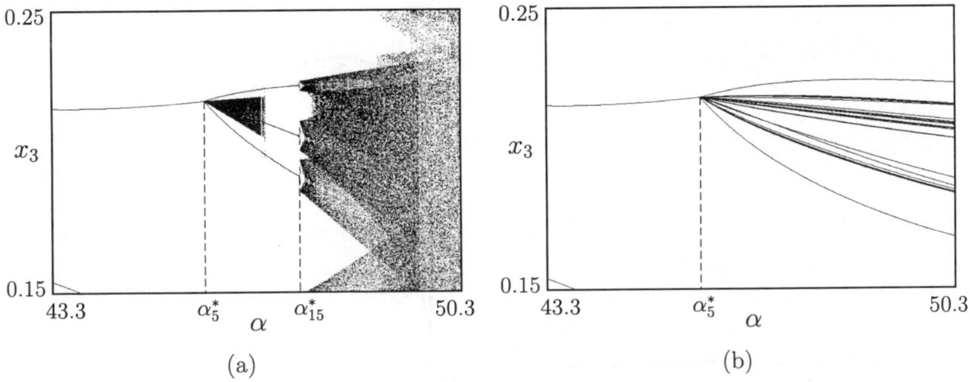

Figure 6.13. Border-collision period-tripling bifurcation of a 5-cycle which is accompanied by the appearance of a chaotic attractor (α_5^* is the bifurcation point). At the point α_{15}^*, the 15-cycle that arises at α_5^* undergoes a local period-doubling bifurcation followed by a transition to a 30-band chaotic attractor. (a) Bifurcation diagram obtained by successive iterations. (b) Results of a calculation of cycles that take part in the transition at α_5^*.

m	j	z_1	z_2	z_3	z_4	z_5	z_6	z_7	z_8	z_9	z_{10}	z_{11}–z_{15}	z_{16}–z_{20}	z_{21}–z_{25}	z_{26}–z_{30}	z_{31}–z_{35}	z_{36}–z_{40}	z_{41}–z_{45}	N
5	1	1	*	0	0	*													4
10	1	1	*	0	0	*	1	*	0	0	1								6
15	1	1	*	0	0	*	1	*	0	0	1	$\mathcal{L}$							8
15	2	1	*	0	0	*	1	*	0	0	*	$\mathcal{L}$							10
20	1	1	*	0	0	*	1	*	0	0	1	$\mathcal{L}$	$\mathcal{L}$						10
25	1	1	*	0	0	*	1	*	0	0	1	$\mathcal{L}$	$\mathcal{L}$	$\mathcal{L}$					12
30	1	1	*	0	0	*	1	*	0	0	1	$\mathcal{L}$	$\mathcal{L}$	$\mathcal{L}$	$\mathcal{L}$				14
35	1	1	*	0	0	*	1	*	0	0	1	$\mathcal{L}$	$\mathcal{L}$	$\mathcal{L}$	$\mathcal{L}$	$\mathcal{L}$			16
40	1	1	*	0	0	*	1	*	0	0	1	$\mathcal{L}$	$\mathcal{L}$	$\mathcal{L}$	$\mathcal{L}$	$\mathcal{L}$	$\mathcal{L}$		18
45	1	1	*	0	0	*	1	*	0	0	1	$\mathcal{L}$	$\mathcal{L}$	$\mathcal{L}$	$\mathcal{L}$	$\mathcal{L}$	$\mathcal{L}$	$\mathcal{L}$	20

Table 6.1. Symbolic characteristics of cycles with periods that are multiple to 5 at $\alpha = 47.3$ and $\chi = 0.62$. These cycles softly branch off a 5-cycle when it loses stability. Here $\mathcal{L}$ denotes the string "1*001" in our symbolic notation.

phase trajectory of the m-cycle is sewed up is indicated in the last column of the table.

With further increase of the parameter α the period of a 15-cycle is doubled at the line 4–19. This bifurcation is local and is connected with the crossing of a multiplier of the 15-cycle through -1. The produced 30-cycle subsequently undergoes two other border-collision bifurcations: The bifurcation of a simple change of the solution type at first, and then a transition at which a cycle of chaotic intervals softly arises from the 30-cycle [see Fig. 6.13(a)].

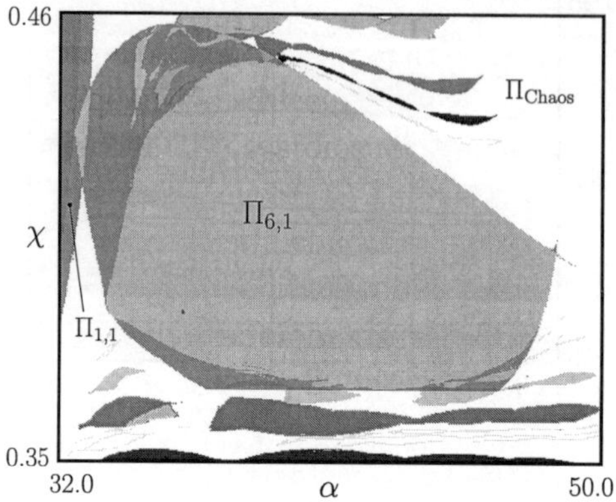

Figure 6.14. Domain of stability of a 6-cycle $\Pi_{6,1}$.

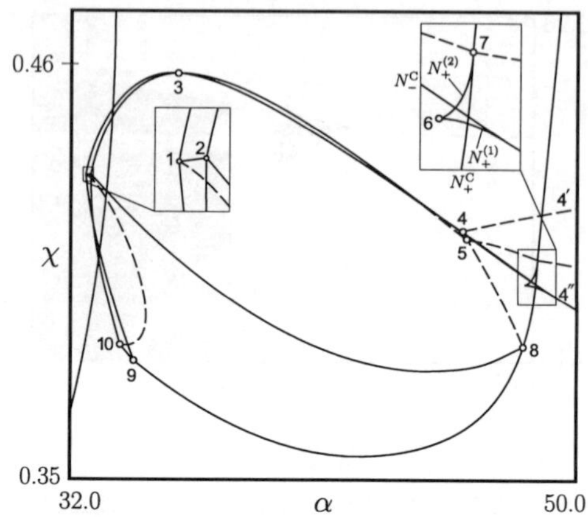

Figure 6.15. Bifurcation boundaries of the domain $\Pi_{6,1}$.

Period-doubling curves for the domains $\Pi_{10,2}$ and $\Pi_{15,1}$ correspond in the diagram to the lines 15–20 and 4–19 (in the points of these curves a multiplier of the corresponding cycle passes through -1). On the right edge the domain $\Pi_{15,1}$ is bounded by a border-collision bifurcation curve of the merging of a stable and an unstable cycle of different types (the line 16–20). The upper parts of the domains $\Pi_{10,2}$ and $\Pi_{15,1}$, which connect to the domains $\Pi_{20,1}$, $\Pi_{20,2}$, $\Pi_{30,3}$, $\Pi_{30,4}$, $\Pi_{30,1}$, and $\Pi_{30,2}$ have a similar structure.

Now let us consider another domain. Figure 6.14 shows a magnified part of the two-parameter diagram of dynamical modes. This part includes the domain of stability of a 6-cycle $\Pi_{6,1}$. Inspection of the figure shows that, similar to the domain $\Pi_{5,1}$, the domain $\Pi_{6,1}$ has non-empty intersections with $\Pi_{1,1}$ as well as with the domain of existence of a two-dimensional ergodic torus. There is a denumerable number of narrow zones of resonance oscillations with higher periods near the upper and lower boundaries of $\Pi_{6,1}$. These domains intersect one another and they also intersect the domain $\Pi_{6,1}$.

The results of a two-parameter bifurcation analysis of $\Pi_{6,1}$ are presented in Fig. 6.15. The external boundary of $\Pi_{6,1}$ corresponds to a border-collision bifurcation curve of the merging of a stable and an unstable 6-cycle of different types with the exception of the part 4–4″ where the solution changes type and loses its local stability. The right part of $\Pi_{6,1}$ has a singularity of fold type. One of the insets of Fig. 6.15 contains a magnified view. This structure contains

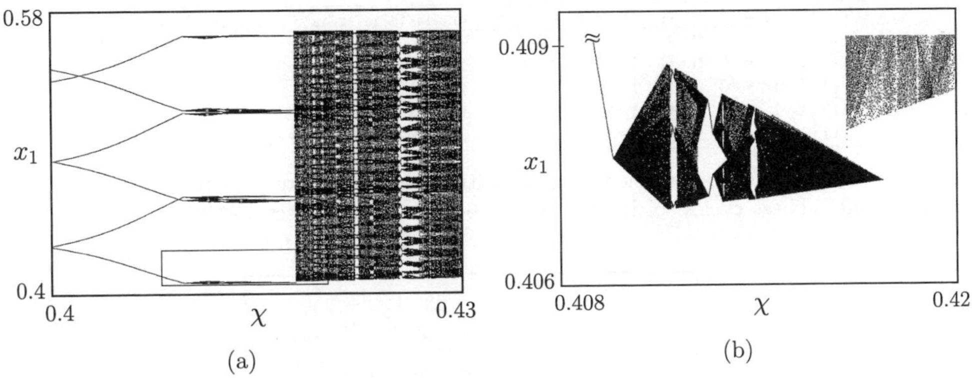

Figure 6.16. (a) Bifurcations of a 6-cycle when intersecting the boundary N_{-}^{C}. (b) Magnification of the lower branch of Fig. 6.16(a).

two saddle-node bifurcation lines (the lines $N_{+}^{(1)}$ and $N_{+}^{(2)}$) supported by the point of codimension two (the cusp point). One of the saddle-node bifurcation lines ($N_{+}^{(1)}$) approaches the border-collision bifurcation line of a change of the solution type N_{-}^{C} in the points of which the loss of local stability of the 6-cycle is accompanied with the soft appearance of a denumerable set of saddle cycles with periods that are multiple to 6, while another line ($N_{+}^{(2)}$) approaches the border-collision bifurcation line N_{+}^{C} where a stable 6-cycle merges with an unstable 6-cycle of another type and disappears.

Both a stable and an unstable 6-cycle can undergo border-collision bifurcations of a simple change of the solution type within the domain $\Pi_{6,1}$. In the diagram the dashed lines mark the curves of border-collision bifurcations of unstable cycles, while the solid curves mark the same transitions for stable cycles.

Figure 6.16(a) is a bifurcation diagram illustrating the system behavior when intersecting the boundary N_{-}^{C}. The magnified part of this diagram, which is outlined by a rectangle around part of the lower branch in Fig. 6.16(a), is shown in Fig. 6.16(b). At the points of the line N_{-}^{C} a 6-cycle loses stability and softly transforms into a saddle 6-cycle of another type. Here a denumerable set of unstable cycles with periods that are multiple to 6 softly branch off the 6-cycle. Part of the data on the unstable cycles that we have succeeded in obtaining numerically, is given in Table 6.2. With further increase of the parameter χ an unstable 6-cycle undergoes a border-collision bifurcation of a simple change of the solution type and then disappears at the line 4–4′ merging with an unstable

m	j	z_1	z_2	z_3	z_4	z_5	z_6	z_7	z_8	z_9	z_{10}	z_{11}	z_{12}	z_{13}	z_{14}	z_{15}	z_{16}	z_{17}	z_{18}	z_{19}	z_{20}	z_{21}	z_{22}	z_{23}	z_{24}	N
6	1	1	1	*	0	0	1																			2
12	1	1	1	*	0	0	*	1	1	*	0	0	1													6
18	1	1	1	*	0	0	1	1	1	*	0	0	*	1	1	*	0	0	1							8
18	2	1	1	*	0	0	*	1	1	*	0	0	*	1	1	*	0	0	1							10
24	1	1	1	*	0	0	1	1	1	*	0	0	1	1	1	*	0	0	*	1	1	*	0	0	1	10

Table 6.2. Symbolic characteristics of cycles with periods that are multiple to 6 at $\alpha = 47.5$ and $\chi = 0.409$. These cycles softly branch off a 6-cycle when it loses stability.

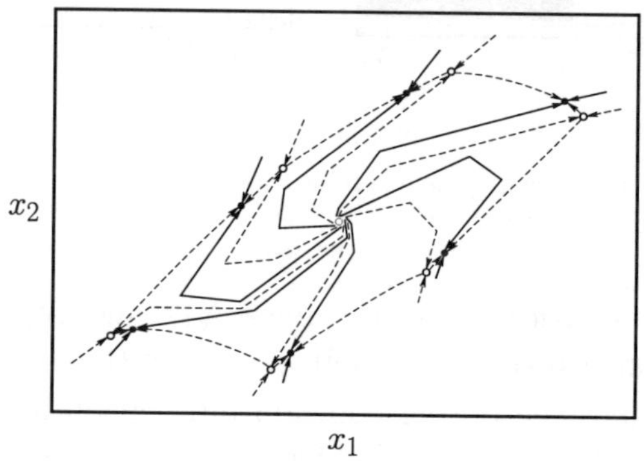

Figure 6.17. Sketch of the resonance torus corresponding to the 6-cycle.

6-cycle of different type that arises at the line N_+^C together with a stable cycle.

A different scenario for the change of oscillation modes is observed when intersecting the line N_+^C. It is worth noticing that the specific character of the behavior of the dynamical system will depend upon the direction of change of the parameters within $\Pi_{6,1}$. If we intersect the line N_+^C in the left part of $\Pi_{6,1}$, a 6-cycle transforms into a 1-cycle or into an ergodic beating mode via a hard bifurcation similar to the domain $\Pi_{5,1}$ considered above (see Figs. 6.9–6.12). In those sections of the line N_+^C that fall in the upper and lower parts of $\Pi_{6,1}$ one can observe hard transitions to resonance oscillations with higher periods or to a chaotic mode based upon the corresponding resonance cycle. The right part of $\Pi_{6,1}$ is bounded by a section of the line N_+^C, where a 6-cycle transforms into a mode of quasiperiodic oscillations without hysteresis (softly). Our numerical experiments show that the transition from a mode of ergodic beating to a 6-cycle in this part of the line N_+^C is connected with a resonance on the two-dimensional torus. Close to this boundary a 6-cycle lies upon the surface of a two-dimensional piecewise-smooth torus formed by the closing of the unstable manifold of the saddle cycle to the stable cycle. The projection of the phase portrait of a resonance torus in the Poincaré section is schematically depicted in Fig. 6.17. With decreasing α the resonance torus loses its smoothness in the points of the 6-cycle when two multipliers of this cycle become complex. However, the torus continues to exist.

6.3 Resonance Tongues and Border-Collision Bifurcations

The domains of resonance on a two-dimensional torus have a peculiar shape for piecewise-smooth systems (see, e.g., [31] and our discussion of the *rotating sawtooth map* in Chapter 1). They are similar in several ways to the Arnol'd tongues described in numerous works for time discrete as well as for flow systems [22–26]. However, resonance tongues for piecewise-smooth systems display essential differences both with respect to their inner structure and when it comes to the detailed bifurcational transitions.

Our numerical analyses show that tongues of synchronization are located everywhere dense in the parameter plane. Nevertheless, their total width close to the curve N_φ is likely to vanish. To the right of N_φ the width of the tongues shows a characteristic non-monotonous variation like a necklace with stones of increasing size or like winding bands (as illustrated by the front-page figure). Moreover, the resonance tongues are bounded not by the curves of saddle-node bifurcation, as in the case of smooth dynamical systems (see, e.g., [8, 20]), but by border-collision bifurcation curves where stable and unstable cycles of different types merge and disappear. Between the synchronization tongues there are domains of quasiperiodicity and chaos. A transition to chaos through a finite sequence of border-collision bifurcations of resonance cycles is observed within each tongue.

Resonance tongues can be ordered in accordance with *the level of complexity* [32]. Tongues of the first level of complexity have rotation numbers of the form $r/q = 1/m$, $m = 5, 6, 7, 8, 9, 10, 11, 12, 13$ ($\Pi_{5,1}$–$\Pi_{13,1}$ in Fig. 6.3). Between every two tongues of the first level of complexity with the rotation numbers $1/m$ and $1/(m+1)$, $m = 5, \ldots, 12$ there are two infinite sequences of tongues of the second level of complexity with the rotation numbers of the form $(s+1)/(sm+m+1)$ and of the form $(s+1)/(s(m+1)+m)$, $s = 1, 2, \ldots$, respectively. The first sequence accumulates to the tongue with the rotation number $1/m$ and the second accumulates to the tongue with the rotation number $1/(m+1)$ [32, 33].

In the general case, between two tongues of the k-th level of complexity, $k \geq 1$ with the rotation numbers r_k^m/q_k^m and r_k^{m+1}/q_k^{m+1} there are two infinite sequences of tongues of the $(k+1)$-th level of complexity with the rotation numbers

$$\frac{r_{k+1}^s}{q_{k+1}^s} = \frac{sr_k^m + r_k^{m+1}}{sq_k^m + q_k^{m+1}} \text{ and } \frac{r_{k+1}^s}{q_{k+1}^s} = \frac{r_k^m + sr_k^{m+1}}{q_k^m + sq_k^{m+1}}.$$

Here, the index $s = 1, 2, \ldots$ is the number of the tongue in the sequence. The first sequence accumulates to the tongue with the rotation number r_k^m/q_k^m and the second accumulates to the tongue with the rotation number r_k^{m+1}/q_k^{m+1}.

For convenience we have divided the domains of synchronization into two groups. The division criterion is the difference in their inner structure and the bifurcations of resonance cycles. We do not include in these groups the domains $\Pi_{5,1}$ and $\Pi_{6,1}$ whose inner structure and bifurcation boundaries were considered in detail in the previous section. The first group comprises such tongues $\Pi_{m,k}$ in which there is a finite number of simply connected domains $\hat{\Pi}_{m,k}$, $\hat{\Pi}_{m,k} \subset \Pi_{m,k}$ where two stable resonance cycles of different type exist. When crossing the boundaries of $\hat{\Pi}_{m,k}$ one can observe hard hysteretic transitions from one stable resonance cycle of period m to another. The second group includes those $\Pi_{m,k}$ where, when changing the parameters within $\Pi_{m,k}$, the transitions from one type of a stable m-cycle to another take place in a smooth way.

Figure 6.18 is a magnified portion of the two-parameter diagram of dynamic modes for the domain of synchronization $\Pi_{11,1}$ that belongs to the first group. The domains of bistability, where two stable period-11 resonance cycles of different types exist, are shaded black. Figure 6.18 shows also the resonance tongues, with a level of complexity up to three, that are situated in the parameter plane near the tongues of the first level of complexity with the rotation number 1/11. These domains are indicated by the periods of the corresponding cycles.

The results of a two-parameter bifurcation analysis for $\Pi_{11,1}$ are presented in Fig. 6.19. The diagram contains four border-collision bifurcation lines L_1, L_2, L_3, and L_4 each being the aggregate of BC-boundaries of two different types

$$L_j = \bigcup_i L_{j,i}^{\pm}, \; j = \overline{1,4}; \; i = 1, 2, \ldots$$

Here the lines $L_{j,i}^{+}$ correspond to the BC-boundaries where a stable and an unstable cycle of different type merge and disappear, and $L_{j,i}^{-}$ correspond to

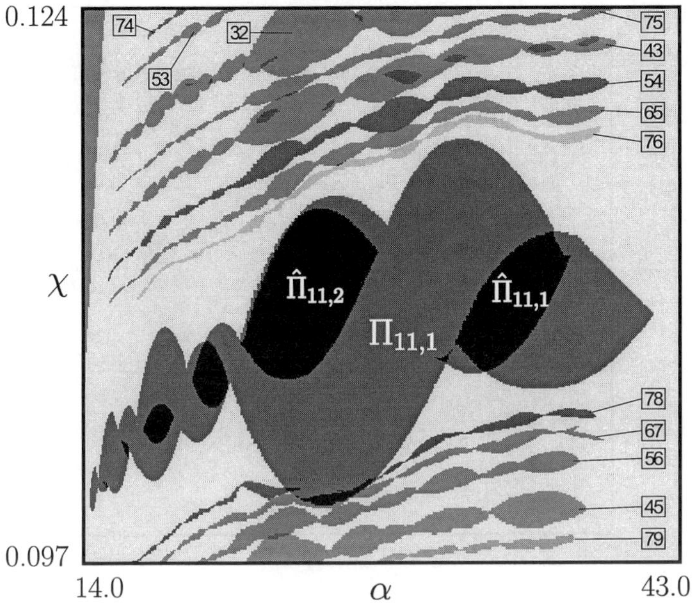

Figure 6.18. Domain of stability $\Pi_{11,1}$ for an 11-cycle. Domains $\Pi_{m,k}$ that correspond to other cycles are indicated by the appropriate periods m.

the BC-boundaries of a simple change of the solution type. The line of the destruction of the two-dimensional resonance torus is not shown in Fig. 6.19. Our investigations indicate that this boundary has a complex structure.

The domain $\Pi_{11,1}$ is bounded by the aggregate of BC-bifurcation lines $L^+_{j,i}$, except for the part NS to the right in the figure which corresponds to a BC-bifurcation line of change of the solution type with the loss of local stability. In Fig. 6.19 $\hat{\Pi}_{11,i}$, $i = 1, 2, \ldots$, $\hat{\Pi}_{11,i} \subset \Pi_{11,1}$ designate those parts of $\Pi_{11,1}$ where, for all the points $P \in \hat{\Pi}_{11,i}$, there are two stable 11-cycles of different types. The domains $\hat{\Pi}_{11,i}$, $i = 1, 2, \ldots$ are bounded by BC-bifurcation lines $L^+_{j,i}$, in the points of which a stable 11-cycle of one type joins an unstable cycle of another type and disappears. Within each $\hat{\Pi}_{11,i}$, $i = 1, 2, \ldots$ the periodic motions vary smoothly with the parameters. When intersecting the BC-boundaries $L^+_{j,i}$, hard transitions occur from a stable 11-cycle of one type to an 11-cycle of another type, with the hysteresis typical for such transitions. Bifurcation lines $L^+_{j,i}$ within $\Pi_{11,1}$ are supported by the points $P_{11,i}$, $i = 1, 2, \ldots$ of codimension two. Figure 6.20 presents a phase-parameter diagram illustrating the peculiarities of

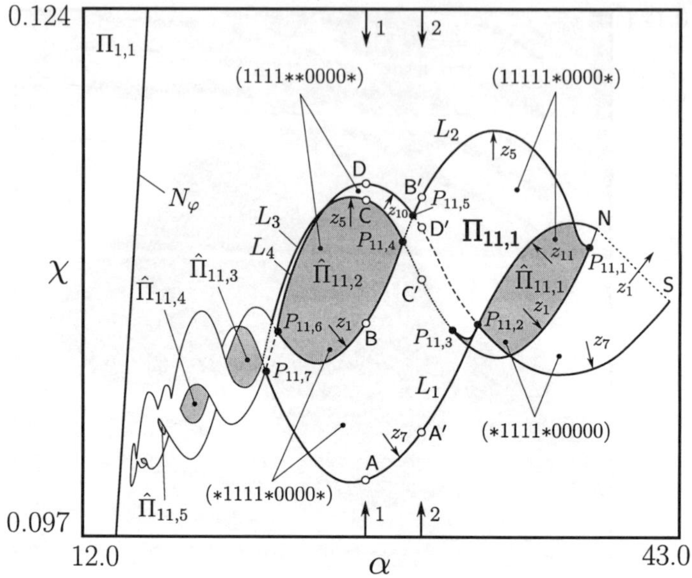

Figure 6.19. Bifurcation boundaries of the domain $\Pi_{11,1}$.

the boundary structure near the points $P_{11,4}$ and $P_{11,5}$. The diagram (Fig. 6.19) also provides the symbolic characteristics of the stable cycles. Moreover, next to each BC-bifurcation curve we have indicated that point z_k^c, $k = 1, 2, \ldots, m$ of the symbolic characteristics that is involved in a change of the solution type by becoming zero or one.

In Fig. 6.19 the dotted and dashed lines mark the BC-boundaries where a simple change in the number of the sewed phase trajectory portions takes place (a simple change of the solution type [34, 35]). These lines continue the BC-boundaries $L_{j,i}^+$ represented in diagram by solid lines. This means that when changing the parameters along L_j, $j = 1, 2, 3, 4$ then, upon passing through the points $P_{11,i}$, $i = 1, 2, \ldots$ of codimension two, the lines $L_{j,i}^+$ will continuously convert to BC-bifurcation lines $L_{j,i}^-$ of a simple change of the solution type. The dotted line shows the BC-boundaries of a change of the type of the stable cycle, and the dashed line is the same for the unstable one.

When moving from the domain $\hat{\Pi}_{11,i}$ into the domain $\hat{\Pi}_{11,i+1}$, the number of parts from which the trajectory of the stable 11-cycle is sewed successively increases exactly by two. For example, the trajectories of the stable cycles existing in $\hat{\Pi}_{11,1}$ is sewed from four parts, in $\hat{\Pi}_{11,2}$ from six parts, etc., and in $\hat{\Pi}_{11,9}$ their number is twenty. The number of parts from which the phase trajectory of the 11-cycle is sewed is limited (it cannot exceed 22), therefore the number of domains $\hat{\Pi}_{11,i}$ is finite.

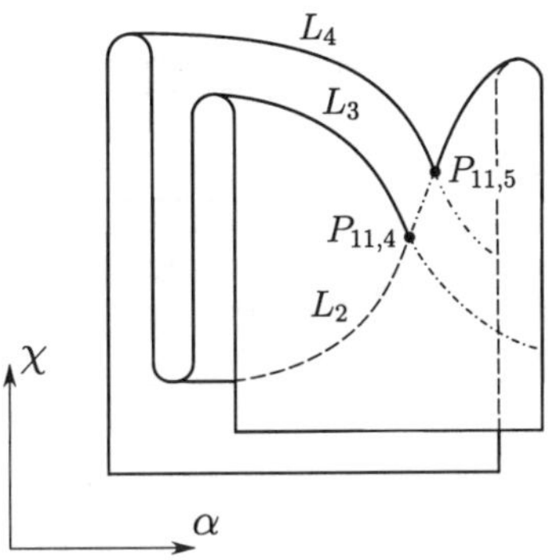

Figure 6.20. Boundary structures near the points $P_{11,4}$ and $P_{11,5}$.

Now let us analyze the bifurcations that occur when intersecting the BC-boundaries. At first we shall study the section $\Omega_3 = \{(\alpha,\chi) : \alpha = 26.7; 0.099 < \chi < 0.116\}$ indicated in Fig. 6.19 by the number 1. This section transverses the lines L_j, $j = 1,2,3,4$ in the points denoted A, B, C and D in Fig. 6.19. The parameter values in these points will be designated by χ_{A}, χ_{B}, χ_{C}, and χ_{D}.

The results of our bifurcation analysis are given in Fig. 6.21. The solid lines in this figure represent the evolution of stable cycles, and dashed lines mark the behavior of unstable cycles. The multiplier dependences of stable and unstable 11-cycles on the parameter χ are shown in Fig. 6.21(b). Note the essential peculiarity of the considered bifurcation. It is easy to see from Fig. 6.21(b), that the multipliers ρ_i, $i = 1,2,3$ of the unstable 11-cycles (i.e., the multipliers of the cycles for which the number of the phase trajectory parts being sewed is changing) depend smoothly on χ, except at the bifurcational values. Near these values, the multipliers display a simple discontinuity.

This feature is easy to prove and typical of border-collision bifurcations [30, 36–39]. Another feature is that border-collision bifurcations are not related with a crossing of one of the multipliers through the unit circle. For example, as it is seen from Fig. 6.21(b), the largest multiplier (in absolute value) of the 11-cycle does not pass through $+1$ at the bifurcational point in contrast to a usual saddle-node bifurcation.

At the point $P_{\mathsf{A}} = (26.7; \chi_{\mathsf{A}})$, $\chi_{\mathsf{A}} \approx 0.09988107$, which is lying on the line L_1, a couple of 11-cycles arise, one being stable and the other unstable. Let us designate them by $X^{\mathrm{S}}_{11,1}$ and $X^{\mathrm{U}}_{11,1}$, respectively. The trajectory of the stable cycle $X^{\mathrm{S}}_{11,1}$ is sewed up from six portions: (*1111*0000*) and that of the unstable one $X^{\mathrm{U}}_{11,1}$ from eight: (*1111**000*). When the parameter χ increases, in the point $P_{\mathsf{C}} = (26.7; \chi_{\mathsf{C}})$, $\chi_{\mathsf{C}} \approx 0.11415795$ located on the line L_3, the stable 11-cycle $X^{\mathrm{S}}_{11,1}$ merges with the unstable 11-cycle $X^{\mathrm{U}}_{11,2}$ and disappears. The un-

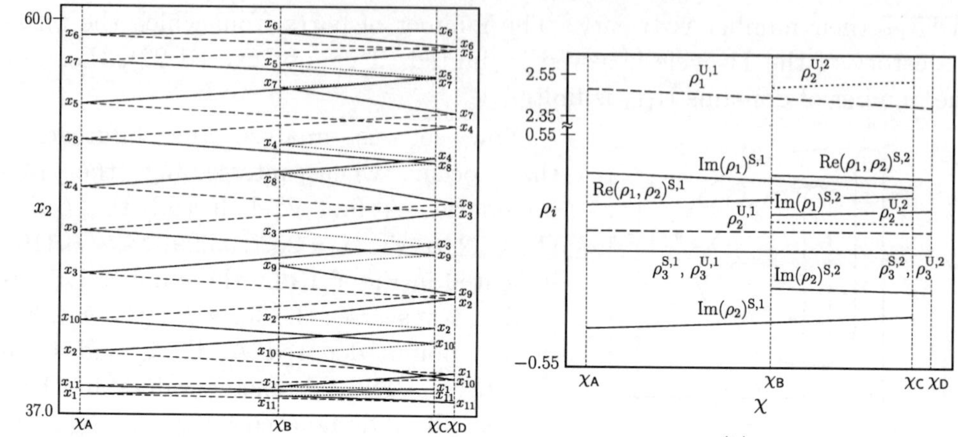

(a) (b)

Figure 6.21. (a) Bifurcation and (b) multiplier diagrams for the section designated in Fig. 6.19 by number 1.

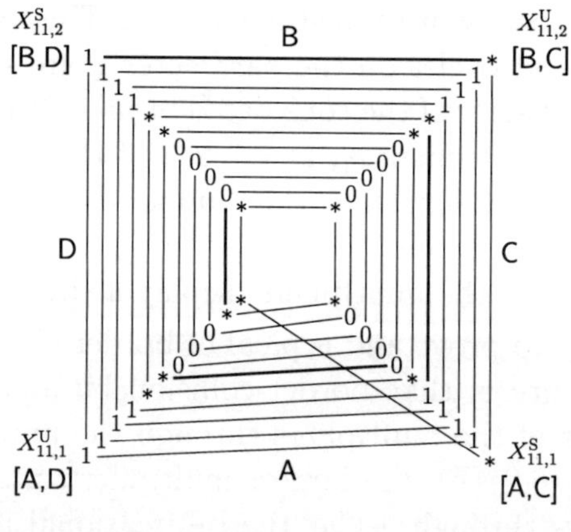

Figure 6.22. Sequence of changes of the solution type for the section designated in Fig. 6.19 by the number 1.

stable 11-cycle $X_{11,2}^{U}$ arises on the line L_2 together with a stable 11-cycle $X_{11,2}^{S}$ of a type different from $X_{11,1}^{S}$. Figure 6.21(a) clearly shows that when changing the parameter χ, the unstable 11-cycle $X_{11,1}^{U}$ disappears as it merges with the stable cycle $X_{11,2}^{S}$ on intersecting the line L_4 in the point $P_D = (26.7; \chi_D)$, $\chi_D \approx 0.11498787$.

This sequence of bifurcations can be expressed symbolically by the diagram

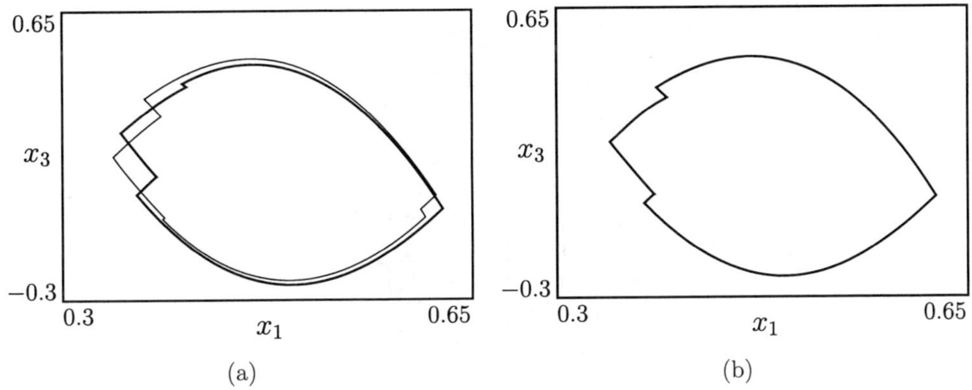

Figure 6.23. Border-collision bifurcation in which a stable focus (thick line) and a saddle cycle (thin line) annihilate with one another. The figures show projected trajectories of the cycles before the bifurcation ((a), $\alpha = 26.7$ and $\chi = 0.109$) and in the bifurcation point $P_{\mathsf{A}} = (26.7; \chi_{\mathsf{A}})$, $\chi_{\mathsf{A}} \approx 0.09988107$ (b).

depicted in Fig. 6.22. The diagonal lines give the symbolic characteristics of the stable $X^{\mathsf{S}}_{11,1}$, $X^{\mathsf{S}}_{11,2}$ and unstable $X^{\mathsf{U}}_{11,1}$, $X^{\mathsf{U}}_{11,2}$ cycles. The brackets near the corners indicate their ranges of existence on the parameter χ. Vertical and horizontal lines correspond to passages through the boundary points where the merging of a stable and an unstable cycles of different types takes place with their subsequent disappearance. Thick solid lines represent the variation of those points z^{c}_k, $k = 1, 2, \ldots, m$, that are responsible for changing the number of sewed portions of the phase trajectories in the border-collision bifurcations.

As we can see from the bifurcation diagram in Fig. 6.21(a), the stable cycles $X^{\mathsf{S}}_{11,1}$ and $X^{\mathsf{S}}_{11,2}$ are sewed from six portions of phase trajectories. The difference between $X^{\mathsf{S}}_{11,1}$ and $X^{\mathsf{S}}_{11,2}$ lies not in the number of the phase trajectories pieces, but in the sewing order. It follows from the fact that the symbolic characteristics of the cycles $X^{\mathsf{S}}_{11,1}$ and $X^{\mathsf{S}}_{11,2}$ can not transform one into the other by a cyclic permutation of their components. A similar picture is observed for the unstable cycles $X^{\mathsf{U}}_{11,1}$, $X^{\mathsf{U}}_{11,2}$ sewed up from eight pieces of phase trajectories.

Figure 6.23 provides an illustration of a border-collision bifurcation in which a stable focus sewed by 6 parts and a saddle cycle sewed by 8 parts meet and are annihilated on the line $P_{11,2}$–$P_{11,7}$ (see Fig. 6.19). Figure 6.24 is a sketch illustrating a border-collision bifurcation when one of the parts of the periodic orbit touches the boundary of the sewing surfaces S_j and S_k. This produces a change in the number of parts from which the trajectory is sewed.

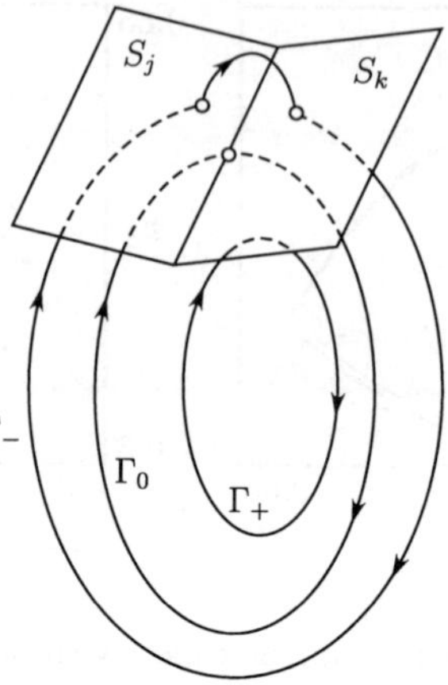

Figure 6.24. Sketch of a border-collision bifurcation leading to a simple change in solution type. Γ_- is a periodic orbit before the bifurcation, Γ_0 in the bifurcation point, and Γ_+ after the bifurcation.

Let us finally examine transitions that occur through points of the BC-boundaries corresponding to a simple change of the solution type. The results of numerical calculations made for the section $\Omega_4 = \{(\alpha, \chi) : \alpha = 29.7; 0.101 < \chi < 0.116\}$, designated in Fig. 6.19 by the number 2, are given in Fig. 6.25. As usual, this figure shows the dependence of the stable and unstable cycles on the parameter χ. The stable 11-cycle annihilates with the unstable 11-cycle in the points A' and B' that belongs to the lines $P_{11,2}$–$P_{11,7}$ and $P_{11,1}$–$P_{11,5}$ in Fig. 6.19, respectively. In the point C' (belonging to the line $P_{11,3}$–$P_{11,4}$ in Fig. 6.19) the stable 11-cycle changes the type. The same transition occurs for the unstable 11-cycle in the point D' (belonging to the line $P_{11,2}$–$P_{11,5}$ in Fig. 6.19). This transition is the border-collision bifurcation of a simple change of the solution type. The parameter values in the points A', B', C', and D' are designated by $\chi_{A'}$, $\chi_{B'}$, $\chi_{C'}$, and $\chi_{D'}$, respectively.

In the right-hand part of $\Pi_{11,1}$ there are two more domains where two stable 11-cycles of different type exist. These domains are bounded by the BC-bifurcation lines of the merging of stable and unstable cycles. The lines of BC-boundaries are supported by the points $P_{11,2}$ and $P_{11,3}$ of codimension two. A phase-parameter diagram illustrating the peculiarities in the organization of the BC-boundaries close to these points is presented in Fig. 6.26.

Now let us analyze the domains of BC-synchronization that belong to the

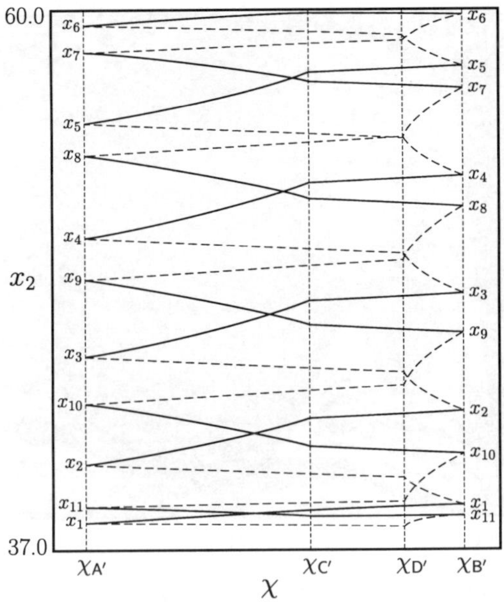

Figure 6.25. Bifurcation diagram for the section designated in Fig. 6.19 by the number 2.

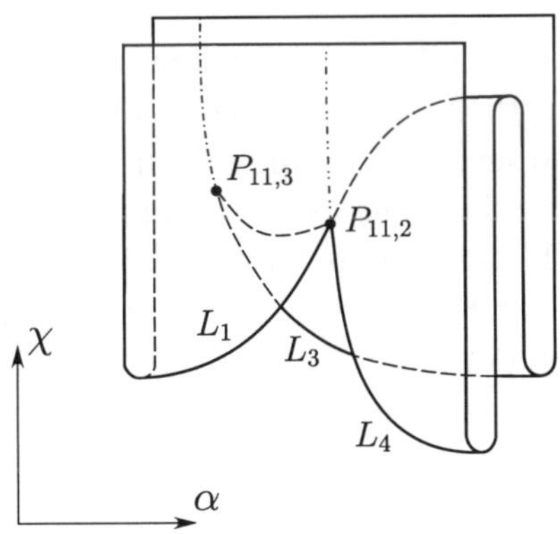

Figure 6.26. Structure of BC-boundaries near the points $P_{11,2}$ and $P_{11,3}$ of codimension two.

second group. Their inner structure, as will be demonstrated later, is relatively simple in comparison with the resonance domains of the first group that have been described above. Let us consider the domain $\Pi_{10,1}$ corresponding to the period-10 resonance cycle as an example. Figure 6.27 is a magnified portion of the two-parameter diagram of dynamic modes

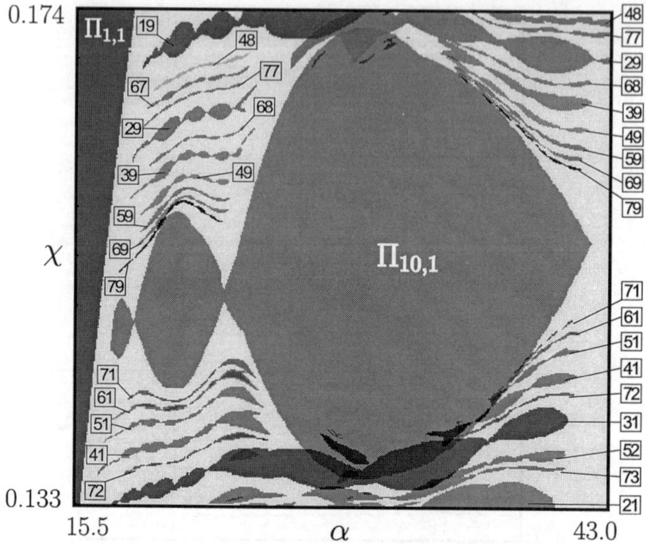

Figure 6.27. Domain of stability for a 10-cycle $\Pi_{10,1}$. Domains $\Pi_{m,k}$ that correspond to other cycles are indicated by the appropriate periods m.

that has been calculated within the range of control parameters $\Pi = \{(\alpha, \chi) : 15.5 < \alpha < 43.0; 0.133 < \chi < 0.174\}$. The diagram also shows the resonance tongues with the level of complexity up to three that are situated in the parameter plane near the tongues of the first level of complexity with the rotation number $1/10$. These domains are indicated by the periods of the corresponding cycles similar to Fig. 6.18. Some of the tongues intersect, and in the areas of intersection, depending upon the initial conditions, different resonance cycles can be attained. Crises, resulting in the merging of chaotic sets that appear on the basis of different resonance cycles, are possible in this part of the diagram.

Figure 6.28 exhibits the results of numerical calculations of bifurcation lines for $\Pi_{10,1}$. Similar to the previous case, the domain $\Pi_{10,1}$ is bounded by parts of four border-collision bifurcation lines L_j, $j = \overline{1,4}$ in the points of which a stable 10-cycle of one type merges with an unstable cycle of another type with the same period and disappears. Each of the lines L_j, $j = \overline{1,4}$ consists of continuously joining segments of BC-boundaries of two different types, just as it has been described in the analysis of the domain $\Pi_{11,1}$.

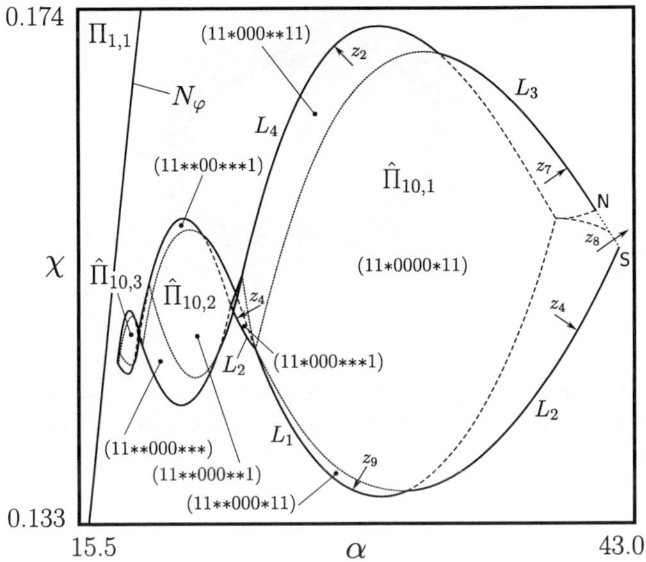

Figure 6.28. Bifurcation boundaries of the domain $\Pi_{10,1}$.

The parts of the lines L_j, $j = \overline{1,4}$ that correspond to border-collision bifurcations of a simple change of type for the stable and unstable 10-cycles divide $\Pi_{10,1}$ into four parts, each being the home for a stable cycle of a particular type. The symbolic characteristics of these cycles are given in the diagram. In every $\hat{\Pi}_{10,i}$, the number of which in $\Pi_{10,1}$ is finite, there are four types of stable cycles. The transition from a stable resonance cycle of one type to a stable cycle of another type when moving along the parameters is continuous in these points of the border-collision bifurcation boundaries that are shown in the diagram by dashed lines.

Let us now consider the behavior of the dynamical system (6.2) when leaving the domain of resonance. When intersecting the boundary N_+^{C}, a stable resonance cycle disappears as it merges with an unstable cycle of a different type. Here there is a possibility for a hard transition to another resonance cycle or to chaotic oscillations developed from another resonance cycle. Such transitions take place within the area of the intersections of the resonances. When leaving the resonance domain in those points of the bifurcation boundaries that are situated between the line of appearance and the line of loss of the smoothness for the two-dimensional torus, a soft transition to a mode of two-frequency

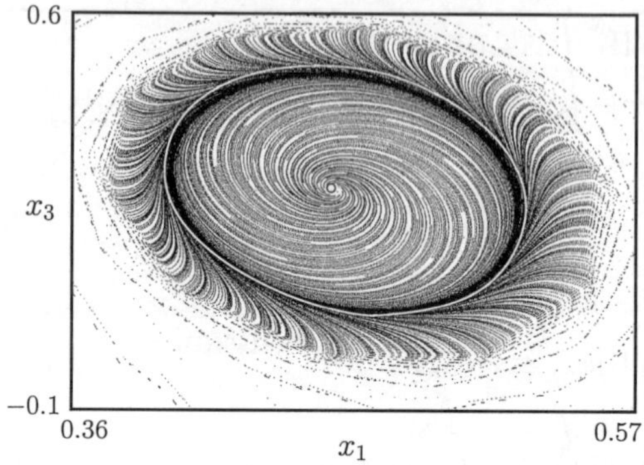

Figure 6.29. Two-dimensional projection of the Poincaré section for two-frequency quasiperiodic oscillations. The figure has been constructed from a great number of transients with different initial conditions.

quasiperiodic oscillations takes place. The two-dimensional projection of the Poincaré section for such a quasiperiodic mode is given in Fig. 6.29. Here the small open circle in the center indicates the location of the unstable fixed point. The phase portrait has been obtained by solving the Cauchy problem with different initial conditions. As it is seen from Fig. 6.29, the Poincaré section for the ergodic beating is a smooth closed curve.

In those part of the boundaries of $\Pi_{10,1}$ and $\Pi_{11,1}$ that correspond to the line N_-^C (the parts NS in Figs. 6.19 and 6.30) a resonance cycle changes its type and becomes a saddle. Here a cycle of chaotic intervals softly arises from a stable cycle.

Within the synchronization tongues resonance cycles can undergo a finite sequence of different border-collision bifurcations. Some of them have already been considered above. At the same time, the two-dimensional torus loses its smoothness and is destroyed. As previously noted, Afraimovich and Shilnikov [20] have described three possible mechanisms for the loss of smoothness and the destruction of a two-dimensional torus. As a conclusion to the present chapter let us consider one of the possible mechanisms for the loss of smoothness in more detail. This is the mechanism that is connected with the appearance of complex multipliers of the stable cycle.

Figure 6.30(a) shows the results of numerical calculations of the two-dimensional projection of the phase Poincaré section for a stable and an unstable period-10 cycle. These cycles lie upon the surface of the torus formed by clos-

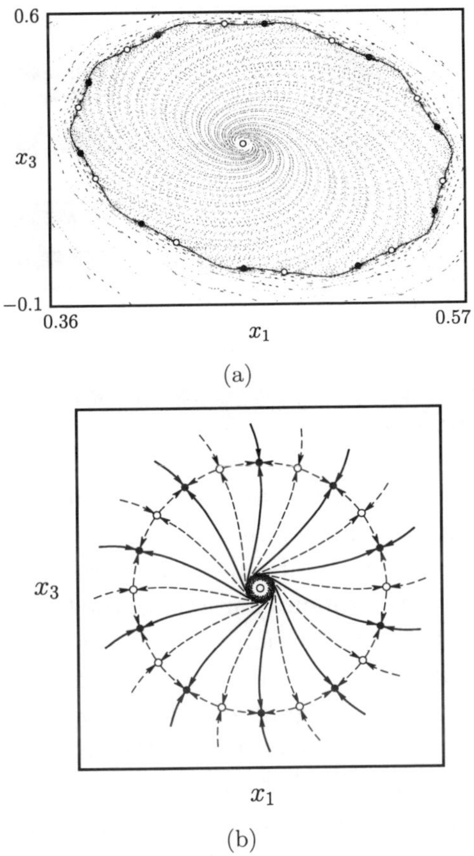

Figure 6.30. (a) Two-dimensional projection of the Poincaré section and (b) sketch of phase trajectories for a resonance 10-cycle with real multipliers.

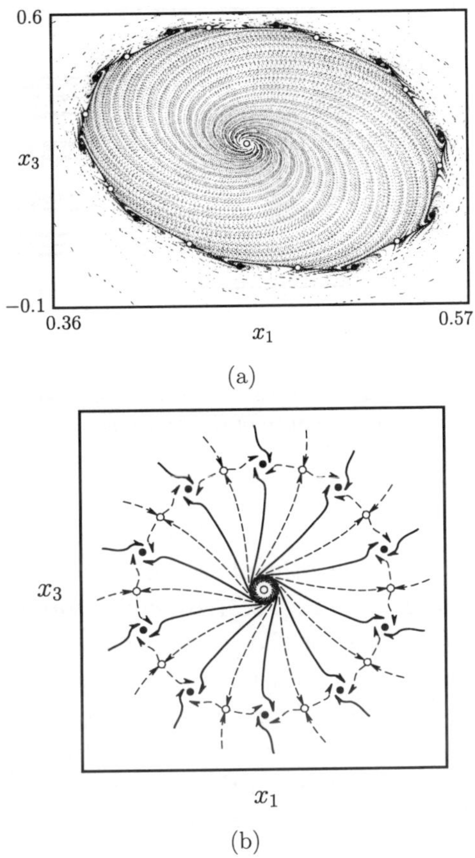

Figure 6.31. (a) Two-dimensional projection of the Poincaré section and (b) sketch of phase trajectories for a resonance 10-cycle with complex multipliers.

ing the unstable manifold of the saddle cycle to the points of the stable cycle [Fig. 6.30(b)]. Here the black circles mark the points of the stable 10-cycle and the open circles indicate the points of the unstable 1- and 10-cycles. With a change of the parameters the two multipliers of the stable 10-cycle may become complex. This will result in the loss of the smoothness of the resonance torus in the points of the stable cycle (see Fig. 6.31). In a smooth system such transitions occur continuously: Two real multipliers become equal to one another and a pair of complex-conjugated multipliers arise. In piecewise smooth systems,

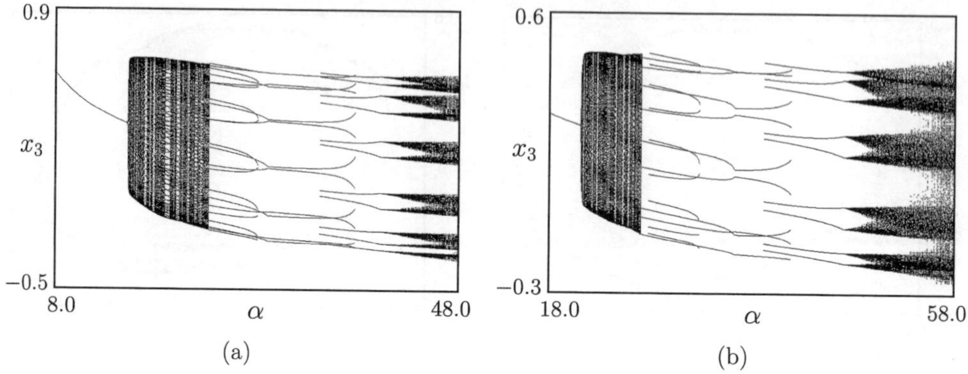

Figure 6.32. Transition to chaos via a sequence of border-collision bifurcations within the domains of resonance of the first kind: (a) $\Pi_{11,1}$; (b) $\Pi_{9,1}$.

however, complex multipliers can arise continuously or they can arise abruptly via border-collision bifurcation of the simple change of the solution type. In the latter case the phase portrait given in Fig. 6.30(b) abruptly transforms into that given in Fig. 6.31(b), which indicates a catastrophic character of the loss of smoothness for the resonance torus. With further variation of the parameters, the complex multipliers of the stable period-10 focus may pass out through the unit circle in a secondary Hopf bifurcation. This results in the destruction of the original torus and the formation of a new 10-band torus. We shall see several examples of this type of torus destruction in Chapter 7.

Figure 6.32(a) presents a bifurcation diagram illustrating the typical transition to chaos via a sequence of border-collision bifurcations within the domains of resonance of the first kind. The diagram shows that when parameters change, hysteretic transitions from one stable cycle to another via border-collision bifurcation are observed together with soft transitions. When intersecting the line NS in Figs. 6.19 and 6.28 a cycle of chaotic intervals softly arises from the stable cycle via a border-collision bifurcation. Another bifurcation diagram is given in Fig. 6.32(b) in order to illustrate the typical nature of such transitions. This diagram is related to the period-9 resonance.

On the whole, bifurcations of resonance cycles and transitions to chaos related with $\Pi_{10,1}$ are similar to those of $\Pi_{11,1}$. The only exception is that hysteretic transitions do not occur.

There are other kinds of resonance domains within Π that display certain differences in the structure of bifurcation boundaries as compared to the examples

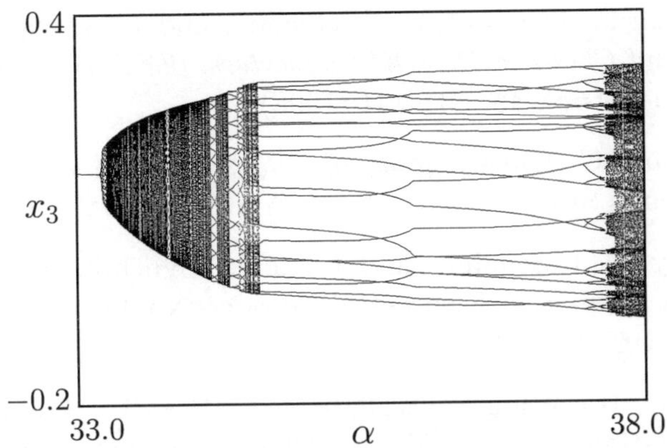

Figure 6.33. Transition to chaos via a sequence of border-collision bifurcations including period-doubling border-collision bifurcation.

we have considered here. These differences lie in the presence of border-collision bifurcation lines of resonance cycle period-doubling and -tripling within the synchronization tongues. The detailed study of bifurcational transitions in the points of such curves was reported in our recent paper [30]. The bifurcation diagram given in Fig. 6.33 is a graphic representation of such transitions. Here the resonance cycle undergoes a finite sequence of border-collision bifurcations of the simple change of the solution type (as within $\Pi_{10,1}$) and then a period-doubling border-collision bifurcation. With further change of the parameter a cycle of chaotic intervals arises from the cycle with doubled period.

Bibliography

[1] Ya. Z. Tsypkin, Relay Control Systems (University Press, Cambridge, 1984).

[2] V. S. Baushev, Zh. T. Zhusubaliyev, Yu. V. Kolokolov, and I. V. Terekhin, *Local Stability of Periodic Solutions in Sample Data Control Systems*, Automation and Remote Control **53**, part 2, 865–871 (1992), Plenum Publishing.

[3] V. S. Baushev, Zh. T. Zhusubaliyev, and S. G. Michal'chenko, *Stochastic Features in the Dynamic Characteristics of a Pulse-Width Controlled Voltage Stabilizer*, Electrical Technology (1), 137–150 (1996), Elsevier Science.

[4] M. di Bernardo, F. Garofalo, L. Glielmo, and F. Vasca, *Switchings, Bifurcation and Chaos in DC/DC Converters*, IEEE Trans. Circuits Syst.-I: Fund. Theory Appl. **45**, 133–141 (1998).

[5] E. Fossas and G. Olivar, *Study of Chaos in the Buck Converter*, IEEE Trans. Circuits Syst.-I: Fund. Theory Appl. **43**, 13–25 (1996).

[6] A. Pikovsky, M. Rosenblum, and J. Kurths, Synchronization: A Universal Concept in Nonlinear Sciences (Cambridge Nonlinear Science Series 12, Cambridge University Press, 2001).

[7] C. Hugenii, Horologium Oscillatorium (Parisiis, France, 1673). English translation: The Pendulum Clock (Iowa State University Press, Ames,1986).

[8] V. S. Anishchenko, Dynamical Chaos — Models and Experiments. Appearance Routes and Structure of Chaos in Simple Dynamical Systems (World Scientific, Singapore, 1995).

[9] E. Mosekilde, Topics in Nonlinear Dynamics. Applications to Physics, Biology and Economic Systems (World Scientific, Singapore, 1996).

[10] M. Togeby and E. Mosekilde, *Mode-Locking and Chaos in a Model of Two Coupled Thermostatically Controlled Radiators*, Open Systems and Information Dynamics **3**, 201–213 (1995).

[11] J. Sturis, E. Van Canter, J. D. Blackman, and K. S. Polonsky, *Entrainment of Pulsatile Insulin Secretion by Oscillatory Glucose Infusion*, J. Clin. Invest. **87**, 439–445 (1991).

[12] C. Schäfer, M. G. Rosenblum, J. Kurths, and H.-H. Abel, *Heartbeat Synchronized with Ventilation*, Nature (London) **392**, 239–240 (1998).

[13] V. I. Arnol'd, *Small Denominators, I: Mappings of the Circumference into Itself*, AMS Trans. Series 2 **46**, 213–284 (1965).

[14] V. I. Arnol'd, Catastrophe Theory, 2nd revised and expanded edition (Springer–Verlag, Berlin, 1986).

[15] D. E. Postnov, T. E. Vadivasova, O. V. Sosnovtseva, A. G. Balanov, V. S. Anishchenko, and E. Mosekilde, *Role of Multistability in the Transition to Chaotic Phase Synchronization*, Chaos **9**, 227–232 (1999).

[16] T. E. Vadivasova, O. V. Sosnovtseva, A. G. Balanov, and V. V. Astakhov, *Phase Multistability of Synchronized Chaotic Oscillators*, Discrete Dynamics in Nature and Society **4**, 231–243 (2000).

[17] M. A. Taylor and I. G. Kevrekidis, *Some Common Dynamic Features of Coupled Reacting Systems*, Physica D **51**, 274–292 (1991).

[18] C. Knudsen, J. Sturis, and J. S. Thomsen, *Generic Bifurcation Structure of Arnol'd Tongues in Forced Oscillators*, Phys. Rev. A **44**, 3503–3510 (1991).

[19] J. Sturis, C. Knudsen, N. M. O'Meara, J. S. Thomsen, E. Mosekilde, E. Van Canter, and K. S. Polonsky, *Phase-Locking Regions in a Forced Model of Slow Insulin and Glucose Oscillations*, Chaos **5**, 193–199 (1995).

[20] V. S. Afraimovich and L. P. Shilnikov, *On Invariant Two-Dimensional Tori, Their Disintegration and Stochasticity*, in Methods of the Qualitative Theory of Differential Equations (Gor'kov. Gos. Univ., Gorkij, 1983), in Russian.

[21] V. I. Arnol'd, V. S. Afraimovich, Yu. S. Il'yashenko, and L. P. Shilnikov, *Bifurcation Theory*, in Dynamical Systems, Encyclopaedia of Mathematical Sciences **5** (Springer–Verlag, Berlin, 1994).

[22] J. Maurer and A. Libchaber, *Rayleigh-Benard Experiment in Liquid Helium: Frequency Locking and the Onset of Turbulence*, J. Phys. Lett. **40**, L419–L423 (1979).

[23] D. G. Aranson, M. A. Chori, G. R. Hall, and R. P. McGenehe, *Bifurcations from an Invariant Circle for Two-Parameter Families of Maps of the Plane: A Computer-Assisted Study*, Comm. Math. Phys. **83**, 303–354 (1982).

[24] V. Franceschini, *Bifurcations of Tori and Phase Locking in a Dissipative System of Differential Equations*, Physica D **6**, 285–304 (1982).

[25] J. Stavans, *Experimental Study of Quasiperiodicity in a Hydrodynamical Systems*, Phys. Rev. A **35**, 4314–4328 (1987).

[26] V. S. Anishchenko and M. A. Safonova, *Mechanism of Destruction of Invariant Curve in a Model Map of the Plane*, Radiotekhnika i Elektronika **32**, 1207–1216 (1987), in Russian.

[27] A. V. Kobzev, G. Ya. Mikhal'chenko, and N. M. Muzhitchenko, Modulating Power Supplies of Radioelectronic Devices (Radio i Svyaz, Tomsk, 1990), in Russian.

[28] A. Kh. Gelig and A. N. Churilov, Stability and Oscillations of Nonlinear Pulse-Modulated Systems (Birkhäuser, Boston, 1998).

[29] Zh. T. Zhusubaliyev, *Investigation of Chaotic Regimes of a Voltage Converter with Pulse-Width Modulation*, Elektrichestvo **6**, 40–46 (1997), in Russian.

[30] Zh. T. Zhusubaliyev, E. A. Soukhoterin, and E. Mosekilde, *Border-Collision Bifurcations and Chaotic Oscillations in a Piecewise-Smooth Dynamical System*, Int. J. Bifurcation and Chaos **11**, 2977–3001 (2001).

[31] D. J. Uherka, *Tongue Art*, CNLS newsletter **78**, 1–23 (1992).

[32] Yu. L. Maistrenko, V. L. Maistrenko, S. I. Vikul, and L. O. Chua, *Bifurcations of Attracting Cycles from Time-Delayed Chua's Circuit*, Int. J. Bifurcation and Chaos **5**, 653–671 (1995).

[33] Zh. T. Zhusubaliyev, E. A. Soukhoterin, and E. Mosekilde, *Quasiperiodicity and Border-Collision Bifurcations in a DC–DC Converter with Pulse-Width Modulation*, IEEE Trans. Circuits Syst.-I: Fund. Theory Appl., submitted.

[34] M. I. Feigin, Forced Oscillations in Systems with Discontinuous Nonlinearities (Nauka Publ., Moscow, 1994), in Russian.

[35] M. di Bernardo, M. I. Feigin, S. J. Hogan, and M. E. Homer, *Local Analysis of C-Bifurcations in n-Dimensional Piecewise-Smooth Dynamical Systems*, Chaos, Solitons and Fractals **10**, 1881–1908 (1999).

[36] H. E. Nusse and J. A. Yorke, *Border-Collision Bifurcations Including "Period Two to Period Three" for Piecewise Smooth Systems*, Physica D **57**, 39–57 (1992).

[37] H. E. Nusse, E. Ott, and J. A. Yorke, *Border-Collision Bifurcations: An Explanation for Observed Bifurcation Phenomena*, Phys. Rev. E **49**, 1073–1076 (1994).

[38] H. E. Nusse and J. A. Yorke, *Border-Collision Bifurcation for Piecewise Smooth One-Dimensional Maps*, Int. J. Bifurcation and Chaos **5**, 189–207 (1995).

[39] Zh. T. Zhusubaliyev, V. S. Titov, E. Yu. Emelyanova, and E. A. Soukhoterin, *C-Bifurcations in the Dynamics of Control System with Pulse-Width Modulation*, in Proc. of Second Int. Conf. "Control of Oscillations and Chaos" (COC'2000), vol. 1 (St.Petersburg, Russia, 2000), pp. 203–204.

Chapter 7

BORDER-COLLISION BIFURCATIONS IN A MANAGEMENT SYSTEM

7.1 Structure Produces Behavior

In the early 1960's, at about the same time as Smale [1] and Shilnikov [2] gained the first more detailed insights into the mathematical intricacies of deterministic chaos, J. W. Forrester of Massachusetts Institute of Technology published his seminal book on Industrial Dynamics [3]. During the preceding two decades, Forrester had established himself as one of the leading figures in the development of computer technology, and his aim was now to create the framework (a management laboratory) for the development of more successful managerial policies by combining dynamic model formulation with computer simulation.

From 1946 to 1951, Forrester had been director of MIT's Digital Computer Laboratory, responsible for the design and construction of Whirlwind I, one of the World's first high-speed digital computers. Whirlwind I was designed with conventional, but rather unreliable, electrostatic memory tubes. In 1949 Forrester happened to see an advertisement for nonlinear magnetic materials and within a few days he had described the principles, and demonstrated the function, of the magnetic core memory. From 1952 to 1956, Forrester headed the Digital Computer Division of MIT Lincoln Laboratory, guiding the planning and technical design of SAGE, the US Air Force's semiautomatic ground environment air defense system.

Forrester came away from the SAGE project with the conviction that no amount of technical expertise can compensate for poor management, and that his engineering background could help him understand important managerial issues. In Forrester's view, existing management science concentrated too much on the criteria for optimal choice in individual decisions and it failed to see management practice as a dynamic process consisting of a nearly continuous and strongly interconnected series of decisions, with each decision influencing the conditions under which the next decisions had to be taken. This led him to emphasize the significance of feedback structures in managerial systems.

In the technical realm, as illustrated by the examples we have discussed in the preceding chapters, computer simulation is used in practically all thinkable situations. The landing procedures for spacecrafts, the steering properties of cars and ships, the energy flows in tileworks and cement factories, the water flows in sewage systems, the information flow in computers, and the operating conditions of relay control systems are simulated before the system is built. The purpose is to experiment with different constructions and to learn about the boundaries for the successful operation of the system. The outcome is enormous savings in time and money combined with improved safety and a dramatically accelerated technical development.

However, management systems are also designable [4]. In Forrester's perspective, the structure of a management system is determined by the flows and accumulations of material quantities (such as products, personnel, capital units, and money) as controlled by economic and technical constraints and by explicit human decisions. Decisions are made in the light of available information and on basis of the pressures, felt by the management, from declining sales, increasing inventories, unrest among the employees, etc. However, the type of information considered and the weight with which a particular information influences the decisions represent gain factors and phase lags (or delays) in the feedback control and, hence, influence the behavior and stability of the system. Controlling the production rate in a firm from information about changes in the inventory of finished goods, rather than about the size of the inventory, may make all the difference between stable and unstable equilibrium behavior [5].

The challenge of introducing dynamic modeling to a softer discipline such as management science led Forrester to sharpen the definition of a *state variable*, to underline the role of *conservation conditions* in *material flows*, and to emphasize the significance of *stability conditions* and dimensional control. At the same time, Forrester recognized the influence of nonlinearities associated, for instance, with transitions from full capacity utilization to underutilization or from surplus inventories to out-of-stock conditions. We encountered an example of this type of nonlinearity in connection with our analysis of the simplified economic long-wave model in Chapter 1.

Until recently, the basic assumption in most economic theory has been that agents act in accordance with *rational expectations* [6, 7]. This implies that all relevant information is available to the decision maker and that he or she has sufficient understanding of the relevant system to always make the optimal decision. The assumption of rational expectations (and the associated assumption of *effective markets*) has allowed the economic sciences to develop an impressive theoretical framework from which a large number of interesting deductions can be made. However, the assumption of rational expectations is not easy to justify from our daily experience. Moreover, dynamic modeling requires a descriptive (rather than a normative) approach to the decision process.

The question also remains open to what extent the concept of effective markets can explain observed economic behavior. The boom-and-bust, boom-and-bust dynamics characteristic of many economic time series bear witness of dynamic phenomena and instabilities that cannot be accounted for within the theory of effective markets. Phase shifts (or lags) among different economic variables show evidence of equilibration processes with surprisingly long time constants [8], and asymmetric temporal variations indicate the presence of significant nonlinear dynamic phenomena. A more detailed discussion of some of these problems may be found in Topics in Nonlinear Dynamics [9].

The *BEER game* [10, 11] was developed at the Sloan School of Management as a role playing experiment to allow students of management to experience the difficulties of controlling a simple commodity distribution chain. The game clearly demonstrates how structure produces behavior. The game is also used to illustrate how *business cycles* arise in the macroeconomic system and to examine how different decision policies influence the dynamics of a distribution system. Mosekilde and Larsen [12] were first to show that a time-continuous version of the BEER model could produce deterministic chaos and other forms of complex behavior. Larsen *et al.* [13] considered the basins of attraction for the case of coexisting solutions, and Towill [14] discussed the problem of *supply chain dynamics* in the light of a changing engineering challenge.

The object for the participants in the game is to minimize the costs of maintaining sufficient inventories of beer at the different stages of a distribution chain while at the same time avoiding out-of-stock conditions that could lead to loss of customers. Many years of experience with running the game has shown that participants systematically misperceive the *feedback structure* and generate instabilities and large scale oscillations. Because of the lags involved in forwarding orders and products from stage to stage, significant inventory fluctuations arise, and costs are typically an order of magnitude larger than necessary. Particularly interesting is the finding that students as well as experienced managers who take part in the game consistently fail to appreciate that the instabilities arise through their own actions [15]. The participants tend to believe that they closely follow a highly variable external demand.

Based on the theory of *bounded rationality* [16, 17], Sterman [15] has formulated a four parameter *anchoring and adjustment heuristics* for the ordering decisions, and he has shown how this heuristics captures the main aspects of the actual decisions made in the game. Thomsen *et al.* [18] have subsequently applied this ordering heuristics to show how the BEER game for realistic parameter values produces extremely complicated modes of behavior, including various forms of hyperchaos and higher-order hyperchaos. Sosnovtseva and Mosekilde [19] have studied torus destruction and various forms of chaos-chaos intermittency in the BEER model.

Chaos-chaos intermittency has previously been described by Ishii *et al.* [20], by Anishchenko and Neiman [21], and by Anishchenko *et al.*[22]. These authors have focused on the $1/f$-character of the power spectrum in the region near the onset of intermittency. This $1/f$-spectrum is associated with the existence of arbitrarily long laminar phases near the transition point.

Chaos-chaos intermittency has also been discussed by Grebogi *et al.* [23] in connection with an analysis of the scaling relations for a particular form of *crisis* in which two symmetric chaotic attractors merge with one another. Immediately after the merging, the characteristic behavior is an intermittent switching between the two types of dynamics observed before the merging. This type of intermittency has also been described by Sturis and Mosekilde [24] for a model of migratory dynamics and by Baier *et al.* [25] for a model of interacting microbiological populations. The purpose of the present chapter is to illustrate some of the results on torus destruction and chaos-chaos intermittency obtained by Sosnovtseva and Mosekilde [19]. At the same time, we shall perform a more detailed bifurcation analysis of one- and two-stage versions of the BEER model.

Here, we observe the same type of bifurcation phenomena that we have met for pulse-width modulated control systems in Chapter 6. The BEER model is an example of a piecewise linear dynamical system because the decision rules that describe the ordering policies for new supplies of beer involve logical functions. As long as the inventory of beer at a given stage is low enough, orders can be placed without restrictions. However, in a situation when the inventory has grown too large, orders cannot be turned negative, even if this might be desirable in order to reduce the costs of carrying excessive stocks.

The chaos-chaos intermittency described by Ishii *et al.* [20], Anishchenko and Neiman [21], Grebogi *et al.* [23], Sturis and Mosekilde [24], and Baier *et al.* [25] typically occur in systems that display some kind of symmetry. Initially, this symmetry may be broken by a pitchfork bifurcation producing two mutually symmetric periodic orbits. Under variation of some parameter, each of these orbits may have developed into a chaotic attractor (through a period-doubling cascade or via torus destabilization) and, as the chaotic attractors continue to grow in size, they may simultaneously collide with their common basin boundary, producing the symmetry-restoring merging transition.

The BEER model displays a different form of chaos-chaos intermittency where the interacting pseudo-attractors are very dissimilar and where, under variation of a parameter, the merging transition is repeatedly turned on and off in a highly irregular manner. In a particular example, one of the pseudo-attractors represents hyperchaotic dynamics with two positive Lyapunov exponents and the other (much larger) represents higher-order hyperchaos with three positive Lyapunov exponents. Chaos-hyperchaos intermittency was also discussed by Anishchenko *et al.* [26] for a system of two coupled Chua circuits.

7.2 A Cascaded Production-Distribution System

It is customary for breweries (and other industries as well) to use a cascaded system with inventory holders at several different levels: A distributor receives the beer from the factory and ships it to the main markets; regional wholesalers receive the beer from the distributor and allocate it to local outlets such as liquor stores and bars. Besides securing availability of beer at the level of individual customers, the cascaded system is meant to facilitate a swift replacement if a dealer runs low in inventory.

The BEER game is a simplified realization of such a production-distribution system developed to illustrate how oscillations can arise in a managerial system [12]. To allow hand simulation there is only one inventory at each level. Each round (to be referred to as a week) customers order beer from the retailer, who supplies the requested quantity out of inventory. Customer demand is exogenous. In most of the simulations to be reported here it consists of a constant demand of four cases of beer per week until week five at which time it is stepped up to eight cases of beer per week and maintained at this level for the rest of the game. A simulation is normally initiated in equilibrium with 12 cases of beer in each inventory. (In Sec. 7.5 we consider the effect of varying initial conditions.)

In response to fluctuations in customer demand, the retailer adjusts the orders placed with the wholesaler. As long as the inventory permits, the wholesaler ships the requested beer. Orders that cannot be met are kept in backlog until delivery can take place. Similarly, the wholesaler orders and receives beer from the distributor, who in turn orders and receives from the brewery. The passing of orders and the production and shipment of beer involve time delays. It is assumed that there is a mail delay of one week from one stage to the next, and similarly it takes a week to ship beer between two sectors.

The objective of the participants is to minimize cumulative costs over a 40 weeks period by keeping inventories as low as possible while avoiding out-of-stock conditions. If the inventory begins to fall below a desired level, extra beer must be ordered. If stocks begin to accumulate, the order rate must be cut down. According to the rules of the game, orders must always be filled if they are covered by available inventory. Orders which have already been placed cannot be canceled, and deliveries cannot be returned.

For all sectors the decision variable is the amount of beer to be ordered in each round. In contrast to the notion of rational expectations underlying much economic theory we do not assume that the participants can maintain a global knowledge of the state of the system, or that they can actually solve the equations of motion and find out in detail how the time delays and nonlinearities influence its dynamics. The participants are assumed to use a relatively simple decision heuristics based on locally available information, i.e., the current value of the inventory or backlog in the individual sector, previous values of these variables, expected orders, anticipated deliveries, etc.

Consistent with *behavioral science* and with the theory of bounded rationality [16] we have proposed an anchoring and adjustment heuristics for the ordering policy. Assuming demand expectations to be formed from incoming orders in an adaptive manner, provision for expected demand is taken to be the main motive (the anchor). Failure to secure sufficient supply to meet this demand would cause the inventory to fall, reducing the sector's ability to cope with subsequent orders. The participants can adjust orders above or below expected demand to keep the inventory and supply line at their desired levels. The main parameters in these adjustments are α and β.

The parameter α ($0 \leq \alpha \leq 1$) is the fraction of the discrepancy between desired and actual inventory ordered in each round, and β ($0 \leq \beta \leq 1$) is the fraction of the supply line taken into account. If $\beta = 1$, the participants fully recognize shipments under way and do not double order. If $\beta = 0$, orders placed are immediately forgotten, and new orders to cover the same lack of inventory are placed in the next round.

Sterman [15] has used a significant number of hand simulations performed at MIT over a period of 4–5 years to determine the parameters of the above ordering policy for each individual participant. According to this study many players operate with α-values around 0.4 and with β-values of the order of 0.1. However, a significant spread from player to player is observed. The more aggressive the inventory adjustment is (i.e., the higher the value of α), the more unstable the system tends to be. On the other hand, the more accurately the supply line is taken into account (i.e., the larger β is), the more stable the system will be.

7.3 The BEER Model

During a game the managers of each of the four sectors week by week perform a sequence of operations: Receive shipments, advance shipments, receive incoming orders, etc. The corresponding simulation model consists of a high-dimensional iterated map that provides a one-to-one representation of these operations. The nonlinearities of the model are associated with obvious non-negativity constraints on orders and shipments.

Simple as the BEER game may appear from a managerial point of view, the actual model consists of a 27-dimensional nonlinear iterated map. Much of this complexity arises from the cascaded nature of the system with four almost identical sectors. A good deal is also associated with the "mechanical" procedures of forwarding orders and shipments. However, the result is a system that is extraordinary complicated from a mathematical point of view [9].

$\mathcal{COR}$ denotes the exogenous customer order rate. Other variables are prescribed by a letter indicating the respective sector. Thus $\mathcal{R}$ stands for retailer, $\mathcal{W}$ for wholesaler, $\mathcal{D}$ for distributor, and $\mathcal{F}$ for factory. In the wholesale sector, $\mathcal{WI}$ is the inventory of beer and $\mathcal{WB}$ the backlog of orders. $\mathcal{WIS}$ and $\mathcal{WOS}$ represent incoming and outgoing shipments, respectively, and $\mathcal{WIO}$ is incoming orders. $\mathcal{WED}$ is the expected demand and $\mathcal{WOP}$ the orders placed by the wholesaler. One time step later, $\mathcal{WOP}$ becomes incoming orders to the distributor $\mathcal{DIO}$. Similarly, as shipments are advanced, $\mathcal{WOS}$ becomes incoming shipments to the retailer $\mathcal{RIS}$. A similar notation is used in the other sectors, except for the factory, where there is a production initiation rate $\mathcal{FPR}$ instead of orders placed. The production time is represented by the two delay variables $\mathcal{FPD}_1$ and $\mathcal{FPD}_2$. In the following, the equations of the model will be presented and explained in terms of the operations conducted in the experiments.

Inventories are updated by adding incoming shipments and subtracting outgoing shipments. To the extent that inventory plus incoming shipments suffice, outgoing shipments are existing backlog plus incoming orders. Otherwise, outgoing shipments are incoming shipments plus inventory, and the new inventory is empty, i.e.

$$\mathcal{WI}_t = \max\{0, \mathcal{WI}_{t-1} + \mathcal{WIS}_{t-1} - \mathcal{WB}_{t-1} - \mathcal{WIO}_{t-1}\}.$$

Similar expressions hold for $\mathcal{RI}$, $\mathcal{DI}$ and $\mathcal{FI}$. In the same operation, distributor outgoing shipments are advanced to wholesaler incoming shipments

$$\mathcal{WIS}_t = \mathcal{DOS}_{t-1}$$

again with similar expressions for $\mathcal{RIS}$, $\mathcal{DIS}$, and $\mathcal{FPD}_2$.

Backlogs are updated by adding incoming orders and subtracting outgoing shipments. If incoming orders plus backlog are completely covered by incoming shipments plus existing inventory, the new backlog is empty, i.e.

$$\mathcal{WB}_t = \max\{0, \mathcal{WB}_{t-1} + \mathcal{WIO}_{t-1} - \mathcal{WI}_{t-1} - \mathcal{WIS}_{t-1}\}.$$

Similar expressions apply for $\mathcal{RB}$, $\mathcal{DB}$ and $\mathcal{FB}$. In the same operation, orders placed by the retail sector last week are advanced to become wholesaler incoming orders of this week

$$\mathcal{WIO}_t = \mathcal{ROP}_{t-1}$$

with similar expressions for $\mathcal{DIO}$, $\mathcal{FIO}$, and $\mathcal{FPD}_1$.

Following the above discussion, outgoing shipments are expressed as

$$\mathcal{WOS}_t = \min\{\mathcal{WI}_{t-1} + \mathcal{WIS}_{t-1}, \mathcal{WB}_{t-1} - \mathcal{WIO}_{t-1}\}.$$

with similar expressions for $\mathcal{DOS}$, $\mathcal{FOS}$, and shipments out of retailer's inventory. Finally, expected demand is updated in accordance with

$$\mathcal{WED}_t = \theta \cdot \mathcal{WIO}_{t-1} + (1-\theta) \cdot \mathcal{WED}_{t-1}$$

where θ $(0 \leq \theta \leq 1)$ is a parameter that describes the rate at which demand expectations are updated. $\theta = 0$ corresponds to *stationary expectations* and $\theta = 1$ describes a situation in which the immediately preceding value of incoming orders is used as an estimate of future demand. Statistical analyses show that θ typically is of the order of 0.25, indicating that expected demand is adjusted with a smoothing time of about four weeks [15].

Orders placed are determined in accordance with the expression

$$\begin{aligned}\mathcal{WOP}_t = \max\{0, WED_t + \alpha(Q - \mathcal{WI}_t + \mathcal{WB}_t) \\ - \alpha\beta(\mathcal{WIS}_t + \mathcal{DIO}_t + \mathcal{DB}_t + \mathcal{DOS}_t)\}.\end{aligned}$$

with $\mathcal{WIS} + \mathcal{DIO} + \mathcal{DB} + \mathcal{DOS}$ representing the supply line for the wholesaler and Q the desired inventory. Again, similar expressions hold for the other sectors. It is unlikely that participants place more emphasis on the supply line than on the inventory itself. The supply line does not directly influence the costs, nor is it as salient as the inventory. Hence, it is assumed that $\beta \leq 1$.

The second term in the expression for orders placed establishes a negative feedback control on the size of the inventory, and the third term establishes a similar control on the size of the supply line. Besides on general concepts from feedback control theory and behavioral science, our formulation of the ordering policy is based on comments (or exclamations) overheard during the game and on detailed discussions with the participants after the experiment. Many participants have a vivid impression about the shipments they expect to receive in the next rounds, and they try to react on this impression. It is also important to note that precise information about variations in the supply line is available, at least in principle. During the game, each manager is required week for week to register and plot the orders placed with the neighboring sector. The supply line variations can then be determined as cumulated orders less incoming shipments.

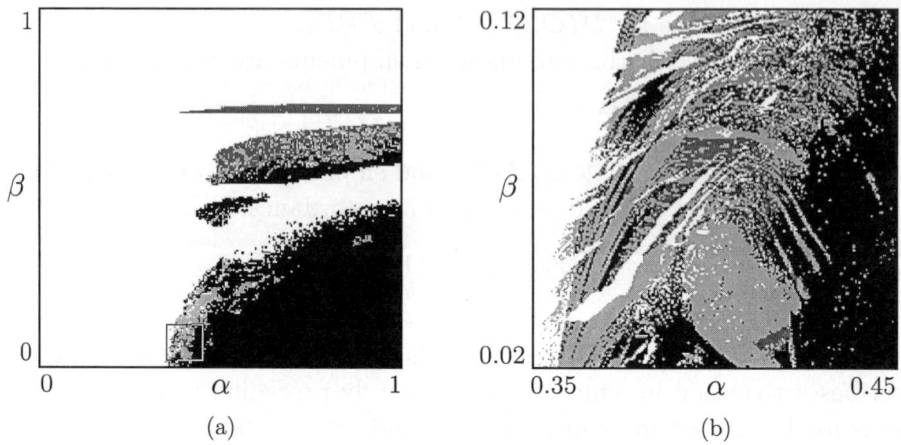

Figure 7.1. 200 × 200 simulations scan over the (α, β) parameter plane. White indicates a stable equilibrium, light gray periodic behavior, dark gray quasiperiodic behavior, and black various forms of chaotic and hyperchaotic dynamics.

In the experiments, each sector manager can apply his own ordering policy. It appears, however, that there is relatively little correlation between the position of a participant in the chain (retailer, wholesaler, distributor, or factory) and the parameters of the applied ordering policy. For simplicity we have therefore performed the computer simulations of the BEER model with the same parameter values for all four sectors. Moreover, in all simulations to be shown we have taken $\theta = 0.25$ and $Q = 17$ units of beer.

7.4 Transitions to Chaos

Further details about the model may be found in our previous publications [9, 12, 15, 18]. Let us briefly recall the main results of these investigations. Computer simulations of the production-distribution system with the estimated order policies show a great variety of behaviors (Fig. 7.1). Detailed analyses of the parameter space reveal an extremely complex distribution with a fractal boundary between stable and unstable solutions and with fingers of periodic solutions protruding deeply into regions of quasiperiodic and chaotic motion. In some regions chaotic transients are observed where, after a randomly distributed time, the system evolves from highly irregular behavior into a stable equilibrium solution. For other parameters an interesting form of intermittency

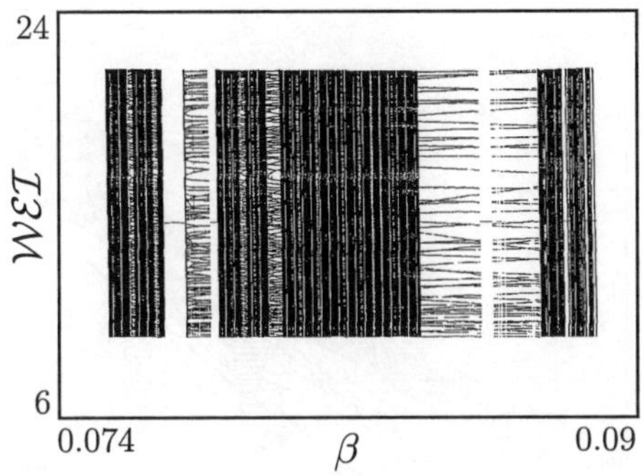

Figure 7.2. One-parameter bifurcation diagram $(\beta, \mathcal{WEI})$ for $\alpha = 0.395$. High-order resonance solutions as well as stable fixed points are clearly observed.

occurs in which the trajectory switches randomly between a hyperchaotic and a higher-order hyperchaotic solution [18]. Figure 7.1(a) shows a 200×200 simulations scan over the *policy plane*. For parameter combinations indicated by white, the model has a stable equilibrium point. For low values of β, the equilibrium inventories are relatively high, while the inventories are small, but the supply lines relatively large, for high values of β. Minimal costs of running the game are typically attained with β-values near 0.7. Light gray points produce periodic behavior, dark gray points give rise to quasiperiodic behavior, and black points to various forms of chaotic and hyperchaotic dynamics. Figure 7.1(b) is a magnification of part of Fig. 7.1(a). Both figures were prepared from calculations of the (four) largest Lyapunov exponents.

Figure 7.2 shows a one-parameter bifurcation diagram $(\beta, \mathcal{WEI})$ for parameter values where only periodic and quasiperiodic motions can be observed. The variable $\mathcal{WEI}$ is the wholesaler effective inventory, i.e., the inventory minus the backlog in the wholesale sector. Similarly, $\mathcal{DEI}$ denotes the distributor effective inventory. We can clearly see parameter values where an ergodic torus exists. In the plane $(\mathcal{DEI}, \mathcal{WEI})$ the two-frequency quasiperiodic motion is represented by an invariant curve. This curve can be considered as the cross section (Poincaré map) of a two-frequency torus in a three-dimensional phase space of a continuous-time dynamical system. Hence, we refer to this invariant curve as a torus. The regions of the ergodic torus are separated by regions of high-order resonance tori. In parameter space these regions often overlap one another. Hence, it is difficult to follow all of the periodic solutions and find the actual routes to chaos.

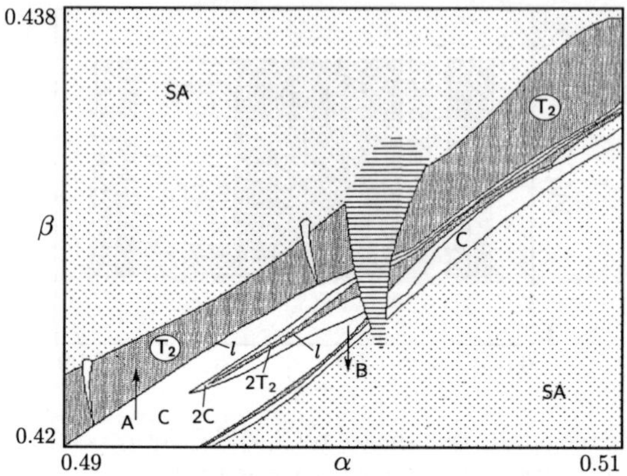

Figure 7.3. Two-parameter bifurcation diagram for a small region of the (α, β) plane.

For a narrow region of parameter space we have been able to construct a two-parameter bifurcation diagram (Fig. 7.3) for the stable period-21 cycle (Fig. 7.4(b)) lying on the torus T_1 (Fig. 7.4(a)). In Fig. 7.3 we have used the following notations: C is the stable limit cycle, 2C is a stable limit cycle with twice the period that arises from C via a period-doubling bifurcation, T_2 is a torus that develops from the cycle C in a torus birth bifurcation (Hopf bifurcation for maps) and $2\mathsf{T}_2$ is a torus that appears from the cycle 2C in the same manner. SA denotes a strange chaotic attractor. The region of chaotic motion in this diagram has a very complicated structure. It includes regimes so-called n-band chaotic attractors and hyperchaotic solutions. Curves corresponding to the transition from periodic motion to quasiperiodic behavior were obtained by calculating the first Lyapunov exponents. On these bifurcation curves the largest Lyapunov exponent vanishes. Along the curves there is an infinite number of resonance regions (Arnol'd tongues) of the torus T_2 with different rational rotation numbers. The bifurcation diagram shows only a few of these tongues. Calculation of Lyapunov exponents was again applied to distinguish the boundary between quasiperiodicity and chaos (the largest Lyapunov exponent turns positive). The hatched region in Fig. 7.3 corresponds to another stable limit cycle (Fig. 7.4(c)) on the torus in Fig. 7.4(a). In this region a structure similar to that observed for the cycle C can be found.

As described above our model demonstrates a great variety of behavioral regimes. In order to investigate the quasiperiodic motion and the transitions from phase-locked regions to chaos let us follow route A marked in Fig. 7.3. A

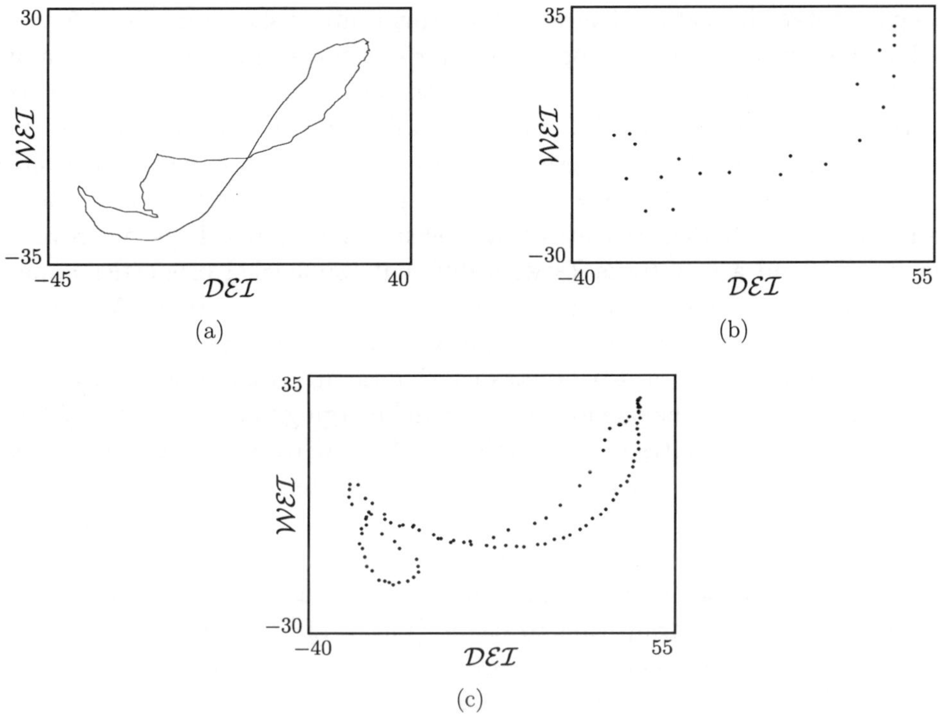

Figure 7.4. Projections of phase trajectories for the following parameter values: (a) $\alpha = 0.5$, $\beta = 0.58$, (b) $\alpha = 0.5$, $\beta = 0.424$, and (c) $\alpha = 0.51$, $\beta = 0.425$. Figures (b) and (c) illustrate some of the high-order resonance behaviors. Figure (a) shows the ergodic torus.

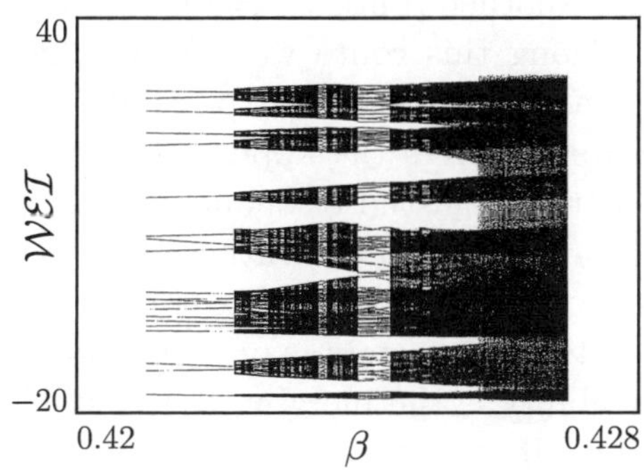

Figure 7.5. One-parameter bifurcation diagram $(\beta, \mathcal{WEI})$ along route A in Fig. 7.3: $\alpha = 0.493$.

one-parameter bifurcation diagram along this route is shown in Fig. 7.5.

The line l in Fig. 7.3 corresponds to a form of torus bifurcation where a new two-dimensional torus develops from the resonance cycle. In terms of maps, this bifurcation looks like a Hopf bifurcation of a fixed point leading to an invariant curve. However, as it is characteristic for piecewise linear systems, the transition is abrupt and without hysteresis. Fig. 7.6 shows the further evolution of this invariant curve. It demonstrates that the invariant torus T_1 is destroyed at the curve l, and a new torus T_2 with different topological properties appears (Fig. 7.6(a)). A further change of the parameter β along route A may lead to chaos, but this transition occurs only via the destruction of this new torus T_2 under formation of an n-band strange chaotic attractor (Fig. 7.6(b)). As β continues to increase, the phenomenon of band merging takes place (Fig. 7.6(c)). Let us describe the routes of destruction of the new torus T_2. Along line l the rotation number varies with changes of the parameter. If the rotation number is irrational, the orbits are everywhere dense on the invariant curve. Whenever the rotation number is rational, a finite number of long-period cycles exists. Inside these resonance regions transitions to chaos as described above can be observed: An abrupt transition to chaos (Fig. 7.7), or a transition to chaos via a sequence of different periodic cycles (Fig. 7.8). Figure 7.8(a) shows a period-14 cycle produced in a period-doubling bifurcation of a period-7 cycle. Figure 7.8(b) shows a stable period-31 cycle and 7.8(c) shows a period-81 cycle produced by a period-tripling of a period-27 cycle. The period-tripling is a typical example of a border-collision bifurcation.

Let us now consider another route to chaos, realized when following the arrow B in Fig. 7.3. Along this route we observe the abrupt emergence of chaos (Fig. 7.9). In smooth systems two possibilities for such a transition are known [27]. The first involves the appearance of a homoclinic structure. This means that the attractor in the resonance region T_1 above the curve on which the homoclinic structure appears rather than being located on a two-dimensional torus is surrounded by a complex non-attractive homoclinic structure. Some variation of the parameters may then lead to an abrupt transition to chaos. Another possibility is the merging of the stable and unstable cycles in a tangent bifurcation. If the torus is not smooth in that region, it is destroyed, and chaos appears. In our model the destruction of the torus and the abrupt transition to chaos are likely to be associated with a border-collision bifurcation.

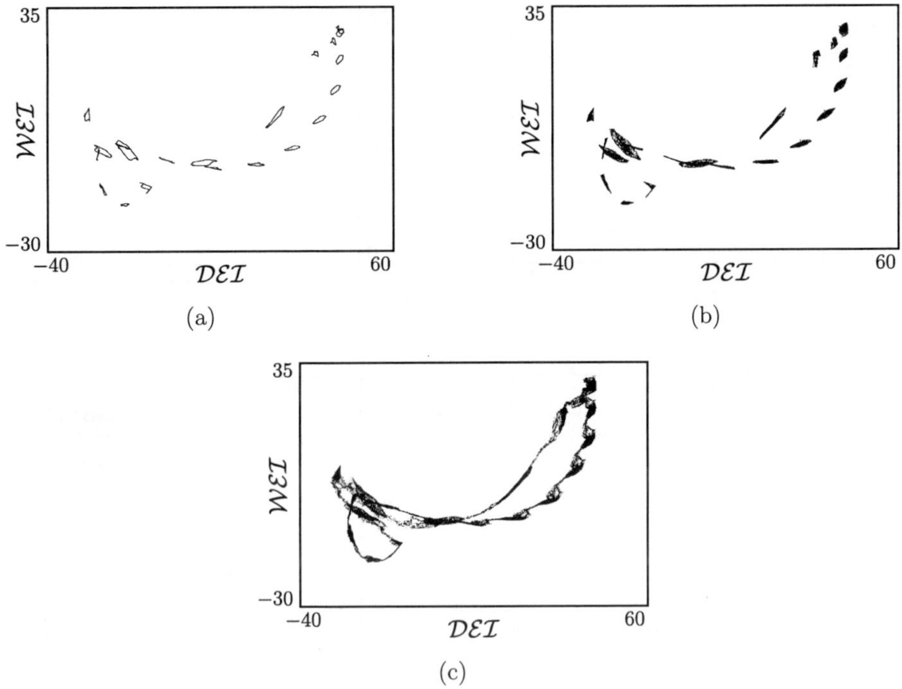

Figure 7.6. Projections of phase trajectories along route A in Fig. 7.3 for $\alpha = 0.493$: (a) $\beta = 0.4233$, (b) $\beta = 0.4255$, and (c) $\beta = 0.426$. Figures (b) and (c) illustrate the development of multiband chaos followed by band merging. Figure (a) shows the formation of a new torus T_2 by destabilization of the 21-cycle in Fig. 7.4(b).

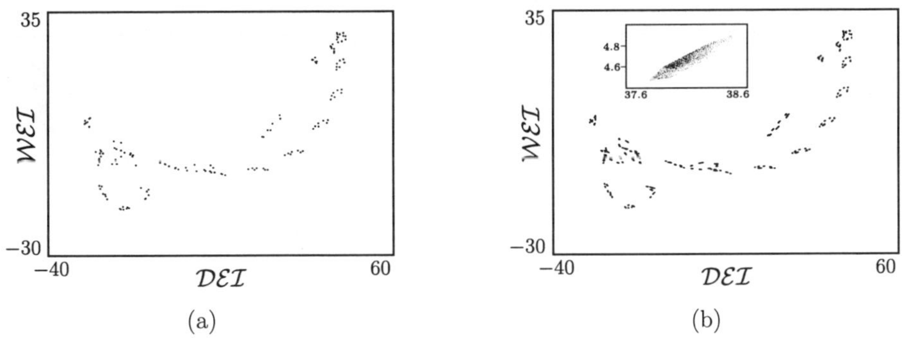

Figure 7.7. Projections of phase trajectories for the destruction of a resonance period-7 cycle on the torus T_2 and the abrupt transition to chaos for $\alpha = 0.49$: (a) $\beta = 0.4217$, and (b) $\beta = 0.4218$. The insert in Fig. (b) illustrates the chaotic structure of each of the spots in the main figure.

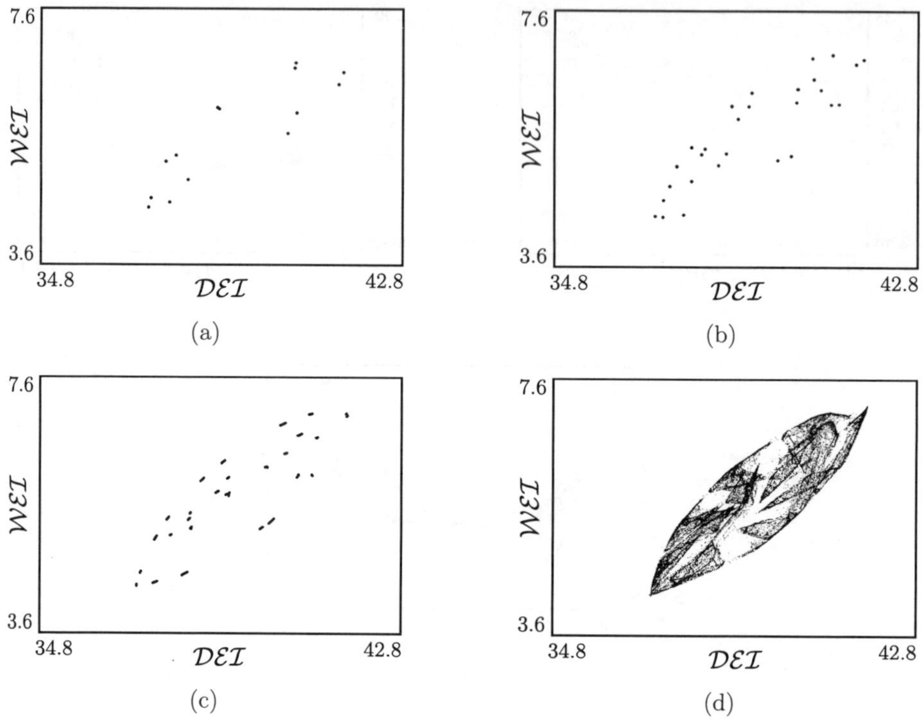

Figure 7.8. Projections of phase trajectories observed in connection with the destruction of the torus T_2 and the transition to chaos via a sequence of periodic cycles for $\alpha = 0.49$: (a) $\beta = 0.4225$, (b) $\beta = 0.4227$, (c) $\beta = 0.42292$, and (d) $\beta = 0.423$. (c) shows a period-81 cycle produced by period-tripling of a period-27 cycle.

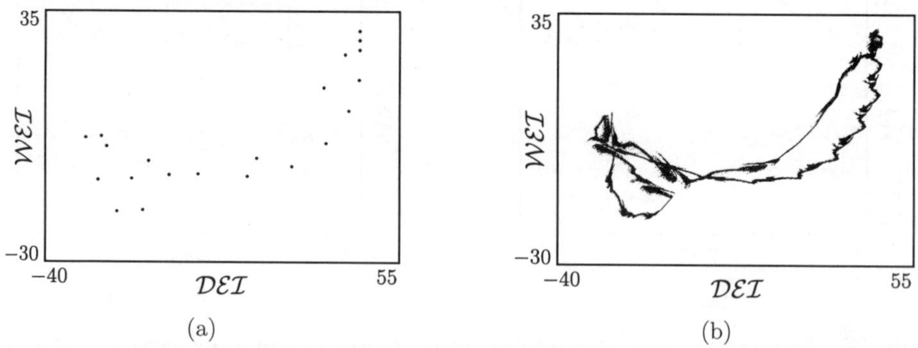

Figure 7.9. Projections of phase trajectories for the destruction of a resonance cycle on torus T_1 along route B in Fig. 7.3 for $\alpha = 0.5$: (a) $\beta = 0.424$, and (b) $\beta = 0.4237$. The chaotic attractor arises abruptly from the period-21 resonance cycle.

7.5 Hyperchaotic Phenomena

As described above, the BEER model contains 27 state variables. Compared with real managerial systems, the model represents a strong simplification. It only considers the flow of a single commodity as opposed to the mutually interacting flows of materials, personnel and capital units in an actual corporation. It supposes that all sectors operate in synchrony, and it provides a clear and simple purpose for the individual manager. Thus personal conflicts and opposing goals should not play a significant role. Compared with most systems investigated in nonlinear dynamics, however, the model is quite complex. In fact, as we shall see in the next section, the model produces chaotic behavior for realistic parameter values with only two sectors. With three sectors, the model produces hyperchaos, and previous investigations [18] have revealed a great variety of complex modes in the four sector model, including intermittency between different hyperchaotic sets.

Figure 7.10(a) shows an example of chaotic solution obtained for $\alpha = 0.695$ and $\beta = 0.2625$. We can clearly identify the structure of the attractor, i.e., the alternating "tube"- and "ball"-like regions associated with the merging bands of an n-band attractor. More detailed simulations show that the largest Lyapunov exponent is positive ($\lambda_1 = 0.98 \cdot 10^{-2}$ /week) while the second exponent is negative, indicating an ordinary form of chaos.

Figure 7.10(b) shows an example of a hyperchaotic attractor obtained for $\alpha = 0.45$ and $\beta = 0.04$. Here, the two largest Lyapunov exponents are positive ($\lambda_1 = 5.03 \cdot 10^{-3}$ /week and $\lambda_2 = 1.19 \cdot 10^{-3}$ /week), but the third exponent is negative.

Finally, Fig. 7.10(c) shows a higher-order hyperchaotic solution for $\alpha = 0.75$ and $\beta = 0.175$. This solution has three positive Lyapunov exponents ($\lambda_1 = 4.64 \cdot 10^{-2}$ /week, $\lambda_2 = 2.73 \cdot 10^{-3}$ /week and $\lambda_3 = 1.24 \cdot 10^{-2}$ /week). At the same time, the excursions in phase space are much larger than observed for the previous solutions, and the structure is also very different. The higher-order hyperchaotic solution is characterized by vertically running lines of points. Such lines arise when the two inventories are both very much larger than desired. Under such circumstances, the wholesaler does not place orders with the distributor, and the distributor does not place orders with the factory. As a result, the distributor inventory remains constant while the wholesaler inventory is gradually reduced by shipments to the retailer. Not until $\mathcal{WEI}$ is close to its desired value will

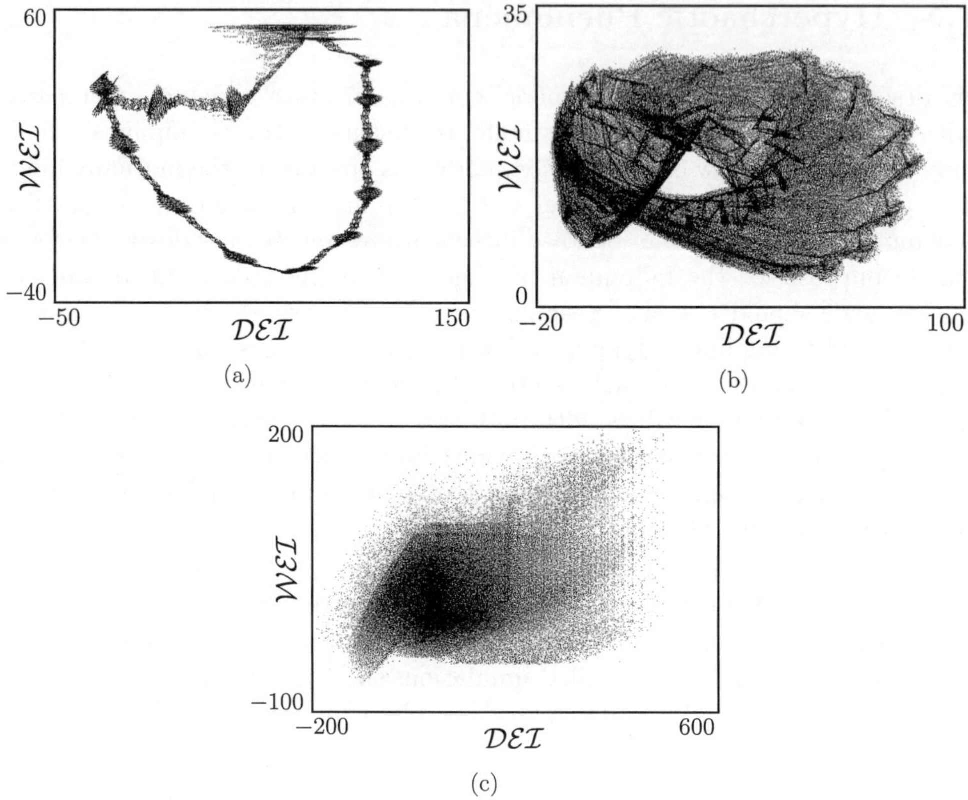

Figure 7.10. Projections of phase trajectories (a) for the chaotic attractor $\alpha = 0.695$, $\beta = 0.2625$; (b) for the hyperchaotic attractor $\alpha = 0.45$, $\beta = 0.04$ and (c) for the higher-order hyperchaotic attractor $\alpha = 0.75$, $\beta = 0.175$.

the wholesaler start to order from the distributor.

As an interesting example of complexity that we meet in the BEER model, Fig. 7.11(a) shows a solution obtained for $\alpha = 0.695$ and $\beta = 0.2635$. By comparing Fig. 7.11(a) with the temporal variation in Fig. 7.11(b) we can see that the system switches between a chaotic set, similar in form to the solution in Fig. 7.10(a), and a hyperchaotic set similar to the solution in Fig. 7.10(c).

To characterize the intermittency phenomenon in detail we have determined the distribution of the residence time N in the smaller chaotic set. This has required very long time series from the model. The distribution $D(N)$ calculated at the vicinity of the critical point is presented in Fig. 7.12(a). It has a

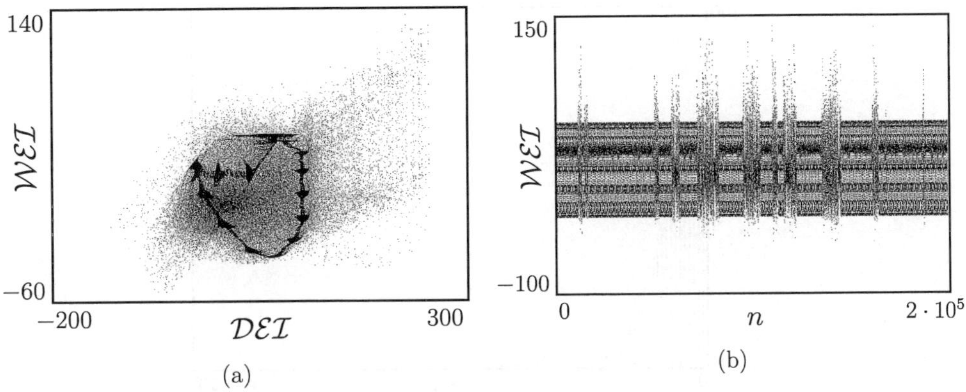

Figure 7.11. Chaos-hyperchaos intermittency observed for $\alpha = 0.695$ and $\beta = 0.2635$: (a) projection of the phase trajectory and (b) time-series.

long tail for $N \to \infty$. With 10^8 iterations at the above parameter values, the maximum residence time in the chaotic set was found to equal 127500 weeks. The distribution has a decreasing trend, and, except for a few points for the shortest residence time it is well approximated by the following exponential law: $D(N) \propto \exp(-\varkappa N)$, $\varkappa \approx 0.000091$/week (Fig. 7.12(b)). This is similar to the results of classical theory [28] which predicts that the residence time of the "laminar" phase in *type-II* and *type-III intermittency* has an exponential distribution for large N: $D(N) \propto exp(-N/\langle N \rangle)$, where $\langle N \rangle$ is the average residence time. On the other hand, such exponential characteristics means that long range correlations are absent in our model, i.e., the "laminar" events are statistically independent.

In Fig. 7.13 the average residence time $\langle N \rangle$ is shown as a function of the control parameter β. For the considered relatively narrow interval of parameter values we can observe regions where the residence time on the smaller chaotic set goes to infinity and regions where intermittency occurs. At some parameter values the bursts into the higher-order hyperchaotic set become more frequent, and at other values transitions to the smaller chaotic set can no longer be distinguished, i.e., the two sets have completely merged.

Pomeau and Manneville [29] have shown that each of the three classical types of intermittency transition displays distinct characteristic behaviors near the critical point β_c. For example, the scaling of the average time between bursts is given by $\langle N \rangle \propto (\beta - \beta_c)^{-\gamma}$, where γ is equal $-1/2$ for type-I intermittency and

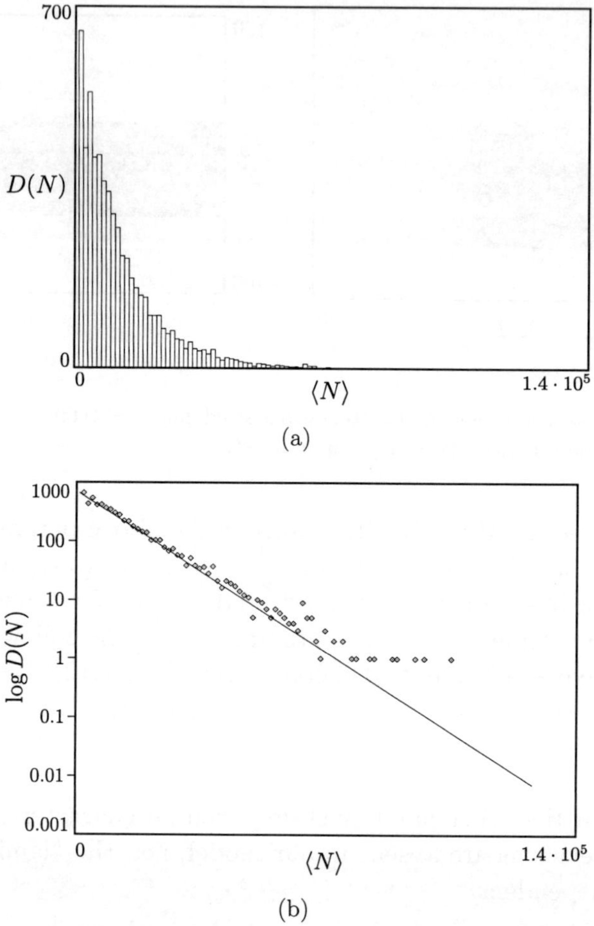

Figure 7.12. Statistical properties of the chaos-hyperchaos intermittency (Fig. 7.11): (a) histogram of the lengths of the residence times on the smaller chaotic set and (b distribution of residence times. The solid line gives the approximation function $\exp(-\varkappa N)$ with $\varkappa \approx 0.000091/\text{week}$.

-1 for type-II and type-III intermittency. In a paper by Anishchenko *et al.* [22] it was shown that such a power law is valid for chaos-chaos intermittency as well. This is contrary to the results in Fig. 7.13 where the characteristics appears to change abruptly.

Whereas the intermittency phenomenon between chaotic attractors observed in the previous investigations [22, 26, 30] took place between sets with simi-

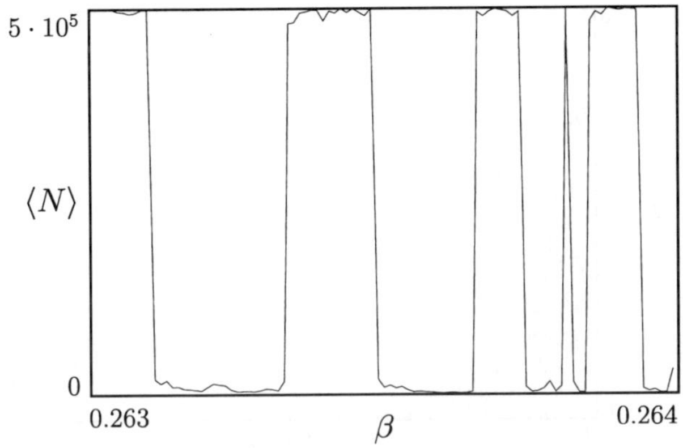

Figure 7.13. Dependence of the average residence time $\langle N \rangle$ on the parameter β. Note the seemingly abrupt shifts from $\langle N \rangle \sim 10,000$ to $\langle N \rangle \sim 500,000$ iterations (or weeks).

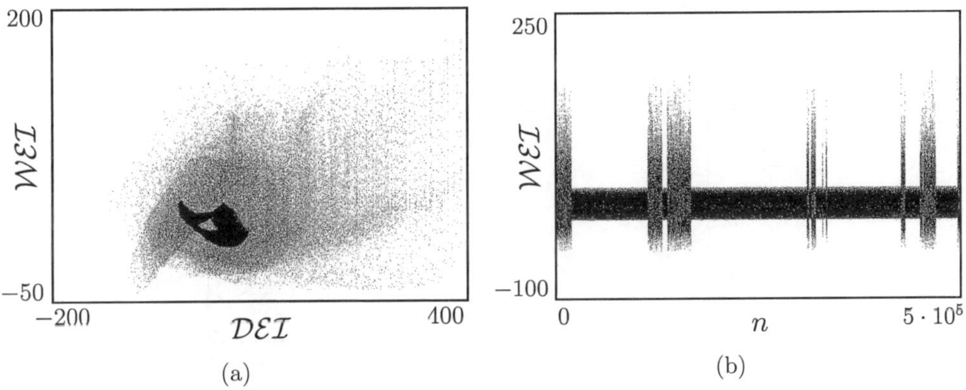

Figure 7.14. Hyperchaos-hyperchaos intermittency observed for $\alpha = 0.45$ and $\beta = 0.0$: (a) projection of the phase trajectory and (b) time-series.

lar topological structures, the BEER model produces intermittency between chaotic sets with widely different structures and properties. Fig. 7.14(a) shows the solution for $\alpha = 0.45$ and $\beta = 0.0$. This combination of parameters is characteristic for subjects who are a little too aggressive in their inventory adjustments and, at the same time, forget to account for orders that have already been placed but not yet supplied. As the system is followed for a long period of time, it is found to switch apparently at random (Fig. 7.14(b)) between two hyperchaotic modes that are similar to the solutions displayed in Figs. 7.10(b) and 7.10(c), respectively. Under these conditions we observe the exponential

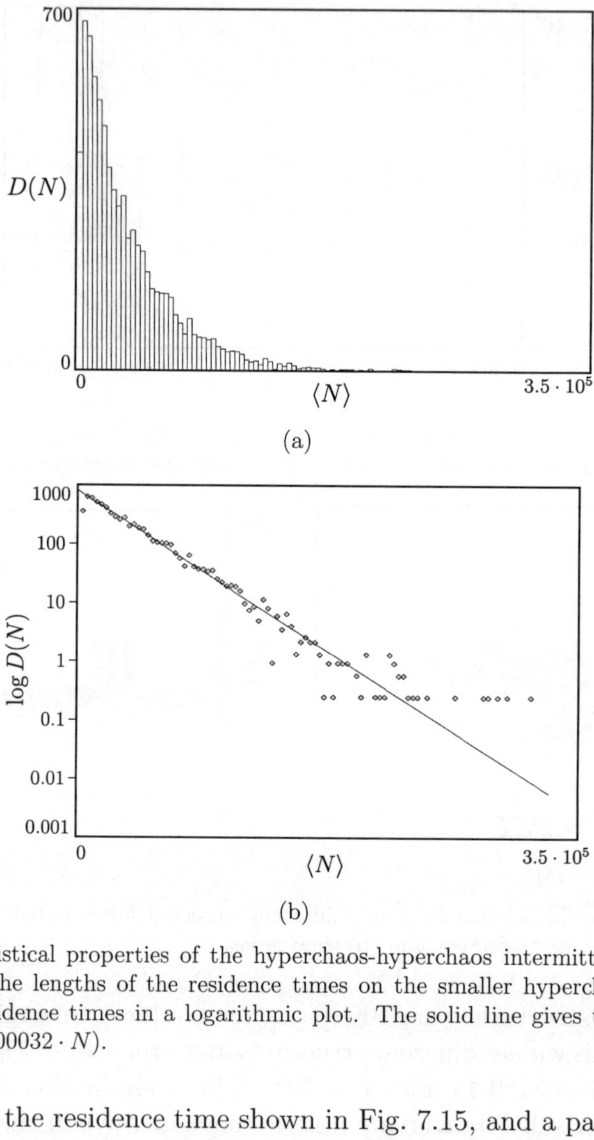

Figure 7.15. Statistical properties of the hyperchaos-hyperchaos intermittency in Fig. 7.14: (a) histogram of the lengths of the residence times on the smaller hyperchaotic set and (b) distribution of residence times in a logarithmic plot. The solid line gives the approximation function $\exp(-0.000032 \cdot N)$.

distribution for the residence time shown in Fig. 7.15, and a parameter dependence of the average residence time in the smaller hyperchaotic set as shown in Fig. 7.16. It is obvious that the statistical properties of this phenomenon are related to the properties of higher-order hyperchaotic solutions.

For some parameter values we can observe (Fig. 7.17(a)) that before reaching the smaller solution (Fig. 7.10(b)) the trajectory first spends an extended time

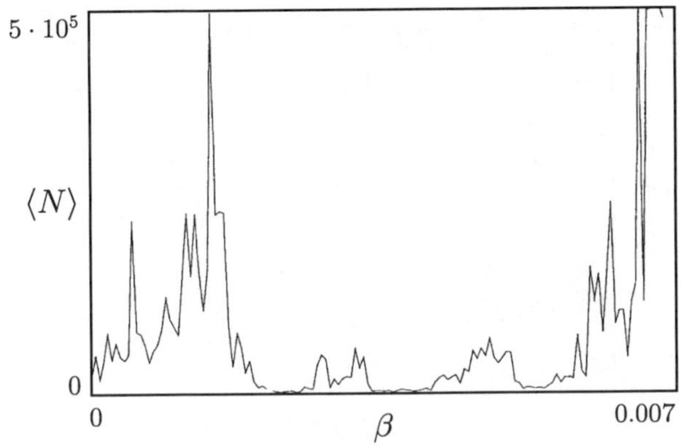

Figure 7.16. Dependence of the average residence time $\langle N \rangle$ on the parameter β for the hyperchaos-hyperchaos intermittency shown in Fig. 7.14.

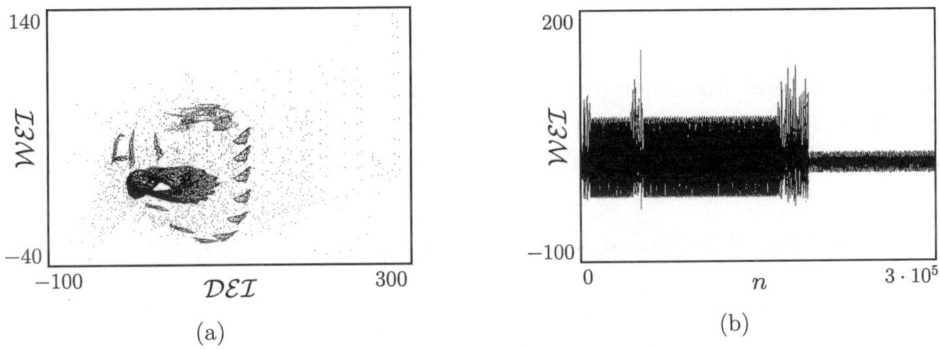

Figure 7.17. Switchings between three different chaotic regimes during the transient period for $\alpha = 0.45$ and $\beta = 0.04$: (a) projection of the phase trajectory and (b) time-series.

on the higher-order hyperchaotic set (Fig. 7.10(c)), and then a period in a region where for other parameter values a chaotic attractor similar in form to the attractor in Fig. 7.10(a) exists. We can see how the system "traces the ghost" of this disappeared attractor. Fig. 7.17(b) shows a time-series for such a transient. Starting from the higher-order hyperchaotic set the trajectory switches into the *ghost solution* and then returns to the higher-order hyperchaotic set to finally approach to the smaller hyperchaotic attractor where it remains "forever".

Whether the ghost solution is visited or not depends in a critical manner on the initial conditions. Fig. 7.18 was constructed for a set of 100×100 initial conditions. For each initial condition the model was iterated a large number of

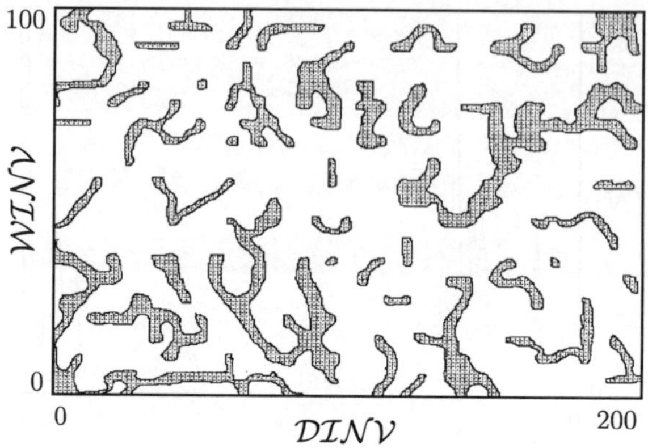

Figure 7.18. Diagram for different initial conditions ($\alpha = 0.45$, $\beta = 0.04$). Gray points correspond to initial conditions starting from which the trajectory spends an appreciable time in the region of the chaotic ghost attractor.

times. It was found that for some initial conditions the trajectory visited all three regions and for other initial conditions only two of them. After a transient time depending on initial conditions the trajectory comes to the smaller attractor which is the stable solution for the considered parameter values.

7.6 Border-Collisions and Resonance Phenomena

In order to perform a more detailed bifurcation analysis of the BEER model we have reduced the original four-stage model to simpler one- and two-stage versions. In the one-stage version of the model we only consider a brewery (factory) subjected to a constant customer order rate $\mathcal{COR}$. The equations of motion then become

$$\begin{aligned}
\mathcal{FI}_t &= \max\{0, \mathcal{FI}_{t-1} + \mathcal{FPD}_{2,t-1} - \mathcal{FB}_{t-1} - \mathcal{COR}\};\\
\mathcal{FPD}_{2,t} &= \mathcal{FPD}_{1,t-1};\\
\mathcal{FPD}_{1,t} &= \mathcal{FPR}_{t-1};\\
\mathcal{FPR}_t &= \max\{0, \mathcal{FED}_t + \alpha(Q - \mathcal{FI}_t + \mathcal{FB}_t) - \alpha\beta(\mathcal{FPD}_{1,t} + \mathcal{FPD}_{2,t})\};\\
\mathcal{FB}_t &= \max\{0, \mathcal{FB}_{t-1} + \mathcal{COR} - \mathcal{FI}_{t-1} - \mathcal{FPD}_{2,t-1}\};\\
\mathcal{FED}_t &= \theta \cdot \mathcal{COR} + (1-\theta) \cdot \mathcal{FED}_{t-1},
\end{aligned}$$

where the same variable designations have been applied as in the full four-stage model described in Sec. 7.3. $\mathcal{FI}$ is the inventory and $\mathcal{FB}$ the backlog of orders in the factory. $\mathcal{FED}$ is the expected demand and $\mathcal{FPR}$ is the production initiation rate. The production time is represented by the two delay variables $\mathcal{FPD}_1$ and $\mathcal{FPD}_2$. Q is the desired inventory and θ is the updating rate for the demand expectations. α and β are our policy parameters.

We do not need an expression for the outgoing shipments $\mathcal{FOS}$ since nothing else depends on this variable. Hence, the one-sector beer model is represented by a 6-dimensional piecewise linear map. This map has a fixed point in which

$$\begin{aligned}\mathcal{FI} &= Q - 2\beta \cdot \mathcal{COR};\\ \mathcal{FB} &= 0,\end{aligned}$$

and all the flow-related variables equal the customer order rate:

$$\mathcal{FPR} = \mathcal{FED} = \mathcal{FPD}_1 = \mathcal{FPD}_2 = \mathcal{COR}.$$

Calculation of the eigenvalues show [40] that the fixed point is stable as long as α is small enough or β is large enough. We may recall that α is a measure of the aggressivity with which inventory differences are corrected. β is a measure of the weight with which supply line variations are taken into account. For

$$\beta = \frac{3\alpha + 1 - \sqrt{\alpha^2 - 2\alpha + 9}}{4\alpha}$$

the fixed point undergoes a Hopf bifurcation as two complex conjugate eigenvalues cross the unit circle, and for higher values of α the system exhibits periodic or quasiperiodic oscillations on a two-dimensional torus.

Figure 7.19 shows the phase diagram for the one-stage BEER model. The Hopf bifurcation curve passes from

$$\alpha - \frac{\sqrt{5} - 1}{2} \approx 0.618 \text{ for } \beta = 0$$

to

$$\beta = \frac{2 - \sqrt{2}}{2} \approx 0.293 \text{ for } \alpha = 1.$$

To the right of this curve we observe the necklace formed resonance zones with shrinking points that are characteristic of piecewise linear maps [41]. Among the main resonance zones we note the $1:9$ and the $1:10$ regions. In the $1:9$ resonance zone the dynamics of the one-stage model repeats itself after 9 iterations. Between the main resonance tongues we observe tongues with $2:19$ and

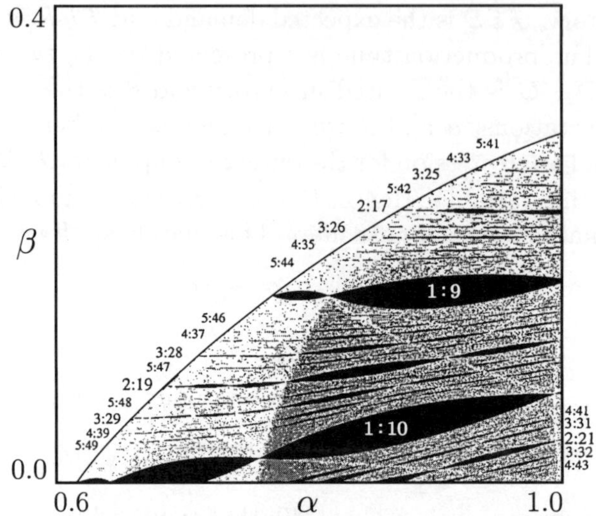

Figure 7.19. Phase diagram for a production-distribution system with only one stage (a factory). The solid curve represents a Hopf bifurcation. To the right of this curve we observe the structure of resonance zones characteristic for piecewise linear systems.

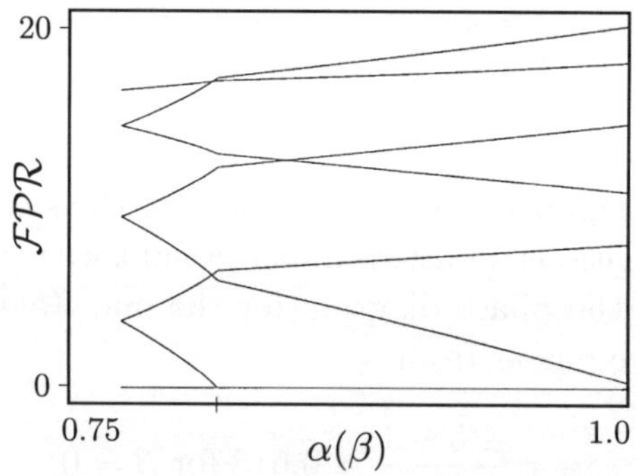

Figure 7.20. Border-collision bifurcation at the edge of the 1 : 9 resonance zone in Fig. 7.19. In symbolic dynamics the solution (00∗∗∗∗∗∗∗) bifurcates into the solution (000∗∗∗∗∗∗).

with 3 : 28 and 3 : 29 dynamics. This is in full agreement with our discussion in Chapter 6 about the organization and levels of complexity for the resonance tongues.

The resonance zones are delineated by border-collision bifurcations. Figure 7.20 illustrates an example of a border-collision bifurcation at the edge of

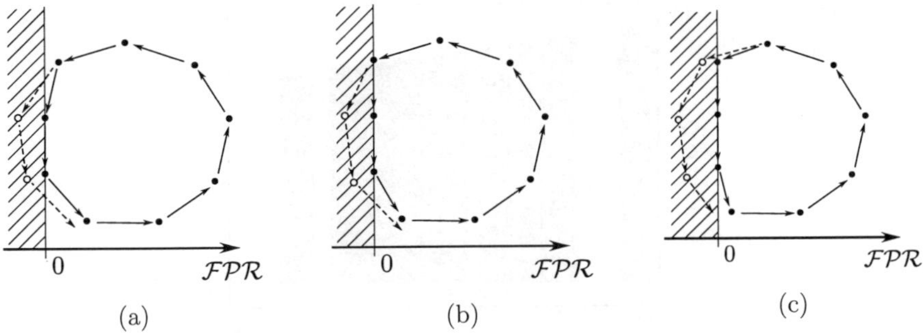

Figure 7.21. Illustration of the border-collision bifurcation in Fig. 7.20. The factory production initiation rate $\mathcal{FPR}$ cannot become negative, even in periods where the inventory of finished goods is very large. During two iterations (a), respectively during three iterations (c) of the period-9 cycle, the factory production vanishes. (b) illustrates the situation in the bifurcation point.

the 1 : 9 resonance zone. For values of α to the left in the figure, the production initiation rate $\mathcal{FPR}$ takes on 8 different values during a period of the system. This period involves 9 iterations, but in two of the points the production rate vanishes because the inventory is too large, and the desired production is negative. To the right in this figure, the production rate assumes 7 different values during a full cycle. Here, $\mathcal{FPR}$ vanishes during three iterations of the map. The border-collision bifurcation takes place for $\alpha \approx 0.82$. Figure 7.21 provides an illustration of the border-collision bifurcation in Fig. 7.20. This bifurcation is associated with the fact that the factory can initiate a positive production of any size. However, it cannot initiate a negative production if the inventory becomes too large. The dashed curves in Fig. 7.21 represent the trajectories that the system would have followed in the absence of the non-negativity restriction.

Figure 7.22 shows a vertical scan through the phase diagram in Fig. 7.19 for $\alpha = 0.8$. We note the abrupt transition to finite amplitude quasiperiodic oscillations at the "Hopf" bifurcation point ($\beta \approx 0.1764$), and the characteristic dense set of resonance zones. By contrast to the border-collision bifurcations that define the edges of the resonance zones, the curves that connect the shrinking points in Fig. 7.19 are saddle-node bifurcations. The one-stage model does not display deterministic chaos.

Figure 7.23 shows the phase diagram for a two-stage version of the BEER model. Besides the brewery we have now included a distributor, and the dimensionality of the map has gone up to 15. The two sectors are not quite identical.

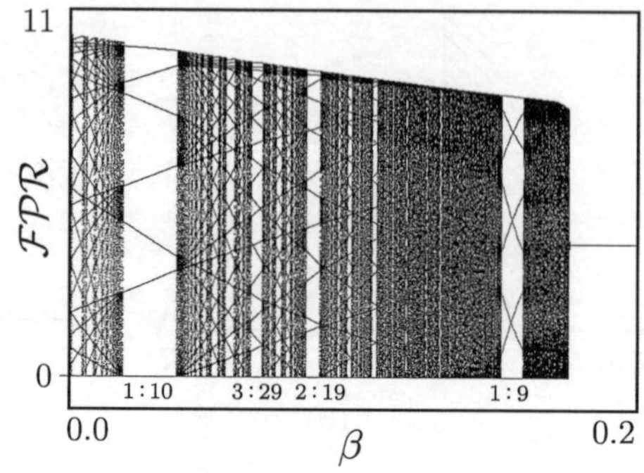

Figure 7.22. One-dimensional bifurcation diagram obtained by scanning Fig. 7.19 vertically for $\alpha = 0.8$. In the manner characteristic of piecewise-linear systems, the "Hopf" bifurcation produces an abrupt transition from a stable equilibrium to a finite-amplitude limit cycle.

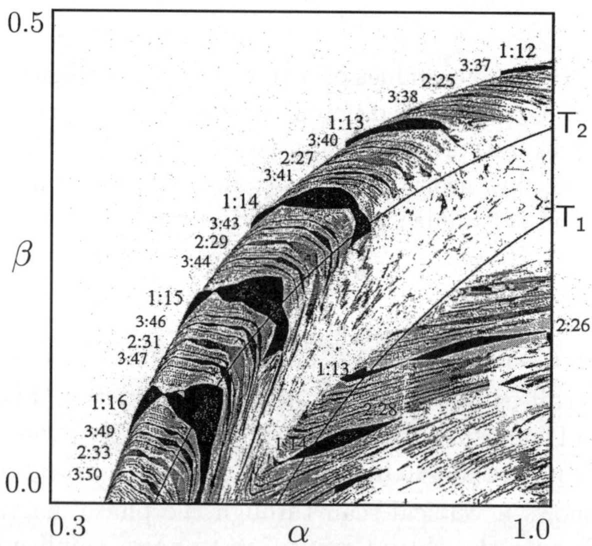

Figure 7.23. Two-dimensional bifurcation diagram for the two-stage production-distribution model illustrating the two torus bifurcation curves T_1 and T_2 for the equilibrium point and the distribution of resonance zones.

The distributor, for instance, does not include a production delay, and the expressions for the supply line also differ between the sectors. This may influence the dynamics of the system. The map now has a fixed point with

$$\mathcal{FB} = 0;$$

$$\mathcal{FI} = Q - 2\beta \cdot \mathcal{COR};$$
$$\mathcal{DB} = 0$$

and

$$\mathcal{DI} = Q - 3\beta \cdot \mathcal{COR}.$$

When analyzing the stability of this fixed point, the characteristic equation splits up into a product of polynomials. There is a third-order polynomial which is identical to the polynomial used to determine the stability of the fixed point for the one-sector model. This gives rise to the same Hopf bifurcation curve for the fixed point as in Fig. 7.19. In Fig. 7.23, this curve is denoted T_1. However, there is also a fourth-order polynomial that produces a Hopf bifurcation along the curve [40]

$$\beta = \frac{2 + 4\alpha - 3\alpha^2 - \sqrt{16 - 16\alpha + 8\alpha^2 - 4\alpha^3 + \alpha^4}}{2\alpha(3 - 2\alpha)}.$$

In the phase diagram (Fig. 7.23), this curve is denoted T_2. It passes from $\alpha \approx 0.46$ for $\beta = 0$ to

$$\beta = \frac{3 - \sqrt{5}}{2} \approx 0.382 \text{ for } \alpha = 1.$$

This implies that, for increasing α, the Hopf bifurcation T_2 is first to occur. It is interesting to note, however, that the numerically determined resonance zones start well before T_2. It is most likely that this is a result of the Hopf bifurcation being subcritical. The main resonance zones are now 1:16, 1:15, 1:14, 1:13, and 1:12. Between the 1:15 and 1:16 zones we observe resonance behavior with 2:31 and with 3:46 and 3:47 dynamics. We note that these zones still maintain some of the necklace structure characteristic for (simple) piecewise linear maps. It is evident, however, that the structure is now more complicated.

Between the resonance zones we can find sets of parameters that produce quasiperiodic dynamics. However, as α increases and we move further to the right in the phase diagram, chaotic dynamics starts to appear. It seems that there is another region of resonant zones associated with the Hopf bifurcation T_1. However, in the lower rightmost corner of the diagram most parameter sets produce chaotic dynamics.

7.7 Conclusions

The instability observed in the BEER model is connected with delays built into the distribution system. Assume, for instance, that a particular sector suddenly experiences a significant increase in incoming orders. Awaiting to find out if this change is of a more permanent character, the sector manager usually hesitates to adjust his orders by the full amount. After all, it is precisely the purpose of the inventory to absorb fluctuations in demand. However, because of this hesitation, and by virtue of the built in mailing and shipping delays, shipments will exceed inventory replacements for several weeks, during which period the inventory will decrease and a backlog of orders may build up. To bring the inventory back to its desired level, the manager must now increase his orders beyond incoming orders, hence amplifying and delaying the original signal.

Similar mechanisms can be identified in the response of real economic sectors to fluctuations in demand, and they are believed to play a major role in the formation of the general business cycle [32]. The nonlinearities of the BEER model are associated with non-negativity conditions on orders and shipments. In the real economy, similar conditions apply to most variables. At the same time, other nonlinearities come into play through limitations in capital and labor availability, and through a variety of bilinear terms involving, for instance, the product of price and quantity variables. For this reason it is of interest to continue to explore the potential of nonlinear dynamics in economics and management science.

Actually, the manner in which our economic systems are controlled has many similarities with the relay control techniques described in the previous chapters, only the feedback structure is less salient. Each time the National Bank or the Government intervenes by lowering or raising the interest rate or by introducing new taxes, the structure of the economic system changes abruptly, and the new trajectory has to be sewed to the previous development by some kind of continuity equations.

A particularly well-defined example of a piecewise smooth macroeconomic system is associated with the so-called intervention thresholds introduced to stabilize certain exchange rates and commodity markets. Typically one defines a target value for a particular exchange rate and negotiates a pair of intervention thresholds. The Danish krone, for instance, is bound to the European Euro in such a *"snake" arrangement.* If the exchange rate moves away from its target value and crosses one of the intervention thresholds, the national banks are committed to intervene by selling or buying the appropriate currencies. We have previously studied a similar problem [36] in connection with UNCTAD's (United Nations Conference on Trade and Development) proposal for a buffer stock scheme to stabilize the prices of some of the World's core commodities [37].

Studies of border-collision bifurcations and synchronization phenomena in piecewise linear economic systems have recently been initiated by Puu and Norin [38] and by Sushko *et al.* [39].

At the same time, there is a great interest in clarifying the scenarios leading from quasiperiodicity to chaos. In the first place, this is important because numerical as well as laboratory investigations of a large number of physical and technical systems show that the birth of stochastic oscillations, when a two-dimensional torus is destroyed, is a common phenomenon. Examples are Couette flow [33], convection in a horizontal layer of a viscous fluid [34], and oscillations in radio and VHF generators [35].

While these systems may all be considered as smooth, the last two chapters of the present book have provided examples of torus breakdown in piecewise-smooth dynamical systems. For such systems, the possible mechanisms of torus destruction may differ qualitatively from the mechanisms described in the classic Afraimovich-Shilnikov scenarios. In the presence of border-collision bifurcations, the eigenvalues of a mode can jump abruptly, and it becomes possible to observe a direct transition from a stable node to an unstable focus point. Hence, in the classic scenario where the resonance torus is destroyed by a secondary Hopf bifurcation, the secondary bifurcation may take place from a node (rather than from a focus).

There is another classic scenario of torus destruction where the unstable manifold from the saddle cycle (defining the resonant torus) starts to develop wrinkles as it approaches the points of the node. The formation of wrinkles is controlled by the eigenvalues of the node, and again one can expect abrupt changes in the dynamics as a consequence of sudden changes in the eigenvalues. The wrinkles produce non-smoothness of the torus in the points of the node, and this non-smoothness spreads to the whole periphery as the system, under variation of a parameter, passes out of the resonance tongue. For piecewise-smooth systems the resonance tongues are generally delineated by border-collision bifurcations (rather than by saddle-node bifurcations), and again one can expect qualitative differences in behavior.

Due to its high dimensionality, the BEER model can produce an interesting sequence of subsequent Hopf bifurcations. Thus we have observed the following scenario: A simple periodic orbit undergoes a Hopf bifurcation and produces a new torus with quasiperiodic dynamics and resonance behavior. A resonance cycle on this secondary torus undergoes a new Hopf bifurcation, etc. We have been able to follow this scenario through three or four subsequent Hopf bifurcations. From a mathematical point of view this problem is important because a general bifurcation theory of quasiperiodic attractors has not yet been developed. Starting from an understanding of the processes through which a two-dimensional torus can break down such a theory will proceed to study the conditions of existence and destruction of n-dimensional tori.

As repeatedly noted throughout the book, one can distinguish between two types of bifurcation for piecewise smooth dynamical systems. The first type is similar to the bifurcations we know for smooth dynamical systems. These include the local bifurcations (saddle-node, period-doubling, and Hopf bifurcations) and the global bifurcations (homoclinic bifurcations). For piecewise-linear systems, however, these bifurcations typically involve abrupt shifts in amplitudes, and both for piecewise-linear and for piecewise-smooth systems, the bifurcation cascades are truncated.

The second type of bifurcations are connected with situations where one of the parts of the periodic trajectory becomes tangent to a sewing surface. These bifurcations are referred to as *border-collision bifurcations*. A simple type of border-collision bifurcation consists in the continuous transformation of a solution from one type into another or the merging and disappearance of solutions of different types. However, more complicated nonlinear phenomena and transitions to chaos are possible.

As our investigations of pulse modulated systems in Chapters 5 and 6 have shown, transitions from periodic to chaotic oscillations can occur through a sequence of border-collision bifurcations including period-doubling, -tripling, -quadrupling, quintupling, etc., bifurcations. Our numerical calculations of the variation of unstable cycles with parameters revealed more subtle features of the border-collision bifurcations. In these transitions a denumerable set of unstable cycles with periods that are multiple to the period of the basic cycle appear together with a stable cycle. The stable cycle can subsequently undergo a sequence of local period-doubling bifurcations connected with the crossing of a multiplier through -1. But the number of such bifurcations is always finite.

Usually, after one or two local period-doubling bifurcations, a finite sequence of border-collision bifurcations occur: A simple change of the solution type first and then a border-collision bifurcation where a cycle of chaotic intervals softly branches off from the stable cycle.

Our investigations of the changes in system behavior as we move along various sections of the parameter plane have revealed bifurcational transitions with even more extraordinary properties.

An example is the soft transition from one family of cycles (the number of which is finite) to another family of cycles (the number of which can be finite or infinite) with multiple periods. In Chapter 5 we considered a border-collision bifurcation where stable and unstable 20-cycles of different types merged with a stable 5-cycle, and a locally stable 15-cycle subsequently arose (together with a denumerable set of unstable cycles with multiple-to-5 periods).

Simpler cases of such behavior were recently described for two-dimensional piecewise-smooth systems by Dutta *et al.* [42] and by Kapitaniak and Maistrenko [43]. Dutta *et al.* refer to these bifurcations as *multiple attractor bifurcations*, while Kapitaniak and Maistrenko call them *multiple choice bifurcations*. We have limited our description to consider only a few examples where such atypical, yet not the most complicated, transitions may occur.

The extraordinary complexity and large variety of nonlinear dynamic phenomena that one can observe in piecewise-smooth systems are associated with the interplay between local and homoclinic bifurcations on one side and different forms of border-collision bifurcations on the other. Understanding of these phenomena is of immediate relevance in engineering and applied science.

We have illustrated this through a number of well-documented and realistic examples of relay and pulse-width modulated control systems. However, it is clear that similar phenomena arise in a great variety of mechanical systems involving finite clearances or stick-slip phenomena.

Bibliography

[1] S. Smale, *Diffeomorphisms with Many Periodic Orbits*, in Differential and Combinatorial Topology, edited by S. S. Cairns (Princeton University Press, NJ, 1963), pp. 63–80.

[2] L. P. Shilnikov, *A Case of the Existence of a Denumerable Set of Periodic Motions*, Sov. Math. Dokl. **6**, 163–166 (1965).

[3] J. W. Forrester, Industrial Dynamics (MIT Press, Cambridge, MA, 1961).

[4] A. V. Ackere, E. R. Larsen, and J. D. W. Morecroft, *Systems Thinking and Business Process Redesign: An Application to the Beer Game*, Eur. Managm. J. **11**, 412–422 (1993).

[5] D. R. Rasmussen and E. Mosekilde, *Bifurcation and Chaos in a Generic Management Model*, Eur. J. Oper. Res. **35**, 80–89 (1988).

[6] D. K. H. Begg, The Rational Expectations Revolution in Macroeconomics: Theories and Evidence (J. Hopkins Univ. Press, Baltimore, 1992).

[7] H. Nagatani and K. Crowley, The Collective Scientific Papers by P. A. Samuelsen (MIT Press, Cambridge, MA, 1977).

[8] S. Rasmussen, E. Mosekilde, and J. Holst, *Empirical Indication of Economic Long Waves in Aggregate Production*, Eur. J. Oper. Res. **42**, 279–293 (1989).

[9] E. Mosekilde, Topics in Nonlinear Dynamics. Applications to Physics, Biology and Economic Systems (World Scientific, Singapore, 1996).

[10] W. E. Jarmain, Problems in Industrial Dynamics (MIT Press, Cambridge, MA, 1963).

[11] J. D. Sterman, *Instructions for Running the Beer Distribution Game*, Working Paper D-3679, Sloan School of Management, MIT, Cambridge, MA (1984).

[12] E. Mosekilde and E. R. Larsen, *Deterministic Chaos in the Beer Production-Distribution Model*, Syst. Dyn. Rev. **4**, 131–147 (1988).

[13] E. R. Larsen, J. D. W. Morecroft, and J. S. Thomsen, *Complex Behavior in a Production-Distribution Model*, Eur. J. Oper. Res. **119**, 61–74 (1999).

[14] D. R. Towill, *Supply Chain Dynamics — The Changed Engineering Challenge of the Mid 1990's*, Proc. Instn. Mech. Engrs. **206**, 233–245 (1992).

[15] J. D. Sterman, *Modeling Managerial Behavior: Misperception of Feedback in a Dynamic Decision Making Experiment*, Managem. Sci. **35**, 321–339 (1989).

[16] H. L. Davis, S. J. Hoch, and E. K. Easton Ragsdale, *An Anchoring and Adjustment Model of Spousal Predictions*, J. Consumer Research **13**, 25–37 (1986).

[17] H. J. Einhorn and R. M. Hogarth, *Ambiguity and Uncertainty in Probabilistic Inference*, Psychological Rev. **92**, 433–461 (1985).

[18] J. S. Thomsen, E. Mosekilde, and J. D. Sterman, *Hyperchaotic Phenomena in Dynamic Decision Making*, J. Syst. Anal. Mod. Sim. **9**, 137–156 (1992).

[19] O. V. Sosnovtseva and E. Mosekilde, *Torus Destruction and Chaos–Chaos Intermittency in a Commodity Distribution Chain*, Int. J. Bifurcation and Chaos **7**, 1225–1242 (1997).

[20] H. Ishii, H. Fujisaka, and M. Inoue, *Breakdown of Chaos Symmetry and Intermittency in the Double-Well Potential*, Phys. Lett. A **116**, 257–263 (1986).

[21] V. S. Anishchenko and A. B. Neiman, *Statistical Properties of Intermittency in a Quasihyperbolic System*, J. Tech. Phys. **17**, 3–14 (1984).

[22] V. S. Anishchenko, A. B. Neiman, and L. O. Chua, *Chaos-Chaos Intermittency and $1/f$-Noise in Chua's Circuit*, Int. J. Bifurcation and Chaos **4**, 99–107 (1994).

[23] C. Grebogi, E. Ott, F. Romeiras, and J. A. Yorke, *Critical Exponents for Crisis-Induced Intermittency*, Phys. Rev. A **36**, 5365–5380 (1987).

[24] J. Sturis and E. Mosekilde, *Bifurcation Sequence in a Simple Model of Migratory Dynamics*, Syst. Dyn. Rev. **4**, 208–217 (1988).

[25] G. Baier, J. S. Thomsen, and E. Mosekilde, *Chaotic Hierarchy in a Model of Competing Populations*, J. Theor. Biol. **165**, 593–607 (1993).

[26] V. S. Anishchenko, T. Kapitaniak, M. A. Safonova, and O. V. Sosnovtseva, *Birth of Double-Double Scroll Attractor in Coupled Chua Circuits*, Phys. Lett. A **192**, 207–214 (1994).

[27] V. I. Arnol'd, V. S. Afraimovich, Yu. S. Il'yashenko, and L. P. Shilnikov, *Bifurcation Theory*, in Dynamical Systems, Encyclopaedia of Mathematical Sciences **5** (Springer–Verlag, Berlin, 1994).

[28] H. G. Schuster, Deterministic Chaos — An Introduction (VCH Verlagsgesellschaft, Weinheim, 1989).

[29] Y. Pomeau and P. Manneville, *Intermittent Transition to Turbulence in Dissipative Dynamical Systems*, Comm. Math. Phys. **74**, 189–193 (1980).

[30] V. S. Anishchenko, Dynamical Chaos — Models and Experiments. Appearance Routes and Structure of Chaos in Simple Dynamical Systems (World Scientific, Singapore, 1995).

[31] C. Grebogi, S. W. McDonald, E. Ott, and J. A. Yorke, *Final State Sensitivity: An Obstruction to Predictability*, Phys. Lett. A **99**, 415–418 (1983).

[32] J. D. Sterman and E. Mosekilde, *Business Cycles and Long Waves: A Behavioral Disequilibrium Perspective*, in Business Cycles: Theory and Empirical Methods, edited by W. Semmler (Kluwer Academic Publisher, 1994), pp. 13–51.

[33] A. Brandstätter, J. Swift, H. L. Swinney, and A. Wolf, *Low-Dimensional Chaos in a Hydrodynamic System*, Phys. Rev. Lett. **51**, 1442–1445 (1983).

[34] J. Stavans, F. Heslot, and A. Libchaber, *Fixed Winding Number and the Quasiperiodic Route to Chaos in a Convective Fluid*, Phys. Rev. Lett. **55**, 596–599 (1985).

[35] V. S. Anishchenko, *Destruction of Quasi-Periodic Oscillations and Chaos in Dissipative Systems*, Sov. Tech. Phys. Lett. **56**, 255–237 (1986).

[36] E. Mosekilde and L. Ranfeld, *Buffer Stock Stabilization of the International Commodity Market*, GRS-report No. 110, Academy of Technical Sciences, Oslo (1977).

[37] *Actions on Commodities in Light of the Need for Change in the World Commodity Economy* (UNCTAD Secretariat, 1976).

[38] T. Puu and A. Norin, *Cournot Duopoly when the Competitors Operate under Capacity Constraints*, Chaos, Solitons, and Fractals, to appear.

[39] I. Sushko, T. Puu and L. Gardini, *The Hicksian Floor-Roof Model for Two Regions Linked by Interregional Trade*, Chaos, Solitons, and Fractals, to appear.

[40] J. Laugesen, Internal report, Department of Physics, DTU (1999).

[41] B. L. Hao, Elementary Symbolic Dynamics and Chaos in Dissipative Systems (World Scientific, Singapore, 1989).

[42] M. Dutta, H. E. Nusse, E. Ott, J. A. Yorke, and G. Yuan, *Multiple Attractor Bifurcations: A Source of Unpredictability in Piecewise Smooth Systems*, Phys. Rev. Lett. **83**, 4281–4284 (1999).

[43] T. Kapitaniak and Yu. L. Maistrenko, *Multiple Choice Bifurcations as a Source of Unpredictability in Dynamical Systems*, Phys. Rev. E **58**, 5161–5163 (1998).

INDEX